Marine Power Systems

Marine Power Systems

Editor

Igor Poljak

MDPI • Basel • Beijing • Wuhan • Barcelona • Belgrade • Manchester • Tokyo • Cluj • Tianjin

Editor
Igor Poljak
University of Zadar
Croatia

Editorial Office
MDPI
St. Alban-Anlage 66
4052 Basel, Switzerland

This is a reprint of articles from the Special Issue published online in the open access journal *Journal of Marine Science and Engineering* (ISSN 2077-1312) (available at: https://www.mdpi.com/journal/jmse/special_issues/Igor_marine_power_systems).

For citation purposes, cite each article independently as indicated on the article page online and as indicated below:

LastName, A.A.; LastName, B.B.; LastName, C.C. Article Title. *Journal Name* **Year**, *Volume Number*, Page Range.

ISBN 978-3-0365-3150-2 (Hbk)
ISBN 978-3-0365-3151-9 (PDF)

Contents

Journal of
Marine Science and Engineering

Editorial

Marine Power Systems

Igor Poljak

Maritime Department, University of Zadar, Mihovila Pavlinovića 1, 23000 Zadar, Croatia; ipoljak1@unizd.hr

1. Introduction

The international seaborne trade by volume is divided into 60% loaded and 70% discharged trade, which means that the marine industry today is still the number one means of transportation for the human kind. As the trade amount is vast, the power included in that transportation field is the same high. Although maritime transport is pointed to as an energy-efficient mode of transportation, its emission of GHGs is still high despite new technologies that have been adopted [1]. Optimization and the use of new fuels with new technologies for power generation of such systems are always tasks that are necessary in order to improve efficiency of the power generation system and to reduce GHG emissions [2–6].

The second important thing is the reliability of such systems. As ships and submarine machines [7,8] are far away from available services, their machinery must be reliable in operation. For example, the failure of turbocharger of the standby generator could cause blackout and loss of propulsion [9]. The probability of avoiding such scenario increases with the number of generators that are on the stand-by mode at the time.

The intent of this Special Issue was to collect recent research in the field that improves such systems. This issue is composed of nine articles and one review paper. The six papers cover energy efficiency with numerical and optimization methods. Two papers are in the field of the submarine machines and two papers are dealing with the system failures. The brief description of each paper are given in the following section.

2. Papers Details

Gospić et al. [1] discusses the possibility of applying the trigeneration energy concept (cogeneration + absorption cooling) on diesel-powered refrigerated ships, based on systematic analyses of variable energy loads during the estimated life of the ship on a predefined navigation route. From a methodological point of view, mathematical modeling of predictable energy interactions of a ship with a realistic environment yields corresponding models of simultaneously occurring energy loads (propulsion, electrical, and thermal), as well as the preferred trigenerational thermal effect (cooling and heating). Special emphasis is placed on the assessment of the upcoming total heat loads (refrigeration and heating) in live cargo air conditioning systems (unfrozen fruits and vegetables) as in ship accommodations. The obtained results indicate beneficiary energy, economic, and environmental effects of the application of diesel engine trigeneration systems on ships intended for cargo transport whose storage temperatures range from -25 to $15\,°C$. Further analysis of trigeneration system application to the passenger ship air conditioning system indicates even greater achievable savings.

Anđelić et al. [2] collected the publicly available dataset for the Combined Diesel-Electric and Gas (CODLAG) propulsion system, which was used to obtain symbolic expressions for estimation of fuel flow, ship speed, starboard propeller torque, port propeller torque, and total propeller torque using genetic programming (GP) algorithm. The dataset consists of 11,934 samples that were divided into training and testing portions in an 80:20 ratio. The training portion of the dataset, which consisted of 9548 samples, was used to train

Citation: Poljak, I. Marine Power Systems. *J. Mar. Sci. Eng.* **2022**, *10*, 195. https://doi.org/10.3390/jmse10020195

Received: 5 January 2022
Accepted: 11 January 2022
Published: 1 February 2022

Publisher's Note: MDPI stays neutral with regard to jurisdictional claims in published maps and institutional affiliations.

the GP algorithm to obtain symbolic expressions for estimation of fuel flow, ship speed, starboard propeller, port propeller, and total propeller torque, respectively. After the symbolic expressions were obtained, the testing portion of the dataset, which consisted of 2386 samples, was used to measure estimation performance in terms of coefficient of correlation (R^2) and Mean Absolute Error (MAE) metric, respectively. Based on the estimation performance in each case, the three best symbolic expressions were selected with and without decay state coefficients. From the conducted investigation, the highest R^2 and lowest MAE values were achieved with symbolic expressions for the estimation of fuel flow, ship speed, starboard propeller torque, port propeller torque, and total propeller torque without decay state coefficients, while symbolic expressions with decay state coefficients had slightly lower estimation performance.

Baress Šegota et al. [3] presented an improvement of marine steam turbine conventional exergy analysis by application of neural networks. The conventional exergy analysis requires numerous measurements in seven different turbine operating points at each load, while the intention of MLP (Multilayer Perceptron) neural-network-based analysis was to investigate the possibilities for reducing measurements. At the same time, the accuracy and precision of the obtained results should be maintained. In MLP analysis, six separate models are trained. Due to a low number of instances within the data set, a 10-fold cross-validation algorithm was performed. The stated goal was achieved, and the best solution suggests that MLP application enables the reduction of measurements to only three turbine operating points. In the best solution, the MLP model errors fall within the desired error ranges: Mean Relative Error (MRE) < 2.0% and Coefficient of Correlation (R^2) > 0.95 for the whole turbine and each of its cylinders.

Pelić et al. [4] discussed the medium-speed diesel engine in diesel-electric propulsion systems, which is increasingly used as the propulsion engine for liquefied natural gas (LNG) ships and passenger ships. The main advantage of such systems is their high reliability, better maneuverability, greater ability to optimize and significant decrease in the engine room volume. Marine propulsion systems are required to be as energy efficient as possible and to meet environmental protection standards. The paper analyzes the impact of split injection on fuel consumption and NO_x emissions of marine medium-speed diesel engines. For the needs of the research, a zero-dimensional, two-zone numerical model of a diesel engine was developed. A model based on the extended Zeldovich mechanism was applied to predict NO_x emissions. The validation of the numerical model was performed by comparing operating parameters of the basic engine with data from engine manufacturers and data from sea trials of a ship with diesel-electric propulsion. The applicability of the numerical model was confirmed by comparing the obtained values for pressure, temperature, and fuel consumption. The operation of the engine that drives the synchronous generator was simulated under stationary conditions for three operating points and nine injection schemes. The values obtained for fuel consumption and NO_x emissions for different fuel injection schemes indicate the possibility of a significant reduction in NO_x emissions but with a reduction in efficiency. The results showed that split injection with a smaller amount of injected pilot fuel and a smaller angle between the two injections allow a moderate reduction in NO_x emissions without a significant reduction in efficiency. The application of split injection schemes that allow significant reductions in NO_x emissions led to a reduction in engine efficiency.

Nirbito et al. [5] explains the performance analysis of a propulsion system engine of an LNG tanker using a combined cycle whose components are gas turbine, steam turbine, and heat-recovery steam generator. The researchers are to determine the total resistance of an LNG tanker with a capacity of 125,000 m^3 by using the Maxsurf Resistance 20 software, as well as to design the propulsion system to meet the required power from the resistance by using the Cycle-Tempo 5.0 software. The simulation results indicate a maximum power of the system of about 28,122.23 kW with a fuel consumption of about 1.173 kg/s and a system efficiency of about 48.49% in fully loaded conditions. The ship speed can reach up to 20.67 knots.

Poljak et al. [6] described and evaluated the atmospheric drain condensate system of a marine steam power plant from the energetic and exergetic points of view at a conventional liquefied natural gas (LNG) carrier. Energy loss and exergy destruction rate were calculated for individual stream flows joined in an atmospheric drain tank with variations of the main turbine's propulsion speed rate. The energy efficiency of joining streams was noted to be above 98% at all observed points as the atmospheric drain tank was the direct heater. The exergy efficiency of the stream flows into the drain tank was in the range of 80% to 90%. The exergy stream flow to the tank was modeled and optimized by the gradient reduced gradient (GRG) method. Optimization variables comprised contaminated and clean condensate temperature of the atmospheric drain tank and distillate water inlet to the atmospheric drain tank with respect to condensate outlet temperature. The optimal temperatures improve the exergy efficiency of the tank as the direct heater to about 5% in the port and 3% to 4% when the LNG carrier was at sea, which is the aim of optimization. Proposals for improvement and recommendations are given for proper plant supervision, which may be implemented in real applications.

Lu et al. [7] studied the safety and stability of the anchoring and hooking of ships, bedrock friction, and biological corrosion of submarine cables. A hydraulic jet submarine cable-laying machine manages to bury the submarine cables deep into the seabed and effectively reduces the occurrence of external damage to the submarine cables. This machine uses a hydraulic jet system to realize trenching on the seabed. However, the hydraulic jet submarine cable-laying machine has a complicated operation and high power consumption with high requirements on the mother ship, and it is not yet the mainstream trenching method. In this paper, a mathematical model for the hydraulic jet nozzle of the submarine cable-laying machine is established, and parameters that affect the trenching efficiency are studied. The effects of jet target distance, flow, angle, and nozzle spacing on the working efficiency of the burying machine are analyzed by setting up a double-nozzle model. The results of the theory, numerical simulation, and experiment show that the operational efficiency of the hydraulic jet submarine cable-laying machine can be distinctly improved by setting proper jet conditions and parameters.

Yang et al. [8] enhance the vibration isolation effectiveness of an underwater vehicle power plant and alleviate the mechanical vibration of the outer housing; initially, discrete vibration isolators were improved, and three new types of ring vibration isolators were designed, i.e., ring metal rubber isolators, magnesium alloy isolators, and modified ultra-high polyethylene isolators (MUHP). A vibrator excitation test was carried out, and the isolation effectiveness of the three types of vibration isolators was evaluated, adopting insertion loss and vibration energy level drop. The results showed that, compared with the initial isolators and the other two new types of isolators, MUHP showed the most significant vibration isolation effectiveness. Furthermore, its effectiveness was verified by a power vibration test of the power plant. To improve the vibration isolation effectiveness, in addition to vibration isolators, it is essential to carry out investigations on high-impedance housings.

Knežević et al. [9] discussed the reliability of marine propulsion systems, which depend on the reliability of several sub-systems of a diesel engine. The scavenge air system is one of the crucial sub-systems of the marine engine with a turbocharger as an essential component. In this paper, the failures of a turbocharger are analyzed through the fault tree analysis (FTA) method to estimate the reliability of the system and to predict the cause of failures. The quantitative method is used to assess the probability of faults occurring in the turbocharger system. The main failures of a scavenge air sub-system, namely air filter blockage, compressor fouling, turbine fouling (exhaust side), cooler tube blockage, and cooler air side blockage, are simulated on a Wärtsilä-Transas engine simulator for a marine two-stroke diesel engine. The results obtained through the simulation can provide improvement in the maintenance plan, reliability of the propulsion system, and optimization of turbocharger operation during exploitation time.

Vizentin et al. [10] explained failures of marine propulsion components or systems that can lead to serious consequences for a vessel, cargo, and the people onboard a ship.

These consequences can be financial losses, delay in delivery time, or a threat to safety of the people onboard. This is why it is necessary to learn about marine propulsion failures in order to prevent worst-case scenarios. This paper aims to provide a review of experimental, analytical, and numerical methods used in the failure analysis of ship propulsion systems. In order to achieve this, the main causes and failure mechanisms are described and summarized. Commonly used experimental, numerical, and analytical tools for failure analysis are given. Most indicative case studies of ship failures describe where the origin of failure lies in the ship propulsion failures (i.e., shaft lines, crankshaft, bearings, and foundations). In order to learn from such failures, a holistic engineering approach is inevitable. This paper tries to give suggestions to improve existing design procedures with a goal of producing more reliable propulsion systems and taking care of operational conditions.

Funding: (3) This research has been supported by the Croatian Science Foundation under the project IP-2018-01-3739, CEEPUS network CIII-HR-0108, European Regional Development Fund under the grant KK.01.1.1.01.0009 (DATACROSS), project CEKOM under the grant KK.01.2.2.03.0004, CEI project "COVIDAi" (305.6019-20), University of Rijeka scientific grant uniri-tehnic-18-275-1447, and University of Rijeka scientific grant uniri-tehnic-18-18-1146. (4) This work was partially supported by the Croatian Science Foundation under the project IP-2018-01-3739. This work was also supported by the University of Rijeka (project no. uniri-tehnic-18-18 1146 and uniri-tehnic-18-266 6469). (6) This research was supported by the Croatian Science Foundation under project IP-2018-01-3739, CEEPUS network CIII-HR-0108, European Regional Development Fund under grant KK.01.1.1.01.0009 (DATACROSS), project CEKOM under grant KK.01.2.2.03.0004, University of Rijeka scientific grant uniri-tehnic-18-275-1447, University of Rijeka scientific grant uniri-tehnic-18-18-1146 and University of Rijeka scientific grant uniri-tehnic-18-14. (7) This research was supported by the Key Research and Development Project of Zhejiang Province (2019C03115). (8) This research was funded by the National Natural Science Foundation of China (62005204) and the Fundamental Research Funds for the Central Universities. (10) This work has been fully supported by the University of Rijeka under the project number uniri-technic-18-200 "Failure analysis of materials in marine environment".

Institutional Review Board Statement: Not applicable.

Informed Consent Statement: Not applicable.

Data Availability Statement: All relevant data and links to that can be found in the presented papers at https://www.mdpi.com/journal/jmse/special_issues/Igor_marine_power_systems (accessed on 6 January 2022).

Acknowledgments: I wish to express my sincere gratitude to all the authors and the reviewers.

Conflicts of Interest: The authors declare no conflict of interest.

References

1. Gospić, I.; Glavan, I.; Poljak, I.; Mrzljak, V. Energy, Economic and Environmental Effects of the Marine Diesel Engine Trigeneration Energy Systems. *J. Mar. Sci. Eng.* **2021**, *9*, 773. [CrossRef]
2. Anđelić, N.; Baressi Šegota, S.; Lorencin, I.; Poljak, I.; Mrzljak, V.; Car, Z. Use of Genetic Programming for the Estimation of CODLAG Propulsion System Parameters. *J. Mar. Sci. Eng.* **2021**, *9*, 612. [CrossRef]
3. Baressi Šegota, S.; Lorencin, I.; Anđelić, N.; Mrzljak, V.; Car, Z. Improvement of Marine Steam Turbine Conventional Exergy Analysis by Neural Network Application. *J. Mar. Sci. Eng.* **2020**, *8*, 884. [CrossRef]
4. Pelić, V.; Mrakovčić, T.; Radonja, R.; Valčić, M. Analysis of the Impact of Split Injection on Fuel Consumption and NO$_x$ Emissions of Marine Medium-Speed Diesel Engine. *J. Mar. Sci. Eng.* **2020**, *8*, 820. [CrossRef]
5. Nirbito, W.; Budiyanto, M.A.; Muliadi, R. Performance Analysis of Combined Cycle with Air Breathing Derivative Gas Turbine, Heat Recovery Steam Generator, and Steam Turbine as LNG Tanker Main Engine Propulsion System. *J. Mar. Sci. Eng.* **2020**, *8*, 726. [CrossRef]
6. Poljak, I.; Bielić, T.; Mrzljak, V.; Orović, J. Analysis and Optimization of Atmospheric Drain Tank of Lng Carrier Steam Power Plant. *J. Mar. Sci. Eng.* **2020**, *8*, 568. [CrossRef]
7. Lu, Z.; Cao, C.; Ge, Y.; He, J.; Yu, Z.; Chen, J.; Zheng, X. Research on Improving the Working Efficiency of Hydraulic Jet Submarine Cable Laying Machine. *J. Mar. Sci. Eng.* **2021**, *9*, 745. [CrossRef]
8. Yang, Y.; Pan, G.; Yin, S.; Yuan, Y.; Huang, Q. Verification of Vibration Isolation Effectiveness of the Underwater Vehicle Power Plant. *J. Mar. Sci. Eng.* **2021**, *9*, 382. [CrossRef]

9. Knežević, V.; Orović, J.; Stazić, L.; Čulin, J. Fault Tree Analysis and Failure Diagnosis of Marine Diesel Engine Turbocharger System. *J. Mar. Sci. Eng.* **2020**, *8*, 1004. [CrossRef]
10. Vizentin, G.; Vukelic, G.; Murawski, L.; Recho, N.; Orovic, J. Marine Propulsion System Failures—A Review. *J. Mar. Sci. Eng.* **2020**, *8*, 662. [CrossRef]

Review

Marine Propulsion System Failures—A Review

Goran Vizentin [1], Goran Vukelic [1,*], Lech Murawski [2], Naman Recho [3,4] and Josip Orovic [5]

[1]   Marine Engineering Department, Faculty of Maritime Studies, University of Rijeka, 51000 Rijeka, Croatia; vizentin@pfri.hr
[2]   Faculty of Marine Engineering, Gdynia Maritime University, 81-225 Gdynia, Poland; l.murawski@wm.am.gdynia.pl
[3]   Institute Pascal CNRS-UMR 6602, University Clermont Auvergne, 63001 Clermont-Ferrand, France; naman.recho@epf.fr
[4]   EPF Engineering School, ERMESS, 92330 Sceaux, France
[5]   Maritime Department, University of Zadar, 23000 Zadar, Croatia; jorovic@unizd.hr
*   Correspondence: gvukelic@pfri.hr; Tel.: +385-51-338411

Received: 11 August 2020; Accepted: 25 August 2020; Published: 27 August 2020

Abstract: Failures of marine propulsion components or systems can lead to serious consequences for a vessel, cargo and the people onboard a ship. These consequences can be financial losses, delay in delivery time or a threat to safety of the people onboard. This is why it is necessary to learn about marine propulsion failures in order to prevent worst-case scenarios. This paper aims to provide a review of experimental, analytical and numerical methods used in the failure analysis of ship propulsion systems. In order to achieve that, the main causes and failure mechanisms are described and summarized. Commonly used experimental, numerical and analytical tools for failure analysis are given. Most indicative case studies of ship failures describe where the origin of failure lies in the ship propulsion failures (i.e., shaft lines, crankshaft, bearings, foundations). In order to learn from such failures, a holistic engineering approach is inevitable. This paper tries to give suggestions to improve existing design procedures with a goal of producing more reliable propulsion systems and taking care of operational conditions.

Keywords: marine propulsion; propulsion failure; propulsion failure analysis; mechanical failure

---

1. Introduction

In order to limit the occurrence of fatalities, environmental damage and economic losses, marine structures are to be designed, built and operated in such manner that the probabilities of overall structural rigid body stability and failures are reduced to a minimum [1]. During the design phase of a specific marine structure, a level of structural safety is chosen by defining individual structural elements, used materials and functional requirements. An important factor that has to be considered is the time dependency of the strength and loads. The strength of a structure decreases with time and true insight into the strength state strongly depends on inspection and maintenance procedures [2]. As for the load, it is very variable through the lifetime of the marine structure.

Previous studies and analysis of marine structure failures had shown that a significant percentage of failures were a consequence of inadequate design due to a lack of operational considerations, incomplete structural element evaluations and incorrect use of calculation methods [3]. Hence, in order to better understand the causes of failures, a failure analysis branch of engineering [4] has developed over the years, serving as a help in the design optimization process. This discipline uses analytical, experimental and numerical tools in order to resolve failure causes. Particular effort has been invested in researching the causes of marine structural failures. Due to recent advances in failure analysis techniques and expected further improvement, it is essential to collect and review current state of the art research in the field and mark paths for future research.

A review of the present state of the scientific and practical development in this field, presented in this paper, should serve as an adequate starting point. The paper will present a brief review of indicative case studies dealing with marine structural failures. Marine structural failures can be divided into three main groups: failures of ships, offshore structures and marine equipment. Here, particular interest will be put on failures of ships, most specifically ship propulsion systems. One part of the paper will summarize experimental, analytical and numerical tools used for failure analysis. The result of this paper will define steps and possible analysis improvement recommendations that will be used as guidelines for future research in failure analysis of ship propulsion systems.

2. General Causes and Mechanisms of Failures

Structural failures occur when the loading exceeds the actual strength of the structure so they can be defined as a loss of the load-carrying capacity of the structure or some of its components [5]. Failures can result in a global catastrophic damage that could easily lead to fatal casualties or partial damage that could lead to pollution or operational delay, but the structure can ultimately be repaired or recovered.

Structural failure is a result of fracture or damage that is initiated when the material is stressed above its strength limit. In particular, structural integrity of marine structures depends, along with the material strength and loading conditions, on material (usually steel) quality, proper manufacturing (usually welding), severity of service conditions (sea, salt, winds, etc.), design quality as well as various human elements that have effects during use of the structure [6].

Causes of failures can be roughly divided in two distinctive groups. The first group is comprised of unforeseeable external or environmental effects which exert additional loading on the structure resulting in overload. Such effects are extreme weather (sea or wind overloads), accidental loads (collisions, explosions, fire, etc.), operational errors or environmental influence (corrosion). The second group comprises causes for failures that occur either during the design and construction phase (dimensioning errors, poor construction workmanship, material imperfections) or due to phenomena growing over time (fatigue, creep), both resulting in reduced actual strength in respect to the design value, Figure 1.

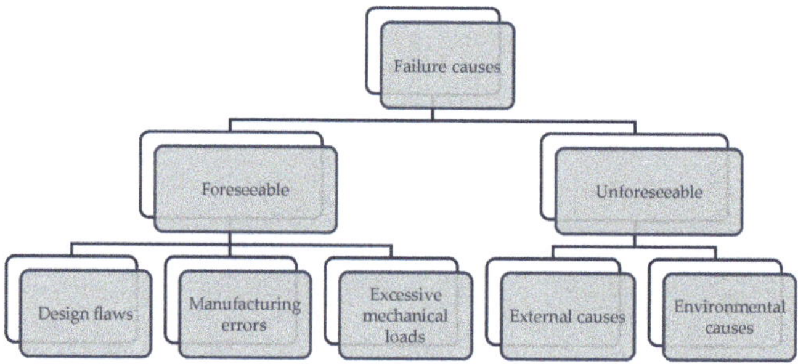

Figure 1. Chart of general failure causes related to marine propulsion systems.

Mechanisms of failures that occur in marine structures can have progressive or sudden natures. Structural designers tend, by all means, to avoid sudden failures like brittle fracture. Progressive failures, which depend on time and specific load conditions, can be monitored and adequate actions can be undertaken to avoid fatal scenarios.

One of such mechanisms, maybe the most important on ships and similar structures, is fatigue. Fatigue can be defined as a process of damage accumulated during each cycle of the dynamic load that the structure is subjected to with an important characteristic of load intensity lower than the values that would cause immediate failure [7]. Fatigue cracks start and evolve in two phases—formation (usually starting on the material surface) of a shear crack on crystallographic slip planes in the first phase, and growth of the crack in a direction normal to the applied stress in the second phase [8]. Cui proposed a division of the failure fatigue process in five stages, namely crack nucleation, microstructurally small crack propagation, physically small crack propagation, long crack propagation and final fracture [9]. The process which occurs before long crack propagation is usually named "fatigue crack initiation", while long crack propagation is called "fatigue crack propagation".

The fatigue failure process is extremely complex in nature and dependent on a large number of parameters like distribution of mean stress, residual stresses, loading history, adequacy of design, environmental effects, manufacturing quality, etc.

Besides fatigue, corrosion effects on marine structures shouldn't be neglected as another aging degradation effect on ship structural integrity [10]. This gradual destruction of materials caused by chemical or electrochemical reaction with their environment weakens the material, opening discontinuities allowing for crack growth and final fracture. Coupled with fatigue, corrosion can indicate lowering of fatigue strength, accelerated initiation of failure at high stresses and elimination of the material's fatigue limit [11]. Furthermore, engineering designers strive to avoid stress corrosion cracking, the formation of microscopic cracks that can remain inconspicuous, but can cause crack formation in a mildly corrosive environment and lead to unexpected failures of ductile metallic materials. However, this paper is primarily concerned with mechanical causes of failure (overloading, fatigue, vibrations, etc.) that affect the structures with reduced strength, even if the reduction is a result of corrosion.

3. Tools Used for Failure Analysis

In order to fully understand the reviewed case studies of marine structural failures, an overview of tools used for failure analysis is desirable. Tools that researchers use are, in most cases, experimental and rely on some non-destructive testing (NDT) technique or microstructural analysis.

NDT plays significant role in failure analysis and control procedures [12]. Classical (eddy-current, magnetic-particle, liquid penetrant, radiographic, ultrasonic and visual testing) or newly developed (e.g., acoustic emission) NDT techniques are used to gain insight into the actual state of the structure, Figure 2. NDT methods must not alter, change or modify the actual condition of the structure, but must survey the failure so that they don't impact, change or further degrade the failure zone. NDT is employed at the beginning of the service life in order to document initial flaws and monitor their progression. Based on these inputs, a structural health monitoring (SHM) strategy can be developed regarding damage detection and characterization [13].

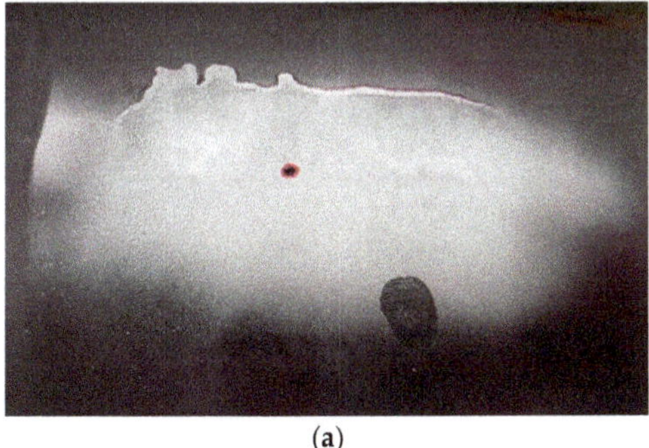
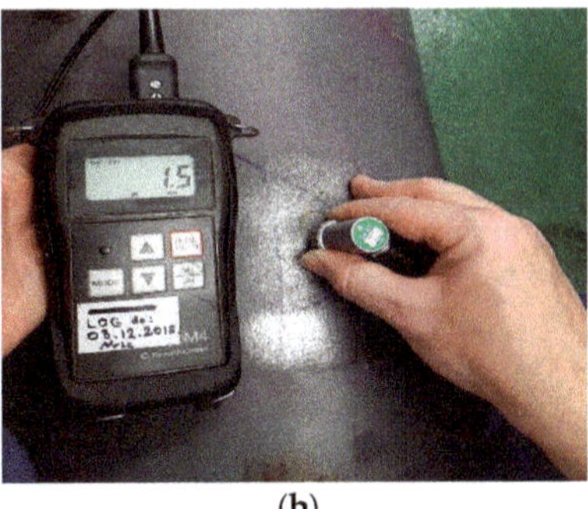

(**a**) (**b**)

Figure 2. Example of non-destructive testing of rotating machinery equipment: (**a**) liquid penetrant testing, (**b**) ultrasonic thickness testing.

If necessary, destructive testing can also be employed, e.g., when the material mechanical parameters are not known and need to be determined. Here, researchers make use of tensile or impact tests performed on specimens extracted from failed structures. Hence, values of the material's ultimate tensile strength, yield strength or Charpy V-notch impact energy, Figure 3, can be determined and used for later numerical modelling [14].

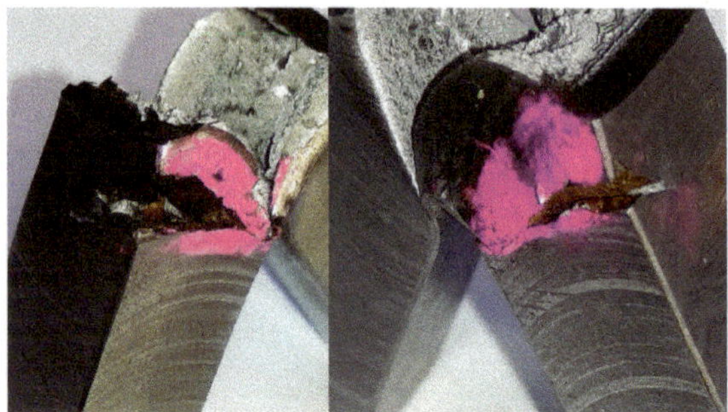

Figure 3. Broken Charpy V-notch specimen machined out of marine shaft to material's fracture toughness. Traces of corrosion near the fracture area can be noted.

Microscopy, optical (OM) or scanning electron (SEM), is a widely used experimental failure analysis method providing insight into the metallurgical state of the fractured zone. This technique is often used in conjunction with micro-sectioning to broaden the application. One of the main disadvantages is the narrow field depth. SEM is an extension of OM and here the use of electrons instead of a light source provides higher magnification, better field depth and the opportunity to perform phase identification. SEM has been extensively used in the analysis of marine structures and equipment [15–18], Figure 4.

Figure 4. SEM image of fractured surface of a rotating shaft at the crack origin showing inclusions that acted as crack initiation points.

Besides experimental, analytical solutions are also being used and further developed to allow fast and reasonably accurate prediction of damage. Analytical procedures are usually based on spectral fatigue analysis, beam theory, fracture mechanics and structural factors. One of the concepts that can be applied is the failure assessment diagram (FAD) approach that spans the entire range from linear elastic to fully plastic behavior of the material preceding the fracture.

The FAD is basically an alternative method for graphically representing the fracture driving force. Depending on the type of the equation used to model the effective stress intensity factors the FAD approach can be sub-divided into the strip-yield based FAD [19], *J*-based FAD [20–22] and approximated FAD. It uses two parameters which are linearly dependent on the applied load. The result is a curve that represents a set of points of predicted failure points and the results fall in acceptable or nonacceptable areas marked by that curve. This method can be applied to analyze and model brittle fracture (from linear elastic to ductile overload), welded component fatigue behavior or ductile tearing.

Another factor that has to be considered are dynamic loads imposed onto the marine structures and their unpredictable, stochastic changes. Probabilistic failure analysis can account for time-dependent crack growth by applying appropriate distribution laws. Most practical situations exhibit randomness and uncertainty of the analysis variables so numerical algorithms for probabilistic analysis may need to be applied. The well-known Monte Carlo method can suit FAD models in most cases of uncertainties [23].

The marine industry relies heavily on standards and regulations set by classification societies that have recently been involved in research and development in order to establish probabilistic methods that are to be used for planning in-service inspection. Det Norske Veritas issued recommendations on how to use probabilistic methods for floating production ships, among others [24,25]. The goal of proposed probabilistic method is to replace conservative inspection planning with mathematical models that consider the influence of exploitation, fatigue causes and crack propagation characteristics on structure lifetime.

Furthermore, with the development of advanced numerical routines and powerful computers, more and more research is done using some kind of numerical analysis. The latest trend in failure analysis development is the unification of analysis methods and procedures [26–28] in order to obtain a comprehensive procedure of structural failure analysis that would cover the main failure modes and enable a safer and more efficient design, manufacturing and maintenance processes. Out of the numerous various methods used, the finite elements (FE) method has been recognized for its universality and efficiency, Figure 5.

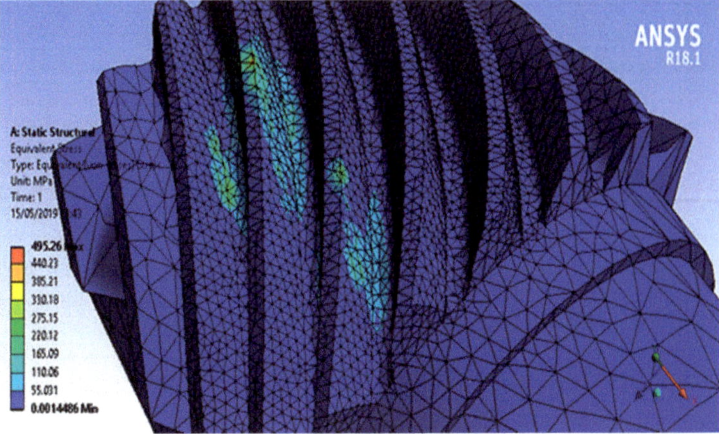

Figure 5. Stress distribution over a gear shaft as a result of finite elements (FE) analysis.

The extended FE method (X-FEM) is the most recent development used mostly for fracture mechanics. It can be applied to solve complex discontinuity issues including fracture, interface and damage problems while proving useful in multi-scale and multi-phase computation [29]. The basic idea is to reduce the re-meshing around the crack so as to enable the crack to be represented independently of the mesh [30], even in 3D applications [31–33].

Various adaptive re-meshing techniques for crack growth modelling have been developed in order to better account for discontinuities and allow time-saving calculations. One of them is the automatic crack box technique (CBT), developed to perform fine fracture mechanics calculations in various structures without global re-meshing [34]. Only the specific crack tip zone has to be re-meshed resulting in quick calculations.

4. Ship Propulsion System Failures

Ship propulsion system failures include failures of shaft lines, crankshafts, bearings, foundations, etc. The causes of ship structure failure can be external (impact, bad weather) or internal (inadequate dimensioning, material grade, fatigue, etc.).

Ship propulsion systems are subjected to vibrational [35], torsional [36], coupled longitudinal (axial) [37] (Figure 6), and lateral [38] loads. Vibrations can cause fracture and failure in system components or on the ship's structure, resulting in complete destruction of the propulsion system, reduction of the service life of shafts and/or their components and fatigue fracture on support brackets and/or engine mountings. The shafts line's misalignment [39] or bend represent one of the most frequent reasons of this kind damages.

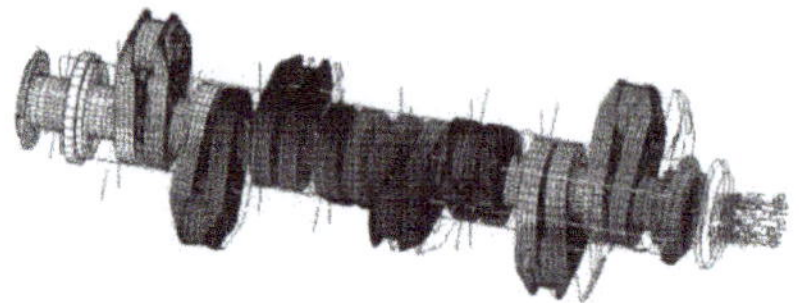

Figure 6. FE model of the MAN B&W 8 S70 MC-C engine crankshaft under longitudinal loading leading to excessive deformations.

Moreover, it has been experimentally proved that frictional losses during power transmission through the universal joints could act as an excitation force for self-excited vibrations [40] of shafting in the propulsion systems of an ocean-going vessel. Research revealed that undamped vibrations will cause failure if coupling connected to the intermediate shaft doesn't have sufficient radial flexibility. Coupling should be designed so that is capable of absorbing the radial shaft displacement, therefore avoiding the effects of the self-excited torsional vibration.

Cracks usually occur in flanges, shaft liners, shaft end and keyways. The causing factors can be grouped in design, workmanship and operation cause groups. A keyway's end design represents a stress concentration point during torque transmission through shaft keys. Poor final processing of key grooves, keyways and keys, inadequate run out radius or material impurities can act as root causes of torsional fatigue failure in shaft keys. This characteristic failure can be recognized as a crack pattern initiating at the keyway end and propagating in a 45° rotational direction marking a helical path, Figure 7. Solved case studies [41,42] have revealed that deficient design against torsional vibrations (i.e., calculations of shaft elements stiffness and damping, natural frequencies, safety factors) causes failures of the shaft's keyway. In the referenced researches root-cause analysis has been performed combining analytical processes set by MIL G 17859D and VDI 3822 standards with FE analysis.

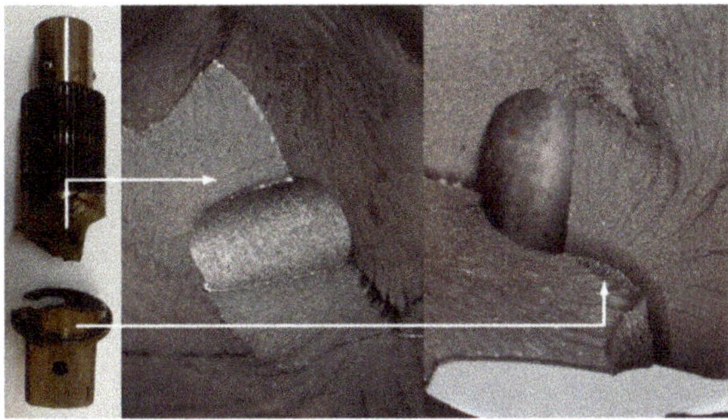

Figure 7. Fracture pattern at 45° to the centerline of the rotating shaft, typical for torsional failure. Fracture area is visible at suitable magnification under optical microscope for both parts.

Propulsion shaft elements can fail while running at low speed due to fatigue caused by torsional stress [43]. The cause of the failure in this particular case was exposure to corrosive environments without any protective coating which resulted in pitting corrosion. The crack grew with multiple starting points due to torsion force (moment) with high stress concentrations, i.e., the failure cause was fatigue and corrosion.

Engine crankshafts are subjected to bending, stretch–compression, and torsional dynamic loads. Thermal displacement (caused by normal engine working conditions) of the crankshaft [44] and thermal interaction between main engine body and ship hull [45] are other sources of variable loads acting on power transmission system. Therefore, the crankshafts are prone to fatigue failures under multiaxial loading. A fatigue analysis for a typical marine crankshaft has shown that a combination of rotating bending with steady torsion stress caused formation of a crack initiated by rotating bending, whilst the effect of the steady torsion became itself significant in the later phases of crack growth. The fact that the propagation was fast in comparison with the total number of the engine work hours indicates that the failure was caused by fatigue [46].

One first indication of failure in a crankshaft is given by the low-pressure value of the lubrication circuit. This is mainly due to the accumulation of debris in the lubrication channels, which causes the oil filters to be clogged. As such, this will cause poor lubrication of the crankshaft, which can consequently cause its catastrophic failure, and frequently originates damage propagation to other components of the engine, namely the crankcase, bearing shells, connecting rods, pistons and other mechanical parts [47].

Gomes at al. [48] performed failure analysis of a maritime V12 diesel engine crankshaft. A series of failures of the crankshafts were reported over a quarter of a century. The authors discussed the influence of material imperfections and applied load to the crankshaft failure but also performed dimensioning assessment using the Soderberg criterion FE crankshaft model. A stress-life equation was used to estimate the fatigue lifetime of the crankshaft so, finally, a modification to the crankshaft's design is suggested to reduce induced stresses. This case study can serve as a showcase for a comprehensive failure analysis that will be discussed in later sections of this paper.

As seen in a previous case study, fillets, tapers and chamfers also represent stress concentration points in shafts and their improper design can lead to fatigue failure. An additional case study of fracture initiated at a fillet [49] shows fatigue failure due to a cyclic torsional-bending load acting on a crack emanating from the fillet shoulders on the shaft. Gradual shaft load bearing reduction led to consequential overloading and final sudden failure. Chemical composition analysis, microstructural characterization, fractography, hardness measurements and FE analysis were incorporated in this research to determine the failure causes.

Spline joints are adequate alternatives to shaft key joints but previous research has shown that the press fitting of the joining elements can cause strains leading to surface crack formation [50]. Spline teeth at the shaft junction zone is the usual crack origin, alternating stress causes crack growth and propagation. Imperfections of the material can further ease crack propagation. This particular case study comprised of a visual and macroscopic inspection, material analysis, hardness measurement, OM and SEM.

Changes of the shaft rotation direction can result in torque moment overloading acting on bolted connections that are used in collar coupling of shaft elements and in propeller blade connections. This can result in fatigue failure of coupling bolts [51]. Fretting creates micro notches that develop into fatigue cracks with a direction of the crack path growing in planes angled from 35° to 60°. As this is not a characteristic of pure torsion fatigue, failure analysis has been performed to shown that the bolts are subjected to an increasing bending moment. Here, experimental findings served as an input for numerical verification of the hypothesis that variable bending stress in the coupling served as the cause of failure.

Damage of one or several blades can cause abnormal performance of the ship's propeller. This can generate uniaxial force which fluctuates once per rotation in a consistent transverse direction across the shaft. This fluctuating force generates a couple which can cause fatigue failure of the propeller hub [52]. A uniaxial type of failure is characterized by a fatigue fracture with a single origination point that progresses across the shaft from the side where the force is being applied and results in the final overload failure occurring on the opposite side from the fluctuating force. Visual inspection, detail axis alignment measurements, microscopic metallurgical examination, hardness measurements and ultrasonic scanning were used in this case study. Numerical modal analysis could prove useful here to determine natural mode shapes and frequencies of a propeller in order to avoid them during operation, Figure 8.

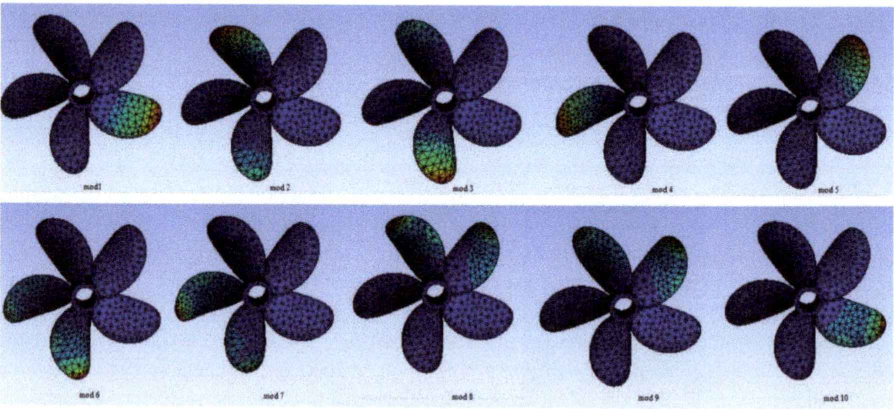

Figure 8. First ten mode shapes of a five-blade marine propeller, obtained by numerical simulation.

AHTS (Anchor Handling/Tug/Supply) ships use ducted azimuth thrusters for propulsion. The integral part of such propulsion system is the gears used for power transmission. Gear failures can occur due to localized stress increase on the teeth surface which is caused by inadequate lubricating and constructional misalignments, i.e., poor maintenance and design [53]. Additionally, gear failures can be initiated at locations with material inclusions, serving as stress raisers, Figure 9.

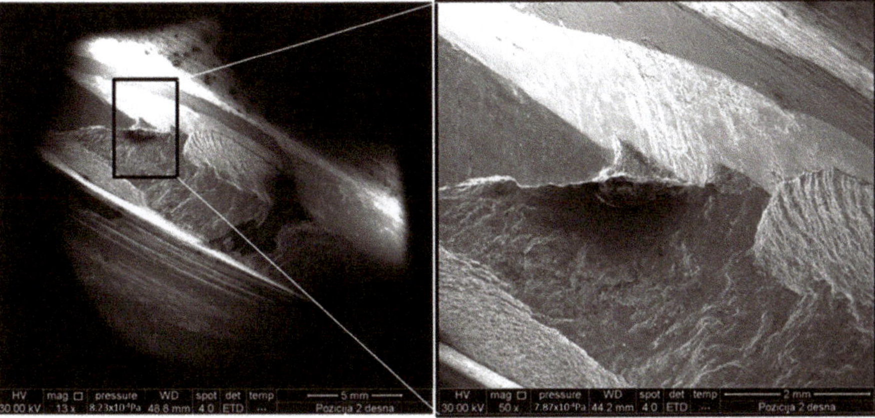

Figure 9. SEM image of a failed gear—detail of a material inclusions just below case-hardened layer serving as a crack initiation point.

The exhaust systems of marine engines and gas turbines are subjected to high service temperatures which can contribute to the reduction of the service life of the system. These structures are usually constructed as welded steel plate structures, with cracks occurring in the welded joints [54]. In conditions of thermal shock and temperature variations, the lifetime of the structure can be influenced significantly as the critical crack size is reduced.

5. Discussion

Two main types of ship structure can be distinguished: ship hull (with superstructure and main engine body) and power transmission system (i.e., crankshaft, shaft line and propeller). Ships operate in aggressive workload and environmental conditions so proper assessment of the technical condition is crucial from the perspective of safety of maritime navigation. Limitation of maritime disasters is of great economic importance and, more importantly, will reduce the negative environmental impact along with human injuries and life losses. Especially the propulsion system of the ship should be subject to thorough assessment, because inoperative propulsion results in a very high probability of disaster in extreme weather conditions.

In order to cope with such requirements, engineering designers rely heavily on the regulations prescribed by the classification societies. Classification societies' rules are based on a wide knowledge collected over hundreds of years and are mostly based on simplified, empirical equations. However, not all the problems occurring on modern ships can be successfully solved using this approach. To properly address issues of marine structural failures, engineers need to turn to failure analysis databases and, learning from the findings, improve procedures for ship designing.

Reviewing case studies in a former section, one can notice that most of them use solely experimental approaches in finding the causes of failures. Techniques like NDT inspection, microscopy or crystallography are used in order to determine the origins of failures. Only a few use numerical analysis as a supplement to traditional experimental techniques used in the field of failure analysis. However, those who do combine experimental and numerical approaches tend to present more reliable results and go a bit further than usual failure analysis does—they suggest modifications to engineering design. So, a combination of failure analysis and design optimization is arising here.

If one goes a step further and tries to identify case studies of failures where experimental and numerical approaches are complemented with analytical analysis, one can find that they are very rare. Only a few case studies (dealing with marine structural failures) can be found that, based on experimental and numerical results, propose an improvement of analytical procedures used in calculations of structures against failures.

So, it is obvious that these separate science disciplines and branches need to bring themselves closer in order to mutually improve. The first step is performing thorough failure analysis—analysis that would incorporate inevitable experimental, numerical and analytical procedures. Experimental, to determine material characteristics and origins of failure. Numerical, to model the structure, analyze its real performance and optimize the design. Analytical, to model complex natural and technical phenomena and then convert them into simple mathematical models. A mathematical model may help to explain a system and to study the effects of different components and to make predictions about behavior. At this stage, failure analysis (or forensic) engineers must work closely with metallurgists, NDT engineers, engineering designers, FEM experts, mechanical engineers, mathematicians, etc. [55,56].

These failure analysis findings should prove valuable in improving analytical procedures defined by rules and regulations that are set by classification societies. The shipping industry is conservative in nature, but all classification societies admit alternatives to their calculation methods, especially FEM. These, more detailed, analyses are usually more expensive but optimization is possible.

Another important aspect, especially in the stage of numerical research, is proper modelling of loads imposed on marine structures. Numerical algorithms used for, e.g., FE analysis of ship mechanical behavior, must account for the randomness and uncertainty of loads coming from sea, wind, operating machinery and moving cargo loads. Using the principles of probabilistic mechanics these problems can be solved successfully and greater safety of navigation can be granted for ships.

Learning from the cases shown in the previous section, several possible research directions can be suggested. These are:

- improved design methodologies,
- condition based monitoring techniques,
- a coupled failure analysis approach.

Improved design methodologies need to take into the account previously acquired practical knowledge about the operation of marine structures and machinery [57], but also need to rely on modern computer-based design and findings from the operational monitoring data and eventual failure analysis. That way, costly and time-consuming experiments can be successfully substituted, the design process can be shortened and safety factors, often too conservative, can be reduced.

Condition based monitoring techniques [58], if introduced for rotating machinery, are most commonly based on vibration and lubrication monitoring [59]. These two techniques prove themselves valuable as they fall into the category of preventive prediction tools, where the monitored machinery can still be satisfactory repaired if unusual values of vibrations or dispensed particles are detected. In addition, ultrasonic detection of failures can also be introduced to ships to detect failures in the early stages, but this technique requires highly skilled operators. Research in this field should find a way to introduce practical and reliable solutions for these techniques to be introduced to ships in order to detect potential failures in the early stages.

The coupled failure analysis approach assumes that adequate failure analysis can no longer be based solely on the techniques such as metallography, microscopy and other experimental methods. Today, experimental methods coupled with big-data acquisition and FE methods provide adequate means of achieving higher degrees of marine machinery safety, suitable operational life prediction and analysis of mechanical failures. Further research in this field should concentrate on blending these two approaches and developing new solutions in FE analysis. These new solutions should seek to close the existing gaps in multi-scale fracture mechanics, transition from damage to fracture, interaction of fracture with heat and moisture transport, dynamic fracture and fatigue prediction [60].

Successfully addressing these research issues could help to reduce the possibility of future failures of marine propulsion systems.

6. Conclusions

In this paper, recent work on the topic of ship failures has been outlined. The list is not exhaustive as literally every day new reports and papers are being published. However, the case studies mentioned here were selected to benchmark the common causes of failures on ships. Further cases can be found, of course, but with the same or similar causes of failures and that is why they were omitted.

Particularly, the failures of ship structures and propulsion systems have been summarized and described. As for the former, it can be noted that failures can be caused either by unfavourable environmental conditions (low temperatures, corrosive surroundings), poor design or workmanship (particularly concerning welds) or fatigue loading that is very often stochastic in nature. As for the latter, causes include inadequate design or assembly (of shaft line) or fatigue very often coupled with fluctuating torsional vibrations.

Some light is shed on the general causes and mechanisms of failures and an overview of the tools used in failure analysis is given. Points for further development of failure analysis are given in the Discussion section mentioning the unification of analysis methods and procedures in order to obtain a comprehensive procedure of structural failure analysis that would cover the main failure modes and enable safer and more efficient design, manufacture and maintenance processes and usage of maritime structures.

This review paper can serve as an introduction to the area of ship failure analysis for new coming engineers, practitioners and researchers or as an initial step in studying structural integrity of rotating machinery [61].

Author Contributions: Conceptualization and methodology, G.V. (Goran Vukelic); investigation and data curation, G.V. (Goran Vizentin), J.O.; writing—original draft preparation, G.V. (Goran Vukelic), G.V. (Goran Vizentin); writing—review and editing, L.M., N.R., J.O.; funding acquisition, G.V. (Goran Vukelic). All authors have read and agreed to the published version of the manuscript.

Funding: This work has been fully supported by the University of Rijeka under the project number uniri-technic-18-200 "Failure analysis of materials in marine environment".

Conflicts of Interest: The authors declare no conflict of interest.

References

1. Bai, Y.; Jin, W.-L. *Marine Structural Design*; Elsevier: Amsterdam, The Netherlands, 2016; ISBN 9780080999975.
2. Gašpar, G.; Poljak, I.; Orović, J. Computerized planned maintenance system software models. *Pomorstvo* **2018**, *32*, 141–145. [CrossRef]
3. Vizentin, G.; Vukelić, G.; Srok, M. Common failures of ship propulsion shafts. *Pomorstvo* **2017**, *31*, 85–90. [CrossRef]
4. Booker, N.K.; Clegg, R.E.; Knights, P.; Gates, J.D. The need for an internationally recognised standard for engineering failure analysis. *Eng. Fail. Anal.* **2020**, *110*, 104357. [CrossRef]
5. Brnic, J. *Analysis of Engineering Structures and Material Behavior*; Wiley: Hoboken, NJ, USA, 2018; ISBN 978-1-119-32907-7.
6. Bužančić Primorac, B.; Parunov, J. Review of statistical data on ship accidents. In *Maritime Technology and Engineering III: Proceedings of the 3rd International Conference on Maritime Technology and Engineering, Lisbon, Portugal, 4–6 July 2016*; CRC Press: Boca Raton, FL, USA; p. 1208.
7. Pastorcic, D.; Vukelic, G.; Bozic, Z. Coil spring failure and fatigue analysis. *Eng. Fail. Anal.* **2019**, *99*, 310–318. [CrossRef]
8. Lukács, J. Fatigue crack propagation limit curves for high strength steels based on two-stage relationship. *Eng. Fail. Anal.* **2019**, *103*, 431–442. [CrossRef]
9. Cui, W. A state-of-the-art review on fatigue life prediction methods for metal structures. *J. Mar. Sci. Technol.* **2002**, *7*, 43–56. [CrossRef]
10. Ivošević, Š.; Meštrović, R.; Kovač, N. Probabilistic estimates of corrosion rate of fuel tank structures of aging bulk carriers. *Int. J. Nav. Archit. Ocean Eng.* **2018**, *30*, 1e13. [CrossRef]

11. Baragetti, S.; Arcieri, E.V. Corrosion fatigue behavior of Ti-6Al-4 V: Chemical and mechanical driving forces. *Int. J. Fatigue* **2018**, *112*, 301–307. [CrossRef]

12. Oktavianus, Y.; Sofi, M.; Lumantarna, E.; Kusuma, G.; Duffield, C. Long-Term Performance of Trestle Bridges: Case Study of an Indonesian Marine Port Structure. *J. Mar. Sci. Eng.* **2020**, *8*, 358. [CrossRef]

13. Lecieux, Y.; Rozière, E.; Gaillard, V.; Lupi, C.; Leduc, D.; Priou, J.; Guyard, R.; Chevreuil, M.; Schoefs, F. Monitoring of a reinforced concrete wharf using structural health monitoring system and material testing. *J. Mar. Sci. Eng.* **2019**, *7*, 84. [CrossRef]

14. Vukelic, G.; Brnic, J. Numerical Prediction of Fracture Behavior for Austenitic and Martensitic Stainless Steels. *Int. J. Appl. Mech.* **2017**, *09*, 1750052. [CrossRef]

15. Vukelic, G.; Pastorcic, D.; Vizentin, G.; Bozic, Z. Failure investigation of a crane gear damage. *Eng. Fail. Anal.* **2020**, *115*, 104613. [CrossRef]

16. Peng, C.; Zhu, W.; Liu, Z.; Wei, X. Perforated mechanism of a water line outlet tee pipe for an oil well drilling rig. *Case Stud. Eng. Fail. Anal.* **2015**, *4*, 39–49. [CrossRef]

17. Harris, W.; Birkitt, K. Analysis of the failure of an offshore compressor crankshaft. *Case Stud. Eng. Fail. Anal.* **2016**, *7*, 50–55. [CrossRef]

18. Ilman, M.N. Kusmono Analysis of internal corrosion in subsea oil pipeline. *Case Stud. Eng. Fail. Anal.* **2014**, *2*, 1–8. [CrossRef]

19. Dowling, A.R.; Townley, C.H.A. The effect of defects on structural failure: A two-criteria approach. *Int. J. Press. Vessel. Pip.* **1975**, *3*, 77–107. [CrossRef]

20. Milne, I.; Ainsworth, R.; Dowling, A.; Stewart, A. Assessment of the integrity of structures containing defects. *Int. J. Press. Vessel. Pip.* **1988**, *32*, 3–104. [CrossRef]

21. Bloom, J.M. Prediction of ductile tearing using a proposed strain hardening failure assessment diagram. *Int. J. Fract.* **1980**, *16*, R73–R77. [CrossRef]

22. Bloom, J.M. Corrections: "Prediction of Ductile Tearing Using A Proposed Strain Hardening Failure Assessment Diagram," by J. M. Bloom. *Int. J. Fract.* **1980**, *16*, R163–R167. [CrossRef]

23. Anderson, T.L. *Fracture Mechanics: Fundamentals and Applications*, 3rd ed.; CRC Press: Boca Raton, FL, USA, 2005; ISBN 9780849316562.

24. Det Norske Veritas, Probabilistic Methods for Planning of Inspection for Fatigue Cracks in Offshore Structures—Recommended Practice DNVGL-RP-C210, 2015, 264. Available online: https://oilgas.standards.dnvgl.com/download/dnvgl-rp-c210-probabilistic-methods-for-planning-of-inspection-for-fatigue-cracks-in-offshore-structures (accessed on 10 August 2020).

25. Det Norske Veritas, Fatigue Design of Offshore Steel Structures—Recommended Practice DNVGL-RP-C203, 2011, 176. Available online: https://oilgas.standards.dnvgl.com/download/dnvgl-rp-c203-fatigue-design-of-offshore-steel-structures (accessed on 10 August 2020).

26. Gutiérrez-Solana, F.; Cicero, S. FITNET FFS procedure: A unified European procedure for structural integrity assessment. *Eng. Fail. Anal.* **2009**, *16*, 559–577. [CrossRef]

27. Cui, W.; Wang, F.; Huang, X. A unified fatigue life prediction method for marine structures. *Mar. Struct.* **2011**, *24*, 153–181. [CrossRef]

28. Yu, S.; Zhang, D.; Yue, Q. Failure analysis of topside facilities on oil/gas platforms in the Bohai Sea. *J. Mar. Sci. Eng.* **2019**, *7*, 86. [CrossRef]

29. Kachanov, L.M. *Introduction to Continuum Damage Mechanics*; Mechanics of Elastic Stability; Springer: Dordrecht, The Netherlands, 1986; Volume 10, ISBN 978-90-481-8296-1.

30. Moës, N.; Dolbow, J.; Belytschko, T. A finite element method for crack growth without remeshing. *Int. J. Numer. Methods Eng.* **1999**, *46*, 131–150. [CrossRef]

31. Sukumar, N.; Moës, N.; Moran, B.; Belytschko, T. Extended finite element method for three-dimensional crack modelling. *Int. J. Numer. Methods Eng.* **2000**, *48*, 1549–1570. [CrossRef]

32. Daux, C.; Moës, N.; Dolbow, J.; Sukumar, N.; Belytschko, T. Arbitrary branched and intersecting cracks with the extended finite element method. *Int. J. Numer. Methods Eng.* **2000**, *48*, 1741–1760. [CrossRef]

33. Marines-Garcia, I.; Paris, P.; Tada, H.; Bathias, C. Fatigue crack growth from small to long cracks in VHCF with surface initiations. *Int. J. Fatigue* **2007**, *29*, 2072–2078. [CrossRef]

34. Lebaillif, D.; Recho, N. Brittle and ductile crack propagation using automatic finite element crack box technique. *Eng. Fract. Mech.* **2007**, *74*, 1810–1824. [CrossRef]

35. Han, H.; Lee, K.; Park, S. Estimate of the fatigue life of the propulsion shaft from torsional vibration measurement and the linear damage summation law in ships. *Ocean Eng.* **2015**, *107*, 212–221. [CrossRef]

36. Murawski, L.; Charchalis, A. Simplified method of torsional vibration calculation of marine power transmission system. *Mar. Struct.* **2014**, *39*, 335–349. [CrossRef]

37. Murawski, L. Axial vibrations of a propulsion system taking into account the couplings and the boundary conditions. *J. Mar. Sci. Technol.* **2004**, *9*, 171–181. [CrossRef]

38. Murawski, L. Shaft Line Whirling Vibrations: Effects of Numerical Assumptions on Analysis Results. *Mar. Technol.* **2005**, *42*, 53–61.

39. Murawski, L. Shaft line alignment analysis taking ship construction flexibility and deformations into consideration. *Mar. Struct.* **2005**, *18*, 62–84. [CrossRef]

40. Song, M.-H.; Nam, T.-K.; Lee, J. Self-Excited Torsional Vibration in the Flexible Coupling of a Marine Propulsion Shafting System Employing Cardan Shafts. *J. Mar. Sci. Eng.* **2020**, *8*, 348. [CrossRef]

41. Han, H.-S. Analysis of fatigue failure on the keyway of the reduction gear input shaft connecting a diesel engine caused by torsional vibration. *Eng. Fail. Anal.* **2014**, *44*, 285–298. [CrossRef]

42. Han, H.S.; Lee, K.H.; Park, S.H. Parametric study to identify the cause of high torsional vibration of the propulsion shaft in the ship. *Eng. Fail. Anal.* **2016**, *59*, 334–346. [CrossRef]

43. Yusoff, N.H.N.; Isa, M.C.; Muhammad, M.M.; Nain, H.; Yati, M.S.D.; Bakar, S.R.S.; Noor, I.M. Failure analysis of a marine vessel shaft coupling. *Def. S T Tech. Bull.* **2012**, *5*, 166–175.

44. Murawski, L. Thermal displacement of crankshaft axis of slow-speed marine engine. *Brodogradnja* **2016**, *67*, 17–29. [CrossRef]

45. Murawski, L. Thermal interaction between main engine body and ship hull. *Ocean Eng.* **2018**, *147*, 107–120. [CrossRef]

46. Fonte, M.; de Freitas, M. Marine main engine crankshaft failure analysis: A case study. *Eng. Fail. Anal.* **2009**, *16*, 1940–1947. [CrossRef]

47. Borras, F.X.; de Rooij, M.B.; Schipper, D.J. Misalignment-induced micro-elastohydrodynamic lubrication in rotary lip seals. *Lubricants* **2020**, *8*, 19. [CrossRef]

48. Gomes, J.; Gaivota, N.; Martins, R.F.; Pires Silva, P. Failure analysis of crankshafts used in maritime V12 diesel engines. *Eng. Fail. Anal.* **2018**, *92*, 466–479. [CrossRef]

49. Sitthipong, S.; Towatana, P.; Sitticharoenchai, A. Failure analysis of metal alloy propeller shafts. *Mater. Today Proc.* **2017**, *4*, 6491–6494. [CrossRef]

50. Arisoy, C.F.; Başman, G.; Şeşen, M.K. Failure of a 17-4 PH stainless steel sailboat propeller shaft. *Eng. Fail. Anal.* **2003**, *10*, 711–717. [CrossRef]

51. Dymarski, C.; Narewski, M. Analysis of Ship Shaft Line Coupling Bolts Failure. *J. Pol. CIMAC* **2009**, *4*, 2.

52. Transport Accident Investigation Commission, New Zealand. Marine Inquiry MO 201 3 203 DEV Aratere, Fracture of STARBOARD Propeller Shaft Resulting in Loss of Starboard Propeller Cook Strait, 5 November 2013, Final Report. Available online: https://www.google.com/url?sa=t&rct=j&q=&esrc=s&source=web&cd=&ved=2ahUKEwi4__rmi7nrAhXIEcAKHaXFAzkQFjACegQIARAB&url=https%3A%2F%2Fwww.iims.org.uk%2Fwp-content%2Fuploads%2F2016%2F12%2FTAIC-Investigation-report-on-loss-of-propeller.pdf&usg=AOvVaw0Vs8mIMXl8SxTAokx3PHtO (accessed on 10 August 2020).

53. Fonte, M.; Reis, L.; Freitas, M. Failure analysis of a gear wheel of a marine azimuth thruster. *Eng. Fail. Anal.* **2011**, *18*, 1884–1888. [CrossRef]

54. Martins, R.F.; Moura Branco, C.; Gonçalves-Coelho, A.M.; Gomes, E.C. A failure analysis of exhaust systems for naval gas turbines. Part I: Fatigue life assessment. *Eng. Fail. Anal.* **2009**, *16*, 1314–1323. [CrossRef]

55. Yang, Q.; Li, G.; Mu, W.; Liu, G.; Sun, H. Identification of crack length and angle at the center weld seam of offshore platforms using a neural network approach. *J. Mar. Sci. Eng.* **2020**, *8*, 40. [CrossRef]

56. Schoefs, F.; Boéro, J.; Capra, B. Long-Term Stochastic Modeling of Sheet Pile Corrosion in Coastal Environment from On-Site Measurements. *J. Mar. Sci. Eng.* **2020**, *8*, 70. [CrossRef]

57. Tilander, J.; Patey, M.; Hirdaris, S. Springing Analysis of a Passenger Ship in Waves. *J. Mar. Sci. Eng.* **2020**, *8*, 492. [CrossRef]

58. Gordelier, T.; Thies, P.R.; Rinaldi, G.; Johanning, L. Investigating polymer fibre optics for condition monitoring of synthetic mooring lines. *J. Mar. Sci. Eng.* **2020**, *8*, 103. [CrossRef]

59. Monkova, K.; Monka, P.P.; Hric, S.; Kozak, D.; Katinič, M.; Pavlenko, I.; Liaposchenko, O. Condition monitoring of Kaplan turbine bearings using vibro-diagnostics. *Int. J. Mech. Eng. Robot. Res.* **2020**, *9*, 1182–1188. [CrossRef]
60. Wang, F.; Cui, W. Recent developments on the unified fatigue life prediction method based on fracture mechanics and its applications. *J. Mar. Sci. Eng.* **2020**, *8*, 427. [CrossRef]
61. Milovanović, N.; Sedmak, A.; Arsic, M.; Sedmak, S.A.; Božić, Ž. Structural integrity and life assessment of rotating equipment. *Eng. Fail. Anal.* **2020**, 104561. [CrossRef]

Article

Analysis and Optimization of Atmospheric Drain Tank of Lng Carrier Steam Power Plant

Igor Poljak [1,*], Toni Bielić [1], Vedran Mrzljak [2] and Josip Orović [1]

1   Department of Maritime Sciences, University of Zadar, Mihovila Pavlinovića 1, 23000 Zadar, Croatia; tbielic@unizd.hr (T.B.); jorovic@unizd.hr (J.O.)
2   Faculty of Engineering, University of Rijeka, Vukovarska 58, 51000 Rijeka, Croatia; vedran.mrzljak@riteh.hr
*   Correspondence: ipoljak1@unizd.hr; Tel.: +385-98-613-848

Received: 3 June 2020; Accepted: 25 July 2020; Published: 28 July 2020

Abstract: The atmospheric drain condensate system of a marine steam power plant is described and evaluated from the energetic and exergetic point of view at a conventional liquefied natural gas (LNG) carrier. Energy loss and exergy destruction rate were calculated for individual stream flows joined in an atmospheric drain tank with variations of the main turbine propulsion speed rate. The energy efficiency of joining streams was noted to be above 98% at all observed points as the atmospheric drain tank was the direct heater. The exergy efficiency of the stream flows into the drain tank was in the range of 80% to 90%. The exergy stream flow to the tank was modeled and optimized by the gradient reduced gradient (GRG) method. Optimization variables comprised contaminated and clean condensate temperature of the atmospheric drain tank and distillate water inlet to the atmospheric drain tank with respect to condensate outlet temperature. The optimal temperatures improves the exergy efficiency of the tank as direct heater, to about 5% in port and 3% to 4% when the LNG carrier was at sea, which is the aim of optimizing. Proposals for improvement and recommendations are given for proper plant supervision, which may be implemented in real applications.

Keywords: atmospheric drain tank; energy analysis; exergy analysis; optimization

---

1. Introduction

There have been a number of studies on stationary steam power plant feed water regenerative groups, their exergy and energy efficiency and possible feed water heater optimization. The importance of the feed water temperature at the entrance of the main boilers is related to fuel consumption, as the feed water temperature is lower, fuel consumption to the main boilers is higher and vice versa. The regenerative feed water cycle usually consists of seven or more regenerative heaters, which may be direct or indirect steam heaters. The selected papers were divided into three groups connected by the same problem.

The first group of authors studied the amount of exergy destruction for the regenerative feed water group, which is relatively low compared to the total exergy destruction of the steam power plant. Aljundi [1] carried out an energy and exergy analysis of the Al-Hussein power plant in Jordan, showing exergy destruction of individual components in the plant. According to the studies, exergy destruction of the feed water heating group, which consists of two low-pressure heaters, a deaerator and two high-pressure heaters, is 0.19% to 0.28% of total exergy destruction. Similarly, Sengupta et al. [2] analyzed a 210 MW thermal power plant and concluded that the contribution of exergy destruction to the regenerative feed water cycle of all feed water heaters and pumps was the lowest of all major components analyzed in the steam cycle. Comprehensive studies of Turkish power plants were made by Erdem et al. [3], in which nine thermal power plants were systematically analyzed with their exergy destruction and efficiency rates of the regenerative feed water groups. In that study, the low-pressure feed water heater group contributed 0.02% to 0.46% to the total exergy destruction of the system. The contribution of the high-pressure feed water group to total exergy destruction was in the range of 0.01% to 0.54%. Conventional analyses of the supercritical 200 MW Shahid Montazeri Power Plant in Iran with installed power capacity of 671 MW gave similar results to previous research according to Wang et al. [4] and Ahmadi and Toghraie [5].

The second group of selected papers is related to the optimization of feed water regeneration. The aim of the optimization is to decrease the fuel consumption of a stationary power plant. Ataei and Yoo [6] optimized ΔT of a thermal plant, combining feed water heaters for exergy and the pinch method by the cycle–tempo simulator, succeeding in decreasing fuel consumption by 5.3%. Modeling of a 312 MW thermal plant showed that increasing the feed water at the steam generator inlet reduced fuel consumption in the steam generators. Toledo et al. [7] conducted an exergy analysis of two 160 MW power plants with six and seven regenerative feed water heaters and the authors concluded that the seventh regenerative feed water heater contributed to decreased specific steam and fuel oil consumption by only 0.5%. Mehrabani et al. [8] optimized a thermal plant for electricity generation in Shahid Rajaei, India, by introducing a feed water heater and new power unit into the system. They used the genetic algorithm to find the optimal amount of turbine extraction steam. The efficiency of the plant with this approach increased by 5%, however a retrofit investment is required for practical realization of that idea. Espatolero et al. [9] optimized a 770 MWe power plant with the addition of one new LP heater, two drain pumps and an indirect flue gas heat recovery system with double-stage integration in the cycle, increasing plant efficiency by 0.7%.

The third group of selected papers is related to research combining the feed water regenerative cycle with solar field collectors and showed the following results. Adibhatla and Kaushik [10] tried to combine feed water regenerative groups to incorporate solar-aided feed water heating for a 500 MWe thermal power plant, but exergy efficiency of such a setup was lower than the classical Rankine regenerative cycle that is associated with exergy destruction in the collector–receiver system. Another study was carried out by Ahmadi et al. [11], integrating a solar field instead of the feed regeneration group [11]. It gained benefits by replacing high-pressure feed water preheaters with a solar farm, resulting in increased energy and exergy efficiencies of the power plant by 18.3%. Following a similar idea, Mohammadi et al. [12] incorporated a solar heating collector upgraded with a thermal storage system, as that system can be used at night, resulting in increased net generated power by 8.14%. The main problem with solar field heat generation is high capital cost, which is a problem with such concepts, but it saves fuel and reduces pollution [13]. The payback time, which varies according to the size and position of the plant, can be about 4.5–5.5 years, according to Bakos and Tsechelidou [14] or 5.13–6.21 years if thermal storage is included in the system [15].

A marine plan analysis carried out by Koroglu and Sogut [16] concluded that a feed water heater's efficiency could be improved externally only as a result of improvements to other components, such as turbine, boiler, condenser and pump equipment. As a marine steam propulsion plant is slightly different compared to a stationary plant, the return of condensate to the feed water system has not yet been evaluated in the scientific literature. Taking this into consideration, a case study of a specific condensate system of a 30 MW marine steam plant is explained and elaborated in this paper. The main difference between the marine condensate system and a stationary steam plant is the condensate cycle loop, which is divided into two groups and is joined together in the atmospheric drain tank. The temperature of the returned condensate in the atmospheric drain tank affects the temperature of the feed water before entering the deaerator as these two streams join before it. As the deaerator is a direct feed water heater, lower feed water temperature will require more steam consumption to heat the feed water to the saturated temperature, which results in higher fuel consumption of the marine power plant.

The paper is divided into two parts. The first part describes the calculation of energy and exergy efficiency of joining condensate water in the atmospheric drain tank as the direct heater, where stream flows from the condensate system are measured. The second part describes the optimization of the obtained exergy results with the adjusted stream flow temperatures in order to improve the efficiency of the joined streams in the direct heater, i.e., the atmospheric drain tank. As the efficiency of the atmospheric drain tank has an impact on the main feed water line temperature before the deaerator since it is mixed with the same. It is important to maintain it at the optimized level. The optimized temperature will save fuel consumption of the plant, which is the motivation for this work.

2. Feed Water and Steam Condensate System

A steam turbine vessel's main condensate system, as part of the closed feed water cycle, allots circulating feed water from the main condenser to the main boilers. Condensed water is taken from the main condenser and passed through the fresh water generator, gland steam condenser and first-stage heater and goes towards the deaerator, main feed water pump and third-stage feed water heater before entering the main boilers (Figure 1). In that feed water line, all heaters are indirect heaters and the deaerator is a direct heater.

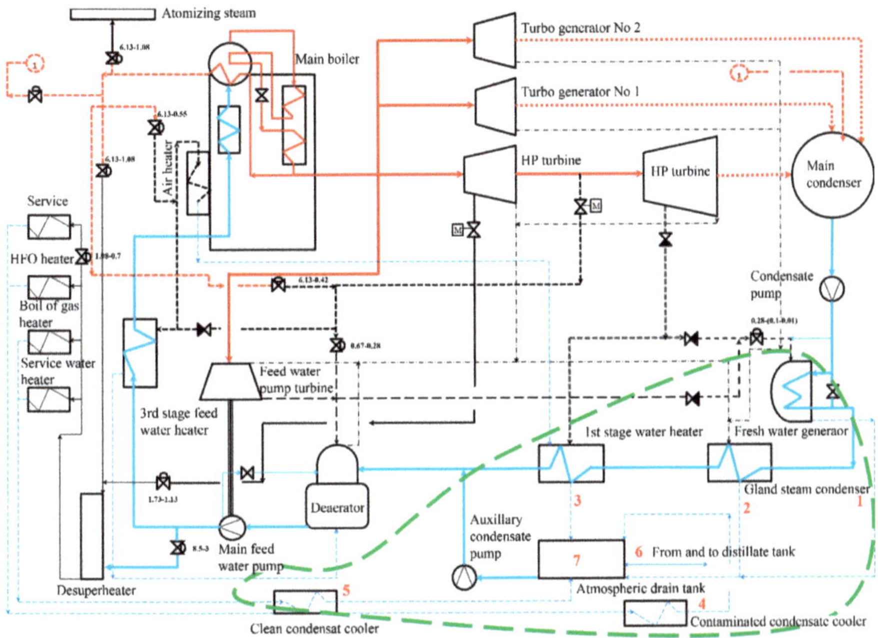

Figure 1. Marine steamship steam plant with highlighted condensate section.

During that process, feed water is taken from the main condenser, where the steam outlet from the main turbine and turbo generators condenses at saturated steam pressure. The temperature of the condensate water depends on the vacuum and seawater cooling temperature and varies from 30 to 40 °C. Condensed water is taken from the main condenser well and passes a group of heaters, whereupon it comes to the main boilers preheated to about 140 °C in the liquid state due to the high pressure of the main boiler water drum, which is maintained at about 6.3 MPa. Section heaters of the mentioned allotment are divided into extraction steam heaters or regenerative heaters and non-extraction or system heaters. Regenerative steam heaters get steam from the main propulsion turbine, which includes first-stage feed water heater, deaerator and third-stage feed water heater, where the fresh water generator, when it is heated from the main turbine extraction, acts as a regenerative feed water heater; otherwise it is a system feed water heater that consumes steam from the system. The condensate section is drawn with a blue line in Figure 1.

Condensate from the service system (5 in Figure 1), fresh water generator (1), gland condenser cooler (2) and first-stage heater (3) collects in an atmospheric drain tank, where it is mixed with distillate makeup water (6) taken from the distillate tanks. Distillate water makes up all water losses in the system, which, if there are no steam leaks, mainly must be refilled to the system due to the atomizing steam and soot blow losses inside the main boilers. Condensate from the service line is drawn with a dashed blue line in Figure 1. In the marine steam turbine plant, the service group is connected to the main propulsion plant system via the atmospheric drain tank. Service steam is used for the various heavy fuel oil (HFO) heaters, boil off gas (BOG) heaters and accommodation service. BOG heaters are used for heating and vaporizing liquefied natural gas (LNG), which is taken from the cargo tank when there is not enough methane vapor from the tank. The amount of steam for the BOG heaters is controlled by the cargo and boiler management system which controls the cargo tank pressure, as described in [17].

The atmospheric drain tank is the collecting node for both contaminated condensate (4) and clean condensate (5) (Figure 2). These two condensate streams arise from the auxiliary steam system, while contaminated condensate is part of various HFO and lube oil heaters. In order to prevent contamination of the system, condensate from these heaters first passes through analyzing and treating units, which set off an alarm of contaminant is detected in the water, such as fuel oil or lube oil, that may destroy the main boiler tubes by depositing into it, causing local overheating [18].

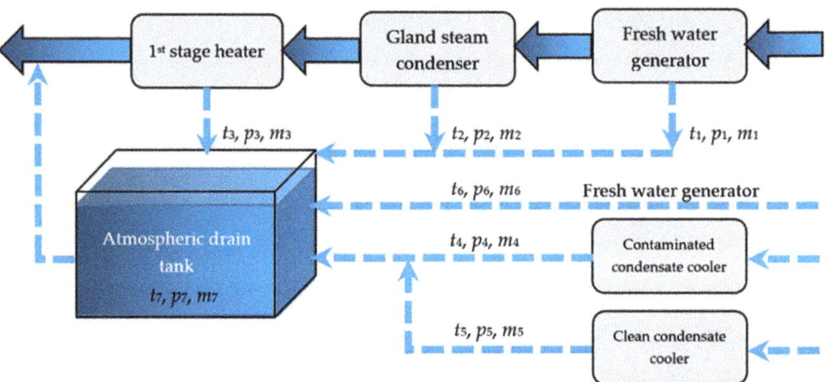

Figure 2. Feed water allotment with atmospheric drain tank.

The clean condensate inlet from the system, which is not in contact with hydrocarbon substances, also enters the atmospheric drain tank, but directly without monitoring and is mixed with the monitored contaminated condensate. Both streams, along with the distillate water stream in the atmospheric drain tank, are again returned back to the system.

As the steam propulsion plant system is dynamic, an additional role of the atmospheric drain tank is to amortize excess and make up the feed water in the system with the spilled feed water back to the distillate tanks or extract it from the distillate tanks again to the system with a change of the plant load. Basic LNG carrier power propulsion plant characteristics for maximal power, vacuum and flow are given in Table 1 [19,20].

Table 1. Steam propulsion plant main characteristics.

Equipment	Size
Main turbine power	29,420 kW
Main condenser vacuum	38 mm Hg, at 27 °C of seawater
Turbo generator	2 × 3850 kW
Feed pump	570 kW
Main boiler, steam generation	2 × 70,000 kg/h

3. Data Collection

A main propulsion turbine run test was carried out with step-by-step increase of the main propulsion shaft revolutions in order to collect the required data. The distilled water inlet to the atmospheric drain tank is the sum of the atomizing steam consumption plus steam losses. Losses are normally calculated for marine steam plants according to the recommendation in [21]. For the purpose of this study, feed water consumption in the tank is listed according to measured consumption from the flow meter described in [22]. Pressure, temperature and main propulsion shaft revolutions taken with standard engines measuring equipment are given in Table 2.

Table 2. Standard marine data collecting equipment.

Component	Measuring Equipment
1. Desuperheating outlet steam pressure	1. Pressure transmitter Yamatake STG940 [23]
2. Steam mass flow	2. Differential pressure transmitter Yamatake JTD960A [24]
3. Desuperheating steam outlet temperature	3. Greisinger GTF 601-Pt100-immersion probe [25]
4. Main propulsion turbine shaft power and rpm	4. Kyma shaft power meter, Model KPM-PFS [26]
5. First-stage feed heater temperature	5. SIKA thermometers for industry and marine sector [27]
6. Gland seal condenser temperature	6. SIKA thermometers for industry and marine sector [27]
7. Gland seal condenser pressure	7. Differential pressure transmitter Yamatake JTD960A [24]
8. Distillate water temperature gauge	8. SIKA thermometers for industry and marine sector [27]
9. Fresh water generator temperature gauge	9. SIKA thermometers for industry and marine sector [27]
10. Contaminated and clean condensate temperature gauge	10. SIKA thermometers for industry and marine sector [27]
11. Atmospheric drain tank temperature	11. SIKA thermometers for industry and marine sector [27]
12. 1st stage feed water pressure gauge	12. SIKA pressure gauges, type MRE-M and MRE-g [28]
13. Fresh water generator distillate flow meter	13. Zenner international GmbH [29]
14. Fresh water generator pressure gauge	14. Type 1259 Process Pressure Gauge—Ashcroft [30]

Measurement results for all fluid streams are presented in Tables A1–A3. All operating parameters were measured by varying the propeller revolutions. Propeller revolutions are increased by a main propulsion turbine which is coupled to the main propeller shaft by reduction gear. As the main turbine is increasing main propeller shaft speed, it is consuming more steam. The consumption of the steam increases the turbine load which must be made up by steam generators and that corresponds to increased steam plant thermal power production. The atmospheric drain condenser was under slight overpressure below ~0.11 MPa.

4. Thermodynamic Analysis

For the presented model, the required enthalpies and entropies were calculated from measured pressures and temperatures for every stream flow by using NIST REFPROP software [31]. Mass, energy and exergy flow stream balances were calculated according to the following [32,33]:

In the steady-state process, the mass balance of control volume is:

$$\sum_{IN} \dot{m}_i = \sum_{OUT} \dot{m}_o. \tag{1}$$

The energy balance of the control volume system is written as:

$$\sum_{IN} \dot{E}_i + \dot{Q} = \sum_{OUT} \dot{E}_o + \dot{W}. \tag{2}$$

In general, energy efficiency is a ratio of useful and used energy rates in the process [34,35]:

$$\eta_I = \frac{\dot{E}_{OUT}}{\dot{E}_{IN}} = 1 - \frac{\dot{E}l}{\dot{E}_{IN}}. \tag{3}$$

Energy loss:

$$\dot{m}_1{\cdot}h_1 + \dot{m}_2{\cdot}h_2 + \dot{m}_3{\cdot}h_3 + \dot{m}_4{\cdot}h_4 + \dot{m}_5{\cdot}h_5 + \dot{m}_6{\cdot}h_6 = \dot{m}_7{\cdot}h_7 + \dot{E}l. \tag{4}$$

Energy efficiency:

$$\eta_I = \frac{\dot{m}_7{\cdot}h_7}{\dot{m}_1{\cdot}h_1 + \dot{m}_2{\cdot}h_2 + \dot{m}_3{\cdot}h_3 + \dot{m}_4{\cdot}h_4 + \dot{m}_5{\cdot}h_5 + \dot{m}_6{\cdot}h_6}. \tag{5}$$

The entropy balance of the control volume system is:

$$\sum_{IN} \dot{S} + \sum_{IN} \frac{\dot{Q}}{T} + \dot{S}_{gen} = \sum_{OUT} \dot{S} + \sum_{OUT} \frac{\dot{Q}}{T}. \tag{6}$$

The exergy balance of the control volume system is written as:

$$\sum_{IN} \dot{E}x_i + \sum_k \left(1 - \frac{T}{T_k}\right){\cdot}\dot{Q}_k = \sum_{OUT} \dot{E}x_o + W + \dot{E}x_d. \tag{7}$$

where exergy rate of the stream is:

$$\dot{E}x = \dot{m}{\cdot}ex. \tag{8}$$

The specific exergy from Equation (8) at standard ambient state of 0.1 MPa and 25 °C is taken as per the recommendations in [36–38]:

$$ex = (h - h_0) - T_0{\cdot}(s - s_0). \tag{9}$$

Exergy efficiency:

$$\eta_{II} = 1 - \frac{\dot{E}xd}{\dot{E}x_{IN}} = \frac{\dot{E}x_{OUT}}{\dot{E}x_{IN}}. \tag{10}$$

Exergy destruction:

$$\dot{m}_1{\cdot}ex_1 + \dot{m}_2{\cdot}ex_2 + \dot{m}_3{\cdot}ex_3 + \dot{m}_4{\cdot}ex_4 + \dot{m}_5{\cdot}ex_5 + \dot{m}_6{\cdot}ex_6 = \dot{m}_7{\cdot}ex_7 + \dot{E}xd. \tag{11}$$

Exergy efficiency:

$$\eta_{II} = \frac{\dot{m}_7{\cdot}ex_7}{\dot{m}_1{\cdot}ex_1 + \dot{m}_2{\cdot}ex_2 + \dot{m}_3{\cdot}ex_3 + \dot{m}_4{\cdot}ex_4 + \dot{m}_5{\cdot}ex_5 + \dot{m}_6{\cdot}ex_6}. \tag{12}$$

5. Energy and Exergy Analysis Results

Atmospheric drain tank energy flow streams show that at lower loads of the main propulsion turbine, total energy loss in the atmospheric drain tank is higher compared to higher load ranges, as seen in Figure 3. The energy loss in the maneuvering range is the result of an accumulating function of the drain tank, where load changes in the system are compensated by adding feed water to the tank. After passing the maneuvering range of the steam propulsion plant, 0.0 to 53.5 min^{-1} and reaching a ship speed of about 13 knots, the main sea water circulating pumps are stopped and cooling of the main condenser is taken over by the scoop system, which collects sea water according to the ship's speed.

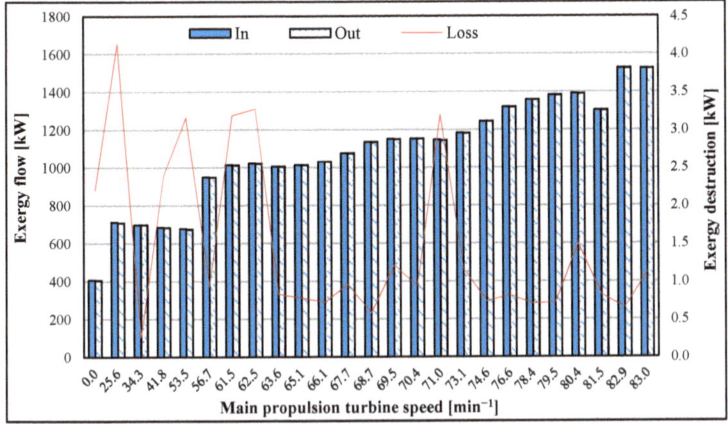

Figure 3. Energy flow of atmospheric drain tank with main turbine load variation.

The opening of the scoop system corresponds to about 61.5 min^{-1}, after which energy losses become lower and are distributed more equally through the upper range of the load range. Exergy flow losses of the atmospheric drain tank have an opposite trend to energy losses and increase even with the increased main propulsion turbine load (Figure 4). Exergy destruction amplitude is about 10 kW at the highest load, which is almost double compared to energy losses. The increment of exergy destruction with increased load is typical for disturbances in the system that may be connected to some equipment, under capacitance or similar construction design failure. The observed shortcoming may be improved by optimizing the respective component flow streams. The moment of decreasing exergy destruction trend at higher loads is at 1.5 min^{-1}, where extraction of steam from the high-pressure turbine begins. The extracted steam is used for ship services. This moment obviously acts positively on the exergy destruction of the atmospheric drain tank.

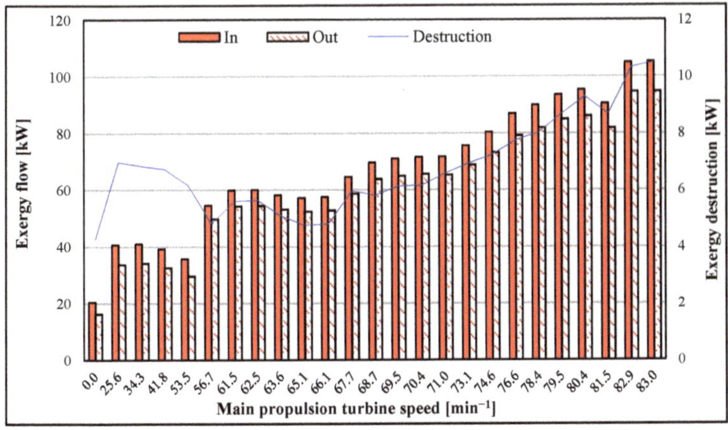

Figure 4. Exergy flow of atmospheric drain tank with main turbine load variation.

A comparison of energy and exergy efficiency is given in Figure 5. Energy efficiency of the joined streams in the atmospheric drain tank was very high, above 98% at all measured ranges. On the other side, results of exergy analysis indicate that exergy efficiency of the joined streams inside the atmospheric drain tank was somewhat worse when the main propulsion turbine was not running and throughout the maneuvering range. The exergy efficiency of the joined streams in the port was below 80% and after passing the maneuvers zone it stabilized to about 90%.

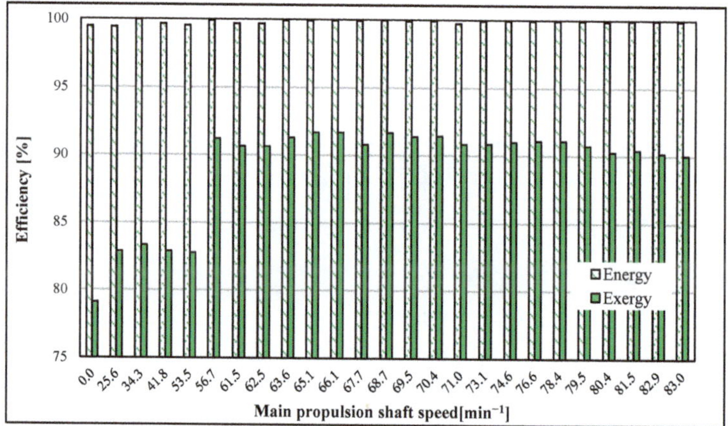

Figure 5. Exergy flow of atmospheric drain tank with main turbine load variation.

Figure 6 shows the condensate mass share from the atmospheric drain tank versus the amount of feed water passing the first-stage heater at their mixing point. When a ship is alongside for a cargo operation, part of the condensate coming from the drain tank to the common feed water line is over 30%, while when maneuvering the vessel and with further increased main propulsion turbine load, that ratio drops down to about 15%. Accordingly, the temperature at atmospheric condenser outlet has an influence on the feed water temperature after the mixing point of the two feed water lines. By decreasing the feed water temperature after the mixing point, deaerator losses are increased, as it will be required to lead more steam onto the deaerator in order to bring feed water to saturation temperature, which is required in order to release various dissolved gasses from the feed water [39].

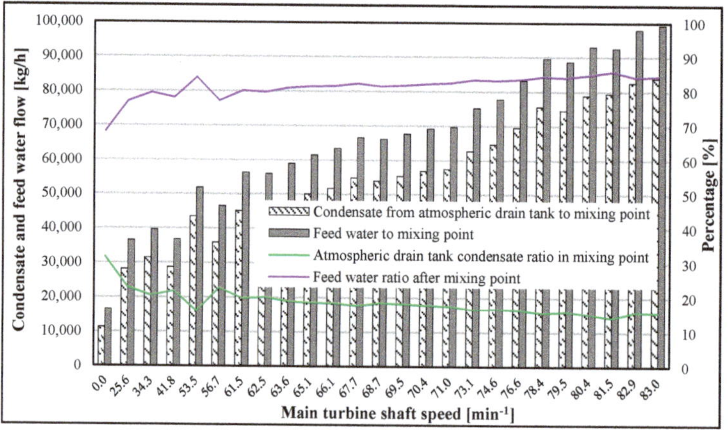

Figure 6. Atmospheric drain tank condensate ratio in mixing point with feed water with main turbine load variation.

Exergy efficiency variation with changing ambient temperature is given in Figure 7 and was assessed according to the recommendations in [40,41]. This measurement gives a good outlook on the effect of ambient temperature on exergy efficiency in various sailing destinations where the LNG carrier is operating. The selected range of exergy variation is surrounding temperature from 10 to 50 °C. The reference temperature is 25 °C and 0.1 MPa. The results of exergy efficiency variation show that it decreases with rising temperature, especially between 40 and 50 °C. Degradation of exergy efficiency is more conspicuous in port. In the upper loads of the steam plant, the difference in exergy efficiency is smaller. Such high discrepancy in exergy efficiency is mainly caused by the condensate temperature from the fresh water generator. The spray water for cooling the steam remains open even when the fresh water generator is not producing the water in the port and lower loads. A cold stream of water reduces the temperature in the atmospheric tank and decreases exergy efficiency.

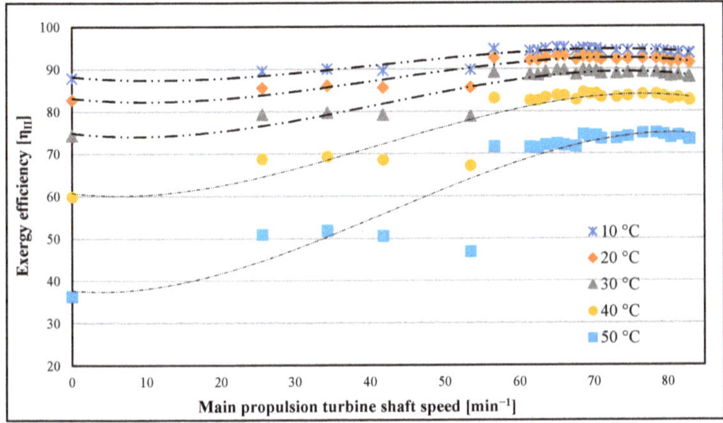

Figure 7. Exergy efficiency with surrounding temperature changes.

6. Mathematical Formulation of Atmospheric Drain Tank Optimization Problem

As an optimizing tool, a fourth-degree polynomial with one variable is used to calculate specific exergy according to data taken from [42] to achieve more accurate optimization results. The common polynomial of *k*th order according to [43,44] is:

$$P(x) = a_0 + a_1 \cdot x + \ldots + a_k \cdot x^k. \tag{13}$$

The sum of residue squares when approximated by value y_i of polynomial $P(x_i)$, $i = 1, \ldots, n$ is:

$$R^2 = \sum_{i=1}^{n} \left[y_i - \left(a_0 + a_1 \cdot x_i + \ldots + a_k \cdot x_i^k \right) \right]^2. \tag{14}$$

A partial differential of Equation (15) yields the following set of equations for extrema:

$$\frac{\partial(R^2)}{\partial a_0} = -2 \sum_{i=1}^{n} \left[y_i - \left(a_0 + a_1 \cdot x_i + \ldots + a_k \cdot x_i^k \right) \right] = 0, \tag{15}$$

$$\frac{\partial(R^2)}{\partial a_1} = -2 \sum_{i=1}^{n} \left[y_i - \left(a_0 + a_1 \cdot x_i + \ldots + a_k \cdot x_i^k \right) \right] \cdot x_i = 0, \tag{16}$$

$$\ldots,$$

$$\frac{\partial(R^2)}{\partial a_k} = -2\sum_{i=1}^{n}\left[y_i - \left(a_0 + a_1{\cdot}x_i + \ldots + a_k{\cdot}x_i^k\right)\right]{\cdot}x_i^k = 0. \tag{17}$$

The previous system of equations is equivalent to the following:

$$a_0{\cdot}n + a_1\sum_{i=1}^{n}x_i + \ldots + a_k\sum_{i=1}^{n}x_i^k = \sum_{i=1}^{n}y_i, \tag{18}$$

$$a_0\sum_{i=1}^{n}x_i + a_1\sum_{i=1}^{n}x_i^2 + \ldots + a_k\sum_{i=1}^{n}x_i^{k+1} = \sum_{i=1}^{n}x_i{\cdot}y_i, \tag{19}$$

$$a_0\sum_{i=1}^{n}x_i^k + a_1\sum_{i=1}^{n}x_i^{k+1} + \ldots + a_k\sum_{i=1}^{n}x_i^{2k} = \sum_{i=1}^{n}x_i^k{\cdot}y_i. \tag{20}$$

The same in matrix notation reads as:

$$\begin{bmatrix} n & \sum_{i=1}^{n}x_i & \cdots & \sum_{i=1}^{n}x_i^k \\ \sum_{i=1}^{n}x_i & \sum_{i=1}^{n}x_i^2 & \cdots & \sum_{i=1}^{n}x_i^{k+1} \\ \vdots & \vdots & \ddots & \vdots \\ \sum_{i=1}^{n}x_i^k & \sum_{i=1}^{n}x_i^{k+1} & \cdots & \sum_{i=1}^{n}x_i^{2k} \end{bmatrix}{\cdot}\begin{bmatrix} a_0 \\ a_1 \\ \vdots \\ a_k \end{bmatrix} = \begin{bmatrix} \sum_{i=1}^{n}y_i \\ \sum_{i=1}^{n}x_i{\cdot}y_i \\ \vdots \\ \sum_{i=1}^{n}x_i^k{\cdot}y_i \end{bmatrix}. \tag{21}$$

The matrix system in (21) is equivalent to the following system with a Vandermonde matrix [45]:

$$\begin{bmatrix} n & \sum_{i=1}^{n}x_i & \cdots & \sum_{i=1}^{n}x_i^k \\ \sum_{i=1}^{n}x_i & \sum_{i=1}^{n}x_i^2 & \cdots & \sum_{i=1}^{n}x_i^{k+1} \\ \vdots & \vdots & \ddots & \vdots \\ \sum_{i=1}^{n}x_i^k & \sum_{i=1}^{n}x_i^{k+1} & \cdots & \sum_{i=1}^{n}x_i^{2k} \end{bmatrix}{\cdot}\begin{bmatrix} a_0 \\ a_1 \\ \vdots \\ a_k \end{bmatrix} = \begin{bmatrix} \sum_{i=1}^{n}y_i \\ \sum_{i=1}^{n}x_i{\cdot}y_i \\ \vdots \\ \sum_{i=1}^{n}x_i^k{\cdot}y_i \end{bmatrix}. \tag{22}$$

Let the Vandermonde matrix from (22) be denoted by X and the vector with the coefficients a_i, $i = 0, \ldots, k$ be denoted by a. The solution to (21) and (22) can be found by multiplying system (23) by the inverse of X:

$$a = X^{-1}{\cdot}y. \tag{23}$$

The procedure described by (14)–(23) was used on data taken from [42], which yielded polynomials for specific exergy at various temperatures and pressures. The polynomials, as a function of temperature at given pressure $f(t, p)$, are listed below; a complete list of the used polynomials is given in the appendix.

Atmospheric tank specific exergy outlet and distillate water inlet to the tank:

$$30 < f(t) < 100\ ^{\circ}\text{C}, \tag{24}$$

$$p = 0.11\ \text{MPa},$$

$$exf(t,p) = 2.8714{\cdot}10^{-8}{\cdot}t^4 - 1.7549625{\cdot}10^{-5}{\cdot}t^3 + 8.1833784{\cdot}10^{-3}{\cdot}t^2 + 0.3776049{\cdot}t + 4.5966608,$$
$$R^2 = 0.999999999461$$

Contaminated condensate cooler specific exergy outlet and clean condensate cooler outlet:

$$30 < f(t) < 100\ ^{\circ}\text{C}, \tag{25}$$

$$p = 0.55 \text{ MPa,}$$

$$exf(t,p) = 2.8752{\cdot}10^{-8}{\cdot}t^4 - 1.7560712{\cdot}10^{-5}{\cdot}t^3 + 8.1832649{\cdot}10^{-3}{\cdot}t^2 + 0.3776048{\cdot}t + 5.0382725,$$
$$R^2 = 0.999999999544$$

$$30 < f(t) < 100\,^{\circ}\text{C,} \tag{26}$$

$$p = 0.65 \text{ MPa,}$$

$$exf(t,p) = 2.8767{\cdot}10^{-8}{\cdot}t^4 - 1.7512137{\cdot}10^{-5}{\cdot}t^3 + 8.1784348{\cdot}10^{-3}{\cdot}t^2 + 0.3774153{\cdot}t + 5.1359903,$$
$$R^2 = 0.999999999434$$

The optimization function is used to achieve maximum exergy efficiency of the joined exergy streams in the atmospheric drain tank by the calculated exergy fourth-degree polynomial functions:

$$\max\eta_{II}(t_4, t_5, t_6) = \frac{\dot{m}_7{\cdot}ex_7}{\dot{m}_1{\cdot}ex_1 + \dot{m}_2{\cdot}ex_2 + \dot{m}_3{\cdot}ex_3 + \dot{m}_4{\cdot}ex_4 + \dot{m}_5{\cdot}ex_5 + \dot{m}_6{\cdot}ex_6}. \tag{27}$$

The optimization variables are:

- Contaminated condensate cooler temperature outlet t_4;
- Clean condensate cooler temperature outlet t_5;
- Distillate temperature t_6.

Fixed conditions are:

- Exergy of stream inlet to atmospheric drain tank from fresh water generator ex_1;
- Contaminated condensate cooler temperature outlet t_4;
- Exergy of stream inlet to atmospheric drain tank from gland steam condenser ex_2;
- Exergy of stream inlet to atmospheric drain tank from first-stage feed water heater ex_3;
- Mass flow inlet to atmospheric drain tank from m_1 to m_6 are fixed;
- Pressure from p_1 to p_6 is fixed.

With following conditions:

- Conservation of mass flow:

$$\dot{m}_1 + \dot{m}_2 + \dot{m}_3 + \dot{m}_4 + \dot{m}_5 + \dot{m}_6 = \dot{m}_7. \tag{28}$$

t_7 is determined by partial temperature ratios of all mass flow participants:

$$\dot{m}_1{\cdot}t_1 + \dot{m}_2{\cdot}t_2 + \dot{m}_3{\cdot}t_3 + \dot{m}_4{\cdot}t_4 + \dot{m}_5{\cdot}t_5 + \dot{m}_6{\cdot}t_6 = \dot{m}_7{\cdot}t_7. \tag{29}$$

Under given constraints:

- Contaminated condensate cooler temperature outlet:

$$30 \leq t_4 \leq 140. \tag{30}$$

- Clean condensate cooler temperature outlet:

$$30 \leq t_5 \leq 140. \tag{31}$$

- Distillate water temperature from the tank:

$$20 \leq t_6 \leq 40. \tag{32}$$

- Energy efficiency of joining streams to atmospheric drain tank:

$$0 \leq \eta_I \leq 1 \text{ or } 0 \leq \frac{\dot{m}_7 \cdot h_7}{\dot{m}_1 \cdot h_1 + \dot{m}_2 \cdot h_2 + \dot{m}_3 \cdot h_3 + \dot{m}_4 \cdot h_4 + \dot{m}_5 \cdot h_5 + \dot{m}_6 \cdot h_6} \leq 1. \tag{33}$$

Optimization was performed with the gradient reduced gradient method (GRG) from Excel's solver analysis packet [46]. The options were adjusted as follows:

- Constraint precision: 0.000001;
- Convergence: 0.0001;
- Derivatives: forward;
- Bounds on the variables: require;

7. Optimization Results

The optimized exergy efficiency of the atmospheric drain tank joining streams is given in Figure 8. The aim of optimization is to achieve the maximum exergy efficiency of the joined exergy streams in the drain tank by calculated exergy fourth-degree polynomial functions. As per the results, better exergy efficiency of the atmospheric drain tank joining streams was achieved in all running ranges of the marine steam propulsion plant. At the maneuvering load of the main propulsion turbine, exergy efficiency increased by about 5%. From 56.7 to 83 min^{-1} at the main propulsion shaft, exergy efficiency increased by 3% to 4%.

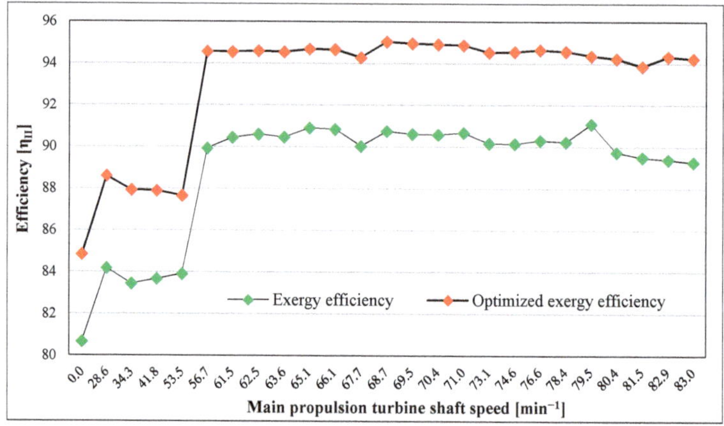

Figure 8. Optimized exergy efficiency of the atmospheric drain tank joining streams.

Optimization results indicate that maintaining distillate temperature at the atmospheric drain tank inlet as high as possible is required (Appendix A, Table A4). Maintaining a higher distillate temperature requires consuming distillate from the tank where it is stored from the fresh water generator, as its temperature from the fresh water generator is ~45 °C. This action is avoided in the operation of the marine steam plant due to safety, which means that if the salinity of the distillate water at the fresh water generator increases and the salinity sensor fails, such higher-salinity distillate will be mixed with the distillate in the tank and could cause damage to the main boiler pipes. According to the Unitor guide [47], for medium-pressure boilers, 3–6 MPa chloride content should be less than 30 ppm. The main boiler maker has even stiffer standards; according to Mitsubishi chloride content should be 20 ppm or less [48]. Normal chloride content when the fresh water generator is producing distillate is below 5 ppm. However, a permanent monitoring system is installed for the boiler water onboard the ship, and there is no harm if boiler water is consumed from the distilled tank, where fresh water is coming directly from the fresh water generator. The recommendation is that this method become the norm.

The steam pressure of the contaminated service system is 0.7 MPa and clean condensate system 1 MPa. The ideal temperature of the contaminated condensate steam is 126 °C and clean condensate system 143 °C after phase change. If the service system is overloaded temperatures are lower due to additional cooling of the steam in the heat exchangers. As the temperature of contaminated and clean condensate outlet flow streams is still higher it should not be cooled at 70 °C but should be maintained at a higher temperature of about 90 °C after cooler. Maintaining the higher temperature increase exergy efficiency of the atmospheric drain tank as a direct heater. The control of the condensate temperature after the cooler is simply achieved by throttling the cooling water inlet to the condensate coolers.

8. Conclusions

According to an exergy analysis in port and at lower loads, it is clear that desuperheating water at the fresh water generator line, which comes from the main condenser feed water line, should be kept closed as fresh water generator is not in service and there is no steam for cooling down the fresh water generator. That part of the feed water is led back to the atmospheric drain tank and cools down the condensate inside the tank.

Optimized parameters clearly show that distillate water should be filled in the atmospheric drain tank from the tank that is in use, as optimized temperature is kept all the way at the upper constraint of 40 °C.

Clean and contaminated condensate temperature follow each other under the proposed optimization setup without regard to condensate mass flow, and they should be kept at the condenser outlet at 90 °C, which can be done simply by throttling the cooling water amount to the condensers or fixing an automatic control valve for temperature control at the condenser outlet.

The benefit of such a procedure is that condensate water will enter into the main feed water line with higher exergy potential, which will increase the efficiency of the power plant and save on fuel consumption. Optimizing the atmospheric condenser drain tank is the first step in the process of optimizing the whole feed water section, which will require investigating the interactions of optimized components in real application conditions, which is planned for future research work.

Author Contributions: Conceptualization, I.P. and V.M.; methodology, I.P.; software, I.P.; validation, I.P.; formal analysis, I.P.; investigation, T.B.; resources, T.B.; data curation, I.P.; writing—original draft preparation, I.P.; writing—review and editing, T.B.; visualization, V.M.; supervision, J.O.; project administration, J.O.; funding acquisition, T.B. All authors have read and agreed to the published version of the manuscript.

Funding: This research was supported by the Croatian Science Foundation under project IP-2018–01–3739, CEEPUS network CIII-HR-0108, European Regional Development Fund under grant KK.01.1.1.01.0009 (DATACROSS), project CEKOM under grant KK.01.2.2.03.0004, University of Rijeka scientific grant uniri-tehnic-18-275-1447, University of Rijeka scientific grant uniri-tehnic-18-18-1146 and University of Rijeka scientific grant uniri-tehnic-18-14.

Conflicts of Interest: The authors declare no conflicts of interest.

Nomenclature

$\dot{E}$	energy flow, kW
$\dot{E}_L$	energy loss, kW
$\dot{S}$	entropy flow rate, kW/K
e_x	specific exergy, kJ/kg
$\dot{E}x$	exergy flow, kW
$\dot{E}x_d$	exergy destruction, kW
h	enthalpy, kJ/k
$\dot{m}$	mass flow rate, kg/s
p	pressure, Pa

$\dot{Q}$	heat flow rate, kW
t	temperature, °C
T	temperature, K
$\dot{W}$	power, kW

Subscript

i	inlet
k	boundary temperature
o	outlet
0	referent temperature

Greek Letter

| η_I | energy efficiency |
| η_{II} | exergy efficiency |

Abbreviations

BOG	boil-off gas
HC	hydrocarbon
HFO	heavy fuel oil

Appendix A

Table A1. Fresh water generator condensate, gland steam condenser and first-stage heater pressure condensate, temperature and mass data.

Main Turbine Propulsion Shaft Speed	Fresh Water Generator			Gland Steam Condenser			1st Stage Feed Water Heater		
n (min^{-1})	t (°C)	p (MPa)	$\dot{m}$ [kg/h]	t (°C)	p (MPa)	$\dot{m}$ (kg/h)	t (°C)	p (MPa)	$\dot{m}$ (kg/h)
0.0	36.8	0.75	720	98.83	0.0973	196	86.0	0.550	1578
25.6	34.3	0.75	720	98.83	0.0973	417	90.0	0.549	3351
34.3	33.3	0.75	720	98.83	0.0973	468	92.0	0.452	3291
41.8	32.5	0.75	720	98.83	0.0973	476	89.0	0.550	3391
53.5	33.3	0.75	720	98.83	0.0973	410	83.0	0.549	3522
56.7	78.7	0.2	2845	98.83	0.0973	410	88.0	0.549	3688
61.5	78.7	0.2	3099	98.83	0.0973	410	90.0	0.548	4083
62.5	78.7	0.2	3060	98.83	0.0973	410	90.0	0.551	4013
63.6	78.7	0.2	3026	98.83	0.0973	410	88.0	0.548	4142
65.1	78.7	0.2	3309	98.83	0.0973	410	85.0	0.547	4197
66.1	78.7	0.2	3342	98.83	0.0973	410	84.0	0.546	4296
67.7	78.7	0.2	3328	98.83	0.0973	410	92.0	0.546	4260
68.7	78.7	0.2	3440	98.83	0.0973	410	94.0	0.082	4699
69.5	78.7	0.2	3500	98.83	0.0973	410	95.0	0.085	4652
70.4	78.7	0.2	3550	98.83	0.0973	410	95.5	0.087	4692
71.0	78.7	0.2	3454	98.83	0.0973	410	96.0	0.088	4699
73.1	78.7	0.2	3570	98.83	0.0973	410	97.8	0.094	4893
74.6	78.7	0.2	3756	98.83	0.0973	410	98.7	0.097	5161
76.6	78.7	0.2	3726	98.83	0.0973	410	99.6	0.100	5712
78.4	78.7	0.2	3906	98.83	0.0973	410	99.8	0.101	5952
79.5	78.7	0.2	3857	98.83	0.0973	410	102.0	0.110	5984
80.4	78.7	0.2	3639	98.83	0.0973	410	103.0	0.114	6083
81.5	78.7	0.2	3813	98.83	0.0973	410	103.0	0.114	5887
82.9	78.7	0.2	3753	98.83	0.0973	410	104.7	0.120	6362
83.0	78.7	0.2	3847	98.83	0.0973	410	105.0	0.121	6336

Table A2. Contaminated condensate cooler, condensate cooler and distillate water, temperature and mass data.

Main Turbine Propulsion Shaft Speed	Contaminated Condensate Cooler Flow			Condensate Cooler Flow			Distillate Water		
n (min^{-1})	t (°C)	p (MPa)	$\dot{m}$ (kg/h)	t (°C)	p (MPa)	$\dot{m}$ (kg/h)	t (°C)	p (MPa)	$\dot{m}$ (kg/h)
0.0	70	0.55	840	70	0.55	1327	29	0.11	561
25.6	70	0.65	1610	70	0.65	1607	29	0.11	663
34.3	70	0.65	1540	70	0.65	1418	29	0.11	695
41.8	70	0.65	1610	70	0.65	1211	29	0.11	653
53.5	70	0.65	1610	70	0.65	1294	29	0.11	745
56.7	70	0.65	1610	70	0.65	1303	29	0.11	764
61.5	70	0.65	1680	70	0.65	1118	29	0.11	793
62.5	70	0.65	1610	70	0.65	1425	29	0.11	789
63.6	70	0.65	1680	70	0.65	1122	29	0.11	815
65.1	70	0.65	1680	70	0.65	1012	29	0.11	822
66.1	70	0.65	1680	70	0.65	1128	29	0.11	852
67.7	70	0.65	1680	70	0.65	1243	29	0.11	876
68.7	70	0.65	1680	70	0.65	1133	29	0.11	865
69.5	70	0.65	1680	70	0.65	1249	29	0.11	868
70.4	70	0.65	1680	70	0.65	1134	29	0.11	867
71.0	70	0.65	1680	70	0.65	1135	29	0.11	861
73.1	70	0.65	1680	70	0.65	1021	29	0.11	922
74.6	70	0.65	1680	70	0.65	1120	29	0.11	933
76.6	70	0.65	1680	70	0.65	1231	29	0.11	939
78.4	70	0.65	1680	70	0.65	1122	29	0.11	977
79.5	70	0.65	1680	70	0.65	1237	29	0.11	978
80.4	70	0.65	1680	70	0.65	1346	29	0.11	1000
81.5	70	0.65	1680	70	0.65	358	29	0.11	1002
82.9	70	0.65	1610	70	0.65	2350	29	0.11	1022
83.0	70	0.65	1610	70	0.65	2244	29	0.11	1032

Table A3. Atmospheric drain tank joining streams.

Main Turbine Propulsion Shaft Speed	Atmospheric Drain Tank Joined Streams		
n (min^{-1})	t (°C)	p (MPa)	$\dot{m}$ (kg/h)
0.0	67	0.11	5223
25.6	73	0.11	8349
34.3	74	0.11	8108
41.8	73	0.11	8043
53.5	70	0.11	8302
56.7	77	0.11	9050
61.5	78	0.11	10,290
62.5	78	0.11	10,448
63.6	78	0.11	10,369
65.1	76	0.11	10,325
66.1	76	0.11	10,566
67.7	78	0.11	10,665
68.7	80	0.11	11,051
69.5	80	0.11	11,113
70.4	81	0.11	11,035
71.0	81	0.11	11,035
73.0	82	0.11	11,171
74.6	82	0.11	11,564

Table A3. *Cont.*

Main Turbine Propulsion Shaft Speed	Atmospheric Drain Tank Joined Streams		
n (min^{-1})	t (°C)	p (MPa)	$\dot{m}$ (kg/h)
76.6	83	0.11	12,220
78.4	84	0.11	12,393
79.5	85	0.11	12,552
80.4	85	0.11	12,777
81.5	86	0.11	11,597
82.9	85	0.11	14,020
83.0	85	0.11	13,961

Table A4. Optimized temperature from contaminated condensate cooler outlet, clean condensate cooler outlet and distillate tank outlet to atmospheric drain tank.

Main Turbine Propulsion Shaft Speed	Contaminated Condensate Cooler Flow			Condensate Cooler Flow			Distillate Water		
n (min^{-1})	t (°C)	p (MPa)	$\dot{m}$ (kg/h)	t (°C)	p (MPa)	$\dot{m}$ (kg/h)	t (°C)	p (MPa)	$\dot{m}$ (kg/h)
0.0	84.64	0.55	840	84.64	0.55	1327	40	0.11	561
25.6	90.83	0.65	1610	90.83	0.65	1607	40	0.11	663
34.3	92.57	0.65	1540	92.57	0.65	1418	40	0.11	695
41.8	90.30	0.65	1610	90.30	0.65	1211	40	0.11	653
53.5	84.82	0.65	1610	84.82	0.65	1294	40	0.11	745
56.7	85.75	0.65	1610	85.75	0.65	1303	40	0.11	764
61.5	86.85	0.65	1680	86.85	0.65	1118	40	0.11	793
62.5	86.90	0.65	1610	86.90	0.65	1425	40	0.11	789
63.6	85.74	0.65	1680	85.74	0.65	1122	40	0.11	815
65.1	83.84	0.65	1680	83.84	0.65	1012	40	0.11	822
66.1	83.29	0.65	1680	83.29	0.65	1128	40	0.11	852
67.7	87.86	0.65	1680	87.86	0.65	1243	40	0.11	876
68.7	88.53	0.65	1680	88.53	0.65	1133	40	0.11	865
69.5	89.09	0.65	1680	89.09	0.65	1249	40	0.11	868
70.4	89.37	0.65	1680	89.37	0.65	1134	40	0.11	867
71.0	89.81	0.65	1680	89.81	0.65	1135	40	0.11	861
73.1	90.93	0.65	1680	90.93	0.65	1021	40	0.11	922
74.6	91.55	0.65	1680	91.55	0.65	1120	40	0.11	933
76.6	92.66	0.65	1680	92.66	0.65	1231	40	0.11	939
78.4	92.73	0.65	1680	92.73	0.65	1122	40	0.11	977
79.5	94.38	0.65	1680	94.38	0.65	1237	40	0.11	978
80.4	95.45	0.65	1680	95.45	0.65	1346	40	0.11	1000
81.5	94.91	0.65	1680	94.91	0.65	358	40	0.11	1002
82.9	96.91	0.65	1610	96.91	0.65	2350	40	0.11	1022
83.0	96.96	0.65	1610	96.96	0.65	2244	40	0.11	1032

Appendix B

$exf(30–100, 0.55) = 0.000000028752 \cdot t^4 - 0.000017560712 \cdot t^3 + 0.008183264951 \cdot t^2 - 0.377604877762 \cdot t + 5.038272554766$

$R^2 = 0.999999999544$

$exf(30–94.151, 0.082) = 0.000000029449 \cdot t^4 - 0.000017724546 \cdot t^3 + 0.008198483260 \cdot t^2 - 0.378158562476 \cdot t + 4.575836547896$

$R^2 = 0.999999999420$

$exf(30–95.444, 0.086) = 0.000000029237 \cdot t^4 - 0.000017674814 \cdot t^3 + 0.008194275370 \cdot t^2 - 0.378007050602 \cdot t + 4.577893162052$

$R^2 = 0.999999999452$

$exf(30\text{--}95.759, \ 0.087) = 0.000000029173 \cdot t^4 - 0.000017659707 \cdot t^3 + 0.008192987671 \cdot t^2 - 0.377960479895 \cdot t + 4.578292957558$

$R^2 = 0.999999999452$

$exf(30\text{--}96.071, \ 0.088) = 0.000000029118 \cdot t^4 - 0.000017646885 \cdot t^3 + 0.008191909080 \cdot t^2 - 0.377921808781 \cdot t + 4.578797824676$

$R^2 = 0.999999999455$

$exf(30\text{--}97.885, \ 0.094) = 0.000000028969 \cdot t^4 - 0.000017612085 \cdot t^3 + 0.008188960776 \cdot t^2 - 0.377816106864 \cdot t + 4.583459267439$

$R^2 = 0.999999999472$

$exf(30\text{--}98.757, \ 0.097) = 0.000000028718 \cdot t^4 - 0.000017550696 \cdot t^3 + 0.008183545317 \cdot t^2 - 0.377613475150 \cdot t + 4.583763019346$

$R^2 = 0.999999999433$

$exf(30\text{--}99.606, 0.1) = 0.000000028618 \cdot t^4 - 0.000017526578 \cdot t^3 + 0.008181456140 \cdot t^2 - 0.377536693237 \cdot t + 4.585760948106$

$R^2 = 0.999999999452$

$exf(30\text{--}99.884, \ 0.101) = 0.000000028648 \cdot t^4 - 0.000017533970 \cdot t^3 + 0.008182090496 \cdot t^2 - 0.377559797341 \cdot t + 4.587064973324$

$R^2 = 0.999999999442$

$exf(30\text{--}100, 0.11) = 0.000000028714 \cdot t^4 - 0.000017549625 \cdot t^3 + 0.008183378488 \cdot t^2 - 0.377604963571 \cdot t + 4.596660891228$

$R^2 = 0.999999999461$

$exf(30\text{--}100, 0.114) = 0.000000028688 \cdot t^4 - 0.000017543499 \cdot t^3 + 0.008182856108 \cdot t^2 - 0.377586423719 \cdot t + 4.600440241904$

$R^2 = 0.999999999449$

$exf(30\text{--}100, 0.12) = 0.000000028701 \cdot t^4 - 0.000017547513 \cdot t^3 + 0.008183249124 \cdot t^2 - 0.377603071989 \cdot t + 4.606707460725$

$R^2 = 0.999999999451$

$exf(30\text{--}100, 0.121) = 0.000000028681 \cdot t^4 - 0.000017542259 \cdot t^3 + 0.008182768307 \cdot t^2 - 0.377584755059 \cdot t + 4.607463940507$

$R^2 = 0.999999999461$

References

1. Aljundi, I.H. Energy and exergy analysis of a steam power plant in Jordan. *Appl. Therm. Eng.* **2009**, *29*, 324–328. [CrossRef]
2. Sengupta, S.; Datta, A.; Duttagupta, S. Exergy analysis of a coal-based 210MW thermal power plant. *Int. J. Energy Res.* **2007**, *31*, 14–28. [CrossRef]
3. Erdema, H.H.; Akkaya, A.V.; Cetin, B.; Dagdas, A.; Sevilgen, S.H.; Sahin, B.; Teke, I.; Gungor, C.; Atas, S. Comparative energetic and exergetic performance analyses for coal-fired thermal power plants in Turkey. *Int. J. Therm. Sci.* **2009**, *48*, 2179–2186. [CrossRef]
4. Wang, L.; Yang, Y.; Morosuk, T.; Tsatsaronis, G. Advanced thermodynamic analysis and evaluation of a supercritical power plant. *Energies* **2012**, *5*, 1850–1863. [CrossRef]
5. Ahmadi, G.R.; Toghraie, D. Energy and exergy analysis of Montazeri Steam Power Plant in Iran. *Renew. Sustain. Energy Rev.* **2016**, *56*, 454–463. [CrossRef]
6. Ataei, A.; Yoo, C.K. Combined pinch and exergy analysis for energy efficiency optimization in a steam power plant. *Int. J. Phys. Sci.* **2010**, *5*, 1110–1123.
7. Toledo, M.; Abugaber, J.; Lugo, R.; Salazar, M.; Rodríguez, A.; Rueda, A. Energetic analysis of two thermal power plants with six and seven heaters. *Open J. Appl. Sci.* **2014**, *4*, 6–12. [CrossRef]
8. Mehrabani, K.M.; Yazdi, S.S.F.; Mehrpanahi, A.; Abad, S.N.N. Optimization of exergy in repowering steam power plant by feed water heating using genetic algorithm. *Indian J. Sci. Res.* **2014**, *1*, 183–198.
9. Espatolero, S.; Romeo, L.M.; Cortes, C. Efficiency improvement strategies for the feed water heaters network designing in supercritical coal-fired power plants. *Appl. Therm. Eng.* **2014**, *73*, 449–460. [CrossRef]

10. Adibhatlaa, S.; Kaushik, S.C. Exergy and thermoeconomic analyses of 500 MWe sub critical thermal power plant with solar aided feed water heating. *Appl. Therm. Eng.* **2017**, *123*, 340–352. [CrossRef]

11. Ahmadia, G.; Toghraieb, D.; Akbari, O.A. Solar parallel feed water heating repowering of a steam power plant: A case study in Iran. *Renew. Sustain. Energy Rev.* **2017**, *77*, 474–485. [CrossRef]

12. Mohammadi, A.; Ahmadi, M.H.; Bidi, M.; Ghazvini, M.; Ming, T. Exergy and economic analyses of replacing feedwater heaters in a Rankine cycle with parabolic trough collectors. *Energy Rep.* **2018**, *4*, 243–251. [CrossRef]

13. Jamel, M.S.; Rahman, A.A.; Shamsuddin, A.H. Advances in the integration of solar thermal energy with conventional and non-conventional power plants. *Renew. Sustain. Energy Rev.* **2013**, *20*, 71–81. [CrossRef]

14. Bakos, G.C.; Tschelidou, C. Solar aided power generation of a 300 MW lignite fired power plant combined with line-focus parabolic trough collectors field. *Renew. Energy* **2013**, *60*, 540–547. [CrossRef]

15. Adibhatla, S.; Kaushik, S.C. Energy, exergy, economic and environmental (4E) analyses of a conceptual solar aided coal fired 500 MWe thermal power plant with thermal energy storage option. *Sustain. Energy Technol.* **2017**, *21*, 89–99. [CrossRef]

16. Koroglu, T.; Sogut, O.S. Conventional and advanced exergy analyses of a marine steam power plant. *Energy J.* **2018**, *163*, 392–403. [CrossRef]

17. Poljak, I.; Glavan, I.; Orović, J.; Mrzljak, V. Three approaches to low-duty turbo compressor efficiency exploitation evaluation. *Appl. Sci.* **2020**, *10*, 3373. [CrossRef]

18. Problems Caused by Oil in Boiler Feed Water System. Available online: http://www.gard.no/web/updates/content/51933/problems-caused-by-oil-in-boiler-feed-water-system (accessed on 30 May 2020).

19. Available online: http://www.duivendijk.net/gas/grace_b1.htm (accessed on 1 April 2020).

20. Poljak, I.; Grace, B. *Machinery Instruction Manual*; Hyundai Heavy Industries Ltd.: Caloocan, Philippines, 2007.

21. Holm, J.T. *Technical and Research Bulletin 3-11, Marine Steam Power Plant Heat Balance Practices*; The Society of Naval Architects and Marine Engineers: Alexandria, VA, USA, 1971; p. 11.

22. ELSTER Messtechnik GmbH, M 100/M 110 Multi-Jet Water Meter Traditional Meter Bodies, for Horizontal Installation, for Risers. Available online: https://www.elster.com/assets/products/products_elster_files/M100_M110 (accessed on 1 April 2020).

23. Available online: http://www.ic72.com/pdf_file/-/87315.pdf (accessed on 30 May 2020).

24. Available online: http://www.krtproduct.com/krt_Picture/sample/1_spare%20part/yamatake/Fi_ss01/SS2-DST100-0100.pdf (accessed on 30 May 2020).

25. Available online: https://www.priggen.com/GTF-601-Pt100-Sheath-Element-Immersion-Probe-for-Liquids-and-Gases-200-to-600C_1 (accessed on 30 May 2020).

26. Available online: http://hwt0346.51software.net/uploadfiles/2011112919581355.pdf (accessed on 30 May 2020).

27. Available online: https://www.sika.net/en/products/sensors-and-measuring-instruments/industrial-thermometers/sika-thermometers-industry-and-marine.html (accessed on 30 May 2020).

28. Available online: https://www.sika.net/en/measuring-by-categories/pressure/mechanical-pressure-gauges/bourdon-tube-pressure-gauges.html (accessed on 30 May 2020).

29. Zenner International GmbH & Co. KG. Available online: https://partners.sigfox.com/products/edc-sigfox-868-dv (accessed on 30 May 2020).

30. Available online: https://ashcroft.com/products/pressure_gauges/process_gauges/1259-duragauge.cfm (accessed on 30 May 2020).

31. Lemmon, E.W.; Huber, M.L.; Mc Linden, M.O. *NIST Reference Fluid Thermodynamic and Transport Properties-REFPROP, Version 8.0*; User's Guide: Boulder, CO, USA, 2007.

32. Moran, M.J.; Shapiro, H.N.; Boettner, D.D.; Bailey, M.B. *Fundamentals of Engineering Thermodynamics*, 7th ed.; John Wiley & Sons: Hoboken, NJ, USA, 2011; pp. 210, 249, 334, 403.

33. Cengel, Y.A.; Boles, M.A. *Thermodynamics an Engineering Approach*, 8th ed.; McGraw-Hill Education: New York, NY, USA, 2015; pp. 251, 311, 395, 396, 467–468.

34. Kaushik, S.C.; Reddy, V.S.; Tyagi, S.K. Energy and exergy analyses of thermal power plants: A review. *Renew. Sustain. Energy Rev.* **2011**, *15*, 1857–1872. [CrossRef]

35. Adibhatla, S.; Kaushik, S.C. Energy and exergy analysis of a super critical thermal power plant at various load conditions under constant and pure sliding pressure operation. *Appl. Therm. Eng.* **2014**, *73*, 51–65. [CrossRef]

36. Mrzljak, V.; Poljak, I.; Medica-Viola, V. Dual fuel consumption and efficiency of marine steam generators for the propulsion of LNG carrier. *Appl. Therm. Eng.* **2017**, *119*, 331–346. [CrossRef]

37. Mrzljak, V.; Poljak, I.; Mrakovčić, T. Energy and exergy analysis of the turbo-generators and steam turbine for the main feed water pump drive on LNG carrier. *Energy Convers. Manag.* **2017**, *140*, 307–323. [CrossRef]

38. Poljak, I.; Orović, J.; Mrzljak, V.; Bernečić, D. Energy and exergy evaluation of a two-stage axial vapour compressor on the LNG carrier. *Entropy* **2020**, *22*, 115. [CrossRef]

39. Tolgyessy, J. (Ed.) Part of volume chemistry and biology water. In *Studies in Environmental Science*; Elsevier Science Publishers: Amsterdam, The Netherlands, 1993; pp. 31–32. [CrossRef]

40. Haseli, Y.; Dincer, I.; Naterer, G.F. Optimum temperatures in a shell and tube condenser with respect to exergy. *Int. J. Heat Mass Transf.* **2008**, *51*, 2462–2470. [CrossRef]

41. Bilgili, M.; Ozbek, A.; Yasar, A.; Simsek, E.; Sahin, B. Effect of atmospheric temperature on exergy efficiency and destruction of a typical residential split air conditioning system. *Int. J. Exergy* **2016**, *20*, 66–84. [CrossRef]

42. Wagner, W.I.; Pruss, A. The IAPWS Formulation 1995 for the thermodynamic properties of ordinary water substance for general and scientific use. *J. Phys. Chem. Ref. Data* **2002**, *31*, 387–535. [CrossRef]

43. Weisstein, E.W. "Least Squares Fitting—Polynomial." From MathWorld—A Wolfram Web Resource. Available online: https://mathworld.wolfram.com/LeastSquaresFittingPolynomial.html (accessed on 1 April 2020).

44. Kenney, J.F.; Keeping, E.S. Linear of statistics and correlation. In *Mathematics of Statistics*, 3rd ed.; Princeton: Princeton, NJ, USA, 1962; Available online: https://mathworld.wolfram.com/LeastSquaresFitting.html (accessed on 1 April 2020).

45. Aldrovandi, R. *Special Matrices of Mathematical Physics: Stochastic, Circulant and Bell Matrices*; World Scientific: Singapore, 2001; p. 193.

46. Fylstra, D.; Lasdon, L.; Watson, J.; Waren, A. Design and use of the microsoft excel solver. *INFORMS J. Appl. Anal.* **1998**, *28*, 29–55. [CrossRef]

47. *Water Treatment Handbook a Practical Application Manual*, 1st ed.; UNITOR: Oslo, Norway, 1996; p. 45. Available online: http://slidepdf.com/reader/full/wilhemsen-water-treatment-handbook (accessed on 1 April 2020).

48. Nakano, K. *Boiler Instruction and Maintenance Manual I*; Mitsubishi Heavy Industries LTD: Tokyo, Japan, 2005; Internal vessel document; p. 28.

Journal of
Marine Science and Engineering

Article

Performance Analysis of Combined Cycle with Air Breathing Derivative Gas Turbine, Heat Recovery Steam Generator, and Steam Turbine as LNG Tanker Main Engine Propulsion System

Wahyu Nirbito *, Muhammad Arif Budiyanto and Robby Muliadi

Department of Mechanical Engineering, Universitas Indonesia, Kampus Baru UI, Jawa Barat 16424, Indonesia;
arif@eng.ui.ac.id (M.A.B.); bitomesin76@gmail.com (R.M.)
* Correspondence: bito@eng.ui.ac.id

Received: 1 September 2020; Accepted: 14 September 2020; Published: 20 September 2020

Abstract: This study explains the performance analysis of a propulsion system engine of an LNG tanker using a combined cycle whose components are gas turbine, steam turbine, and heat recovery steam generator. The researches are to determine the total resistance of an LNG tanker with a capacity of 125,000 m^3 by using the Maxsurf Resistance 20 software, as well as to design the propulsion system to meet the required power from the resistance by using the Cycle-Tempo 5.0 software. The simulation results indicate a maximum power of the system of about 28,122.23 kW with a fuel consumption of about 1.173 kg/s and a system efficiency of about 48.49% in fully loaded conditions. The ship speed can reach up to 20.67 knots.

Keywords: LNG tanker; combined cycle; propulsion main engine

1. Introduction

Transportation of natural gas between islands can be done in various ways, such as through transmission pipes or by using sea transportation modes [1]. The transportation of natural gas using pipes has several limitations; namely, limited mobility requires a large investment; handling the compressor system is quite complicated, i.e. the further the supplied distance, the bigger compressor must be used; and the environmental safety management is quite difficult considering that the pressure in the pipeline is very high so that a little leak can be fatal to the environment [2–4]. Therefore, for cross-sea transportation with long distances, ships are chosen as the mode of transportation [5]. In its development, natural gas transportation using ships is divided into two broad lines, namely, transporting natural gas in the gas phase/compressed natural gas (CNG) and in the liquid phase/liquefied natural gas (LNG). The disadvantage of transporting in the gas phase is the need for pressure vessels that are able to withstand high pressures and large volumes, so in general, the transportation of natural gas through ships is done by the liquid/LNG phase, namely, by maintaining a charged temperature that causes the natural gas to be in the liquid phase [6–10]. LNG tankers are an option for transporting large amounts of natural gas for long distances [11]. At present, LNG tankers of various types and sizes are widely available in the world. Based on data from the International Gas Union (IGN), there were 373 active ships with capacities above 30,000 m^3 in 2015, and as many as 28 ships are under construction [12]. The value of charter vessels dropping to $40,000/day in the third quarter of 2014 due to the decreased number of cross-Pacific–Atlantic shipments and the construction of large vessels in 2015 caused the LNG freight market share to decline [13]. Old ships that still use the inefficient steam turbine propulsion system must be able to compete with ships that use new propulsion systems that are far more efficient [14]. This condition encourages owners to build ships with more efficient propulsion systems. One of the factors that influence the level of efficiency of gas transportation using LNG tankers is the propulsion system used. Currently, LNG tankers in the world generally still use a steam turbine with boilers that use fuel from boil-off gas [15–18]. Several alternative propulsion systems have been developed by engineers to increase the efficiency value of the main engine driving LNG tankers, such as the dual-fuel diesel engine (DFDE) [19] and the combined cycle [20–22]. The majority of active LNG ships in the world currently use a steam turbine propulsion system with a low level of efficiency; therefore, today engineers continue to develop an efficient propulsion system at the range of the power required by LNG ships. Propulsion systems with diesel engines and ones with combined cycles have different dimensions due to the different equipment components are used to support the performance of the engines [16,23]. Large and heavy engines cause the ship to lose volume and weight that could otherwise be used to transport cargo. This can be circumvented by designing machines that also produce the right power but are smaller and lighter with an efficient arrangement of the engine room [23,24]. Combined cycle is an alternative propulsion system that can be applied to LNG tankers engine of power between 20 to 50 MW, by considering its overall efficiency. It can be seen that the combined cycle systems have higher total efficiency than other propulsion system engines. Currently, large-capacity LNG tankers require at least 25 MW for propulsion and auxiliary systems on board [25–30]. Therefore, further studies are needed for a marine combined-cycle gas steam turbine power plant. This paper has two main objectives, namely, designing the propulsion system in an LNG tanker with a combined-cycle propulsion system and calculating the performance of the combined-cycle system with an air-breathing derivative gas turbine, heat recovery steam generator (HRSG), and steam turbine as the LNG tanker's main engine propulsion system. Based on these objectives, the formulated problem is to design combined gas–electric steam (COGES) propulsion systems on LNG tankers with engine power requirements related to ship resistance at a certain speed, and to determine the tools needed to support the performance of the propulsion system.

2. Methodology

2.1. Research Stages

The following is a design methodology that was carried out in this design:
1. Field Study
Conduct a search and study of LNG tankers that would be implemented using the combined-cycle gas steam turbine propulsion system for the power requirements of the propulsion system and the ship's auxiliary systems.
2. Literature Study
Learn the basics of propulsion systems using a combined-cycle gas steam turbine and the tools needed to support the operation of the propulsion system.
3. Problem Statement
Identify the propulsion power requirements of the ship to be designed for the propulsion system. Determine the tools needed to support the performance of the propulsion system.
4. Design and Analysis
Design and calculate the design of the propulsion system with the parameters available from the ship to be applied to the propulsion system, to get the power in accordance with the needs of the ship. In this design study, the Maxsurf Resistance 20 software was used to obtain the resistance value using this method for each desired value of the ship's speed. This software can estimate resistance and power requirements for ships designed using industry-standard prediction techniques. The data input required from this software is the shape of the hull of the ship. In this case, the main dimensions of the hull are shown in Table 1. Apart from determining the ship's power requirements, the analysis of the thermodynamic calculations of the propulsion system was designed using the Cycle-Tempo 5.1 software. In simulations using the software, several parameters are needed to simulate a steam turbine,

such as pressure, temperature, and mass flow rate of the steam, for the calculation of the isentropic efficiency of the steam turbine component is used based on relevant references.

Table 1. LNG tankers with a capacity of 125,000 m^3.

Ship Name	LNG Aquarius
LOA	285.3 m
LBP	273.4 m
Beam	43.74 m
Depth to main deck	24.99 m
Full-load draft	10.97 m
Scantling draft	11.53 m
Engine type	Steam turbine
Number of propellers	1
Trial speed	20.4 knots
Service speed	19.5 knots
Cargo tank capacity ($-160\,^\circ$C)	126,400 m^3
Tank design	Spherical aluminum
Crew number	31

2.2. Data Used

In this design study, some reference data needed to get the appropriate design of the system were as follows:

1. LNG Tanker

Table 1 shows the specification of the LNG tanker with a capacity of 125,000 m^3 [31].

2. Ship Load Data

Data of the shipload were obtained from the operation report of the comparison ship, LNG Aquarius, as shown in Table 2 [32].

Table 2. LNG cargo handling averages.

Cargo Aboard	
After loading	125,400 m^3
Before discharge	123,400 m^3
Heel aboard	
After loading	1900 m^3
Before discharge	600 m^3
Cargo loaded	124,800 m^3
Cargo discharged	121,500 m^3
Boil-off	
Loaded leg	2,000 m^3
Ballast leg	1,300 m^3

3. Shipping Conditions

The LNG tanker propulsion system in this study is designed for LNG shipping between Bontang, East Kalimantan to Japan. The assumption of the environmental conditions of the journey are shown in Table 3 [33]:

Table 3. Assumed environmental shipping conditions.

Air Temperature	27 °C (Average Daily Temperature at Bontang)
Sea water temperature	28 °C
Environmental pressure	1.01 bar
Mileage	2,400 nautical miles
Total sailing hours	316.2
Total sailing days	13.2

2.3. Calculation of Ship Resistance

In this design, we needed some data to get an appropriate design, including the amount of power needed by the ship to go at the required speed, so the calculation of the value of the obstacle for 125,000 m^3 LNG tankers was needed. In this design study, the application of Maxsurf Resistance 20 software provided the steps to be followed for determining the value of the ship's resistance [34]. Figure 1 shows a graph of the results of a resistance simulation using the Holtrop method with a speed range between 0 and 22.5 knots.

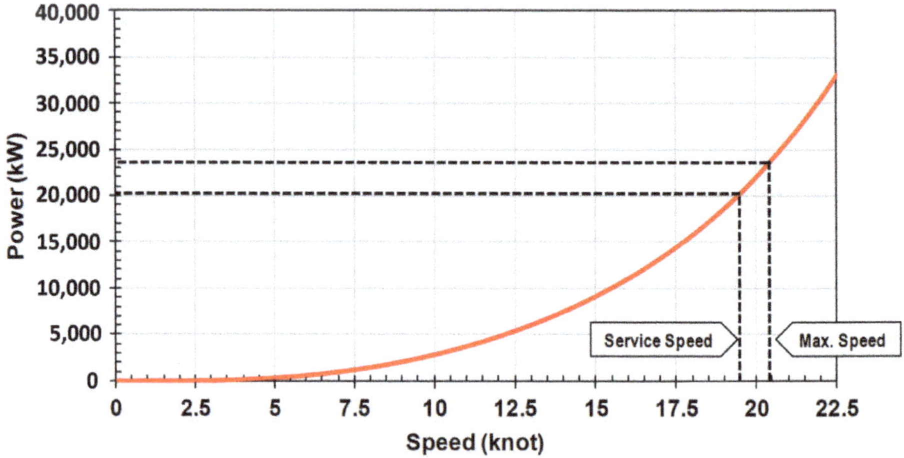

Figure 1. Power prediction using Holtrop methods.

2.4. Design of Combined-Cycle Propulsion System

After getting the data resistance of the LNG tanker, which would be designed to be able to determine the amount of power for ship propulsion, the next step was to design a propulsion system to meet the power requirements of the ship. The design of the combined-cycle system was done using the Cycle-Tempo software so that each component of the designed system could be connected and in accordance with the desired results. Figure 2 presents a scheme of this propulsion system. The processes in the systems are shown in Figure 2 as well. The design of this system consisted of gas turbine components, which consisted of (3) compressor, (4) combustion chamber, and (5) turbine, using fuel sourced from (1) the fuel source. Then this component was directly connected by the shaft to rotate the electric generator.

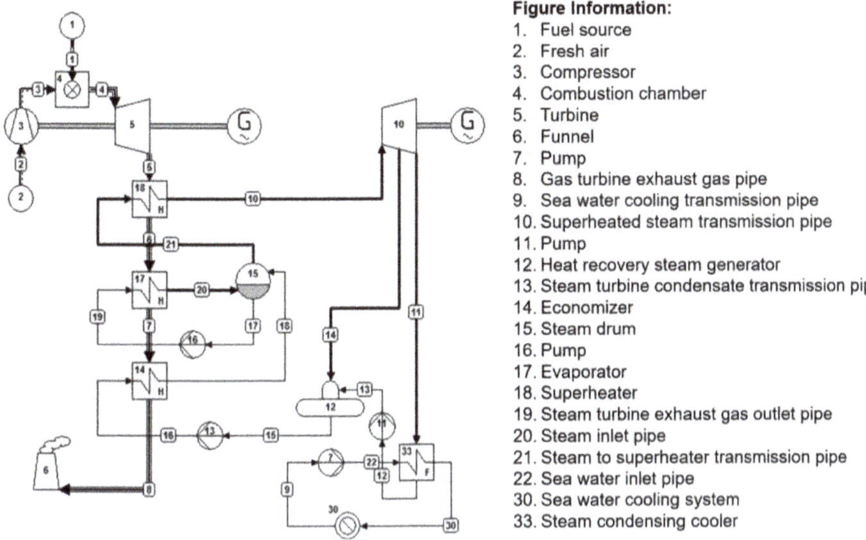

Figure 2. Schematic of a combined-cycle propulsion system.

The working fluid, after coming out of the steam turbine, was still in the saturated vapor phase (saturated steam). Then this fluid entered (33) the condenser, which then turned into saturated liquid phase. To change the phase of the working fluid, the condenser got the cooling medium from (30) the sea water pump. Then the water that had been used as a cooler was discharged into the environment. HRSG is a device that utilizes heat energy from gas turbine flue gas to be used as a source of steam turbine cycle energy. The working fluid was pumped after it exited the condenser and then entered (14) the economizer, where the working fluid temperature was increased. Then it was forwarded to (17) the evaporator where the working fluid phase was changed into saturated steam. Then the steam entered (18) the superheater where the saturated steam phase changed to superheated steam. After that the working fluid, i.e. superheated steam, entered the steam turbine to be converted into mechanical energy to turn the electric generator. After leaving the steam turbine, the working fluid returned to the condenser.

To be able to simulate the design of a combined-cycle propulsion system, several parameter values of the system components to be designed were needed, including:

1. Fuel

The fuel used was natural gas from boil-off cargo gas that was transported. The carried cargo was natural gas from a liquefaction plant in Bontang, East Kalimantan, whose composition and characteristics are shown in Table 4.

Table 4. The composition and characteristics of natural gas [35].

CH_4	91.20%
C_2H_6	5.50%
C_3H_8	2.40%
C_4H_{10}	0.90%
O_2	0.10%
N_2	0.00%
Lower heating value (LHV)	49,426.97 kJ/kg
Liquid density (LNG)	456 kg/m^3
Gas density (CNG)	0.801 kg/m^3 (0 °C, 1 atm)
Expansion ratio (Gas/Liq)	568 m^3 (gas)/m^3 (liq) (0 °C, 1 atm)

2. Gas Turbine

The basis of the selection of the gas turbine specifications used in this design study was the output power, based on the availability of gas turbine types on the market for the required power range. The appropriate gas turbine model was obtained as shown in Table 5. Data supporting the simulations in this design study used sources from a datasheet from the manufacturer, as well as data sourced from research results. In operation, the gas turbine must be within the operating range of the gas turbine to prevent the condition of chocking and surging. Therefore, in this simulation, the gas turbine operation was adjusted to the performance characteristic map issued by the manufacturer.

Table 5. Gas turbine specifications [36].

Model	LM 2500
Manufacturer	General Electric
Power ISO condition (100% load)	25 MW
Power ambient condition (27 °C, 1.01 bar)	22.8 MW
Fuel	Natural gas
Exhaust gas temperature (100% load)	530 °C
Isentropic compressor efficiency	70–87%

3. Steam Turbine

In simulations using the Cycle-Tempo software, several parameters were needed to do a steam turbine simulation, such as pressure, temperature, and mass vapor flow rate, for the calculation of the isentropic efficiency of the steam turbine component used based on the method set by general electric vapor pressure, which was chosen at 25 bar based on the recommendations of the results of an optimization study conducted by Følgesvold [37,38]. The specifications of the steam turbine used were an inlet vapor pressure of 25 bar and isentropic efficiency in the range of 0.8–0.88%.

4. HRSG

The once-through steam generator (OTSG) HRSG was chosen as the most suitable choice for this simulation. This simulation included three main components of the HRSG, namely, superheater, evaporator, and economizer. In its operating conditions, the economizer is useful as a preheater to raise the temperature of the working fluid into the saturated liquid phase; then the working fluid enters the evaporator, then exits the phase change into saturated steam, then enters the superheater so it enters the boiler in the superheated steam phase. The parameters used in the simulation were the superheater temperature (ΔT_{hi}) was 30 °C and the evaporator temperature ($\Delta T_{pinchpoint}$) was 25 °C.

An upper terminal difference temperature of 30 °C and a pinch point of 25 °C were chosen to provide enough energy to move the steam turbine cycle. According to Saravanamuttoo, this difference in temperature values was chosen to maintain the size and weight of the HRSG [39].

5. Supporting Parameters

Other supporting data used in the simulation were the isentropic pump efficiency used was 85%, the deaerator pressure (P_{in}) was 2 bar, the pressure drop in the condenser was 0.1 bar, and the mechanical efficiency in the generator was 97.5%.

2.5. The Equation Used

To be able to find out the appropriateness of the calculations using the Cycle-Tempo software with manual calculations, a manual calculation was made to load the system at maximum (100%). The calculation of T_2 and T_3 under isentropic conditions in gas turbines is as follows:

$$\frac{T_2}{T_1} = \left(\frac{P_2}{P_1}\right)^{\frac{k-1}{k}} \tag{1}$$

where T, P, and k are the temperature, pressure, and heat capacity ratio of the gas, respectively, and subscripts 1 and 2 denote the state before and after the isentropic compression process, respectively. The equations for the calculation of air mass flow rate (m_{air}) and processes (3–4), are the following:

$$Qin = m_{fuel} \times LHV \tag{2}$$

$$Q_{in} = m_{air+fuel}h_3 - m_{air}h_2 \tag{3}$$

where m_{fiel} and LHV are the mass flow rate and lower heating value of the fuel, respectively, whereas $m_{air+fuel}$, m_{air}, h_3, and h_2 are the mass flow rate of the air and fuel mixture, mass flow rate of the air, specific enthalpy of the air after the isobaric combustion process, and specific enthalpy of the air before the combustion, respectively. The values for enthalpies h_2 and h_3 were determined by interpolation of the relevance values in the table of water properties. The actual work of a gas turbine is:

$$W_{Gas\ Turbine} = W_{Steam\ Turbine} - W_{Compressor} \tag{4}$$

where $W_{Gas\ Turbine}$ and $W_{Steam\ Turbine}$ are the work output of the gas turbine and steam turbine, respectively, and $W_{Compressor}$ is the compressor work. The work of a steam turbine is:

$$W_{Steam\ Turbine} = m\ (h_4 - h_3) - v\Delta P \tag{5}$$

where m, h_4, h_3, v, and ΔP are the mass flow rate of the working fluid, fluid enthalpy before entering the steam turbine, fluid enthalpy after exiting the turbine, specific volume of the working fluid, and pressure drop before and after the fluid enters the turbine, respectively. Looking for the value of the mass flow rate of the working fluid in the vapor cycle, the heating value (Q_{in}) was obtained from the heat recovery steam generator (HRSG), where this equation applied:

$$Q_{Exhaust\ gas} = Q_{in} \tag{6}$$

$$m_{Exhaust\ gas}c_p\Delta T = m_{fluid}(h_4 - h_3) \tag{7}$$

where $Q_{Exhaust\ gas}$, $m_{Exhaust\ gas}$, c_p, and ΔT are the heat output, mass flow rate, isobaric specific heat, and temperature difference of the exhaust gas from the gas turbine, respectively, whereas Q_{in}, m_{fluid}, h_4, and h_3 are the heat input from the exhaust gas, mass flow rate, and specific enthalpies before and after the heating of the working fluid. The h_4 and h_3 values were obtained from the interpolation of the relevance values in the properties table for saturated water and superheated steam. Pump work calculation used the following equation:

$$W_{pumps} = \frac{v\Delta P}{\eta} \tag{8}$$

where W_{pump} and η are the work output and efficiency of the pump, respectively. The recommended pressure was 25 bar or 2500 kPa. Based on the literature, the η (efficiency) of the pump was 87.5%. Then the total work of the Rankine cycle is:

$$W_{Rankine} = m_{fluid}(3460 - 2680) - W_{pumps} \tag{9}$$

Therefore, we get the total system efficiency as:

$$\eta_{total} = \frac{Total\ Power\ (kW)}{m_{fuel}\left(\frac{kg}{s}\right) \times LHV\left(\frac{kJ}{kg}\right)} \tag{10}$$

There were result differences between the manual calculations and the calculations using the Cycle-Tempo software due to the less accuracy in manual calculations. The plotted values of the fluid properties, i.e. the enthalpy, specific heat capacities c_p and c_v, were determined by interpolation in manual calculation whereas more accurate digital calculations were performed in the application software. Therefore, a slight difference in the values of the system power and system efficiency under the maximum loading conditions which result the total efficiency of the systems using empirical equations by manual calculation is 52.7%, and based on the simulation software is 48.49%.

3. Results and Discussions

3.1. Thermodynamic Cycle Analysis

Figure 3 is a T-s diagram of the actual gas turbine cycle with a dual shaft configuration. Point 1 is the environmental condition. Actual compressor work is illustrated by 1–2a. Then actual work of the gas turbine generator is illustrated by line 3–4a, and the work of the power turbine is represented by points 4a–5a. The temperature of this cycle at the time of maximum loading (100% load) was 1215.87 °C, while the inlet air temperature was 27 °C. Heat discharged Q_{out}, i.e. at point 5a–1 in the graph of Figure 3, was then used for the next cycle below (bottoming cycle) as the heat input in the HRSG unit.

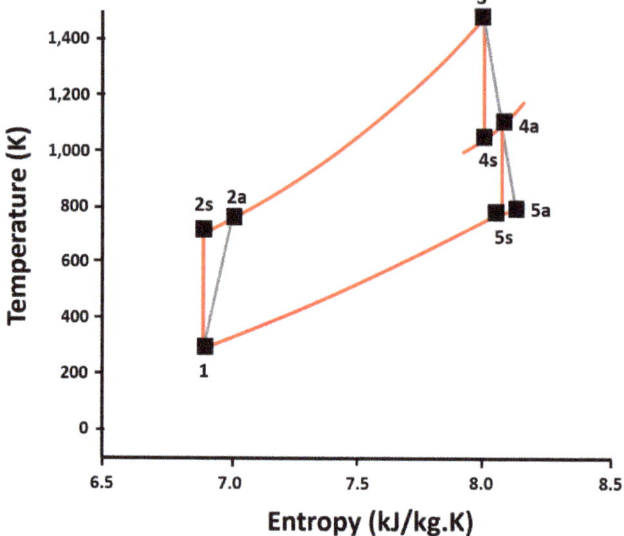

Figure 3. T-s diagram for the actual gas turbine cycle of the designed system.

In Figure 4 can be seen the Q-T diagram for the heat recovery steam generator (HRSG) at maximum loading. The inlet temperature of the working fluid was 120 °C, while the exhaust gas temperature of the gas turbine was 530 °C. The upper terminal difference was 30 °C, and the value of the pinch point temperature was 25 °C. The exhaust gas was released into the environment through a stack at a temperature of 198 °C.

In this design power generation cycle, as can be seen in Figure 5, the incoming heat (Q_{in}) was heat taken from the gas turbine exhaust gas. Point 1–2 is the work of the pump (feed water pump). Then the water was pumped into the HRSG system. Point 2–3 is an economizer component that acts as a preheater. Water increased in temperature but was still in the saturated liquid phase. Then point 3–4 is the heat transfer process in an evaporator component. Here, the working fluid changed its phase into saturated steam. Then point 4–5 is a superheater. This system worked at a pressure of 25 bar. Point 5–6 is the un-isentropic expansion or work of the steam turbine. The inlet temperature in the steam turbine was 500 °C at the maximum loading of the gas turbine.

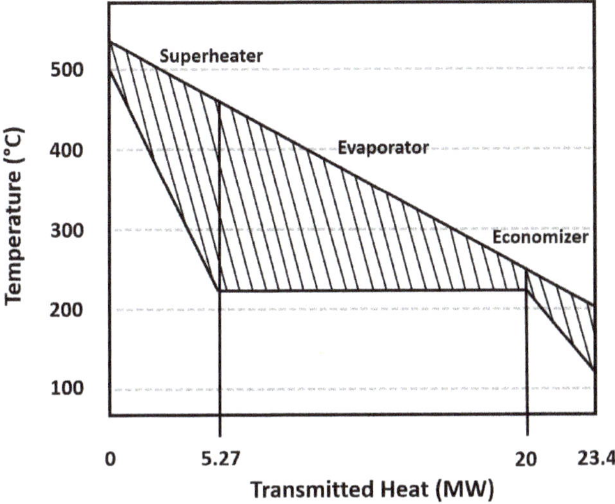

Figure 4. Q-T diagram for HRSG.

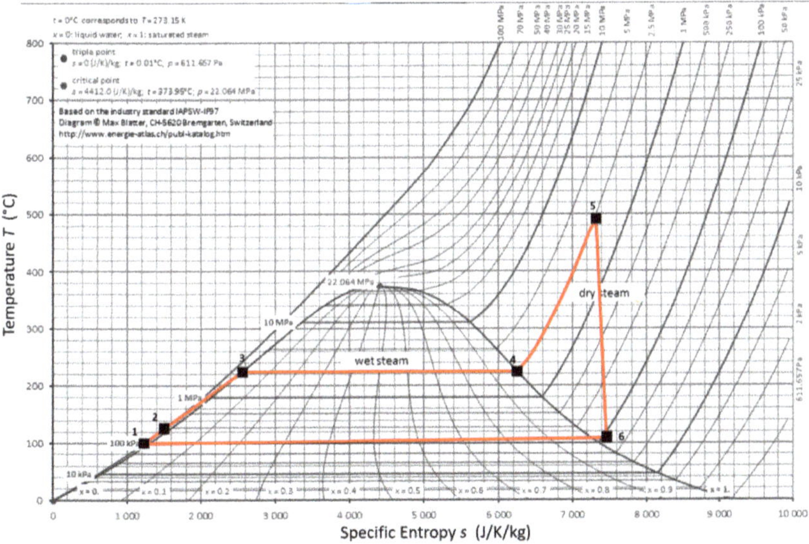

Figure 5. T-s diagram for the actual steam turbine cycle of the designed system.

3.2. Analysis of System Performance

In designing this propulsion system, energy supply only came from gas turbines, because there was no further combustion in the HRSG. Therefore, it is important in determining the performance limits of gas turbines because this can affect the bottom generation cycle. Gas turbines work in a relatively narrow range of performance that can be described in the performance curve of the axial compressor or axial turbine. Figure 6 is a graph plot of the variations of loading with pressure ratio to corrected mass flow of the LM 2500 gas turbine compressor. Gas turbines experience the phenomenon of surging when operated at a loading below 22.8%, so this limit is a reference of gas turbine operations in the design of this system. In plotting the operating points in the performance curve, the designer must consider the magnitude of the mass flow rate and pressure ratio. The blue line represents the constant isentropic efficiency line so that keeping the operating point at a high-efficiency value can result in a more efficient system.

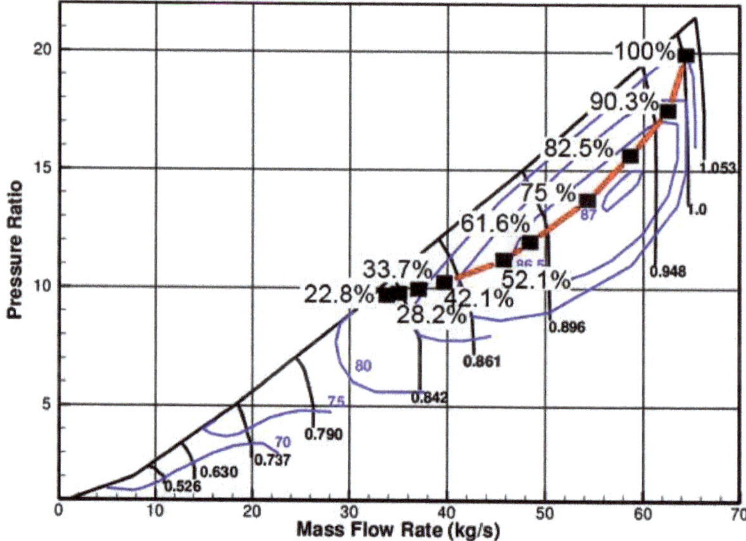

Figure 6. Plot curve performance of the LM 2500 gas turbine compressor against variations in loading from the system.

3.3. Data Analysis of Simulation Results

Based on data obtained from software simulation results, a graph is made to facilitate the drawing of conclusions from the observed data. Figure 7 presents a graph that shows the propulsion power at every percentage of the load of the engines, which can be seen the gas turbine engine become more dominant as the load increases. The gas turbine was taken as the reference for loading since it was the source of the overall heat system. The load or load percentage is the generated power of the gas turbine compared to the maximum power of the gas turbine. The total propulsion power of the system is the summation of the steam turbine power and the gas turbine power so that the value of the maximum total power of the system was 28,122.23 kW and the minimum power of the system was 6990.9 kW. The limits of minimum loading were discussed in the previous subchapter. From the relationship between power and load, it could be seen that the power of the steam turbine decreased significantly at 33.77% loading. So, also be seen that the overall system power was decreased slightly. However, the rest of the total power needed was fulfilled by the power generated from the gas turbine.

Figure 8 shows that the total power efficiency follows on the gas turbine efficiency since the power of the steam turbine unit were relative constant at higher power load. The power efficiency is the total power of the gas turbine and steam turbine divided by the LHV fuel combustion power which is occur only in the gas turbine combustion chamber. Also, the maximum total propulsion system efficiency is at 48.49% at the maximum loading. This total efficiency will decrease as the loading percentage decrease. At minimum load, the total system efficiency is just 34.03%. Gas turbine characteristics have best performance at the high design load. So, therefore part loading of the propulsion system causes the efficiency system to significantly decrease, along with the decrease of the performances of both the gas turbine and the steam turbine engines. The steam cycle, which gets heat from the gas turbine exhaust gas, experiences a decrease in performance due to reducing mass flow rate and the exhaust gas temperature of the gas turbine low load.

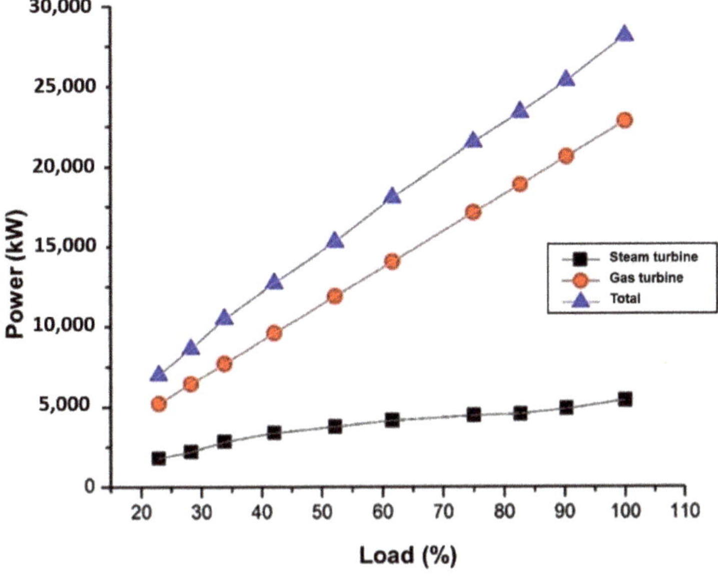

Figure 7. Power to the loading of the propulsion system.

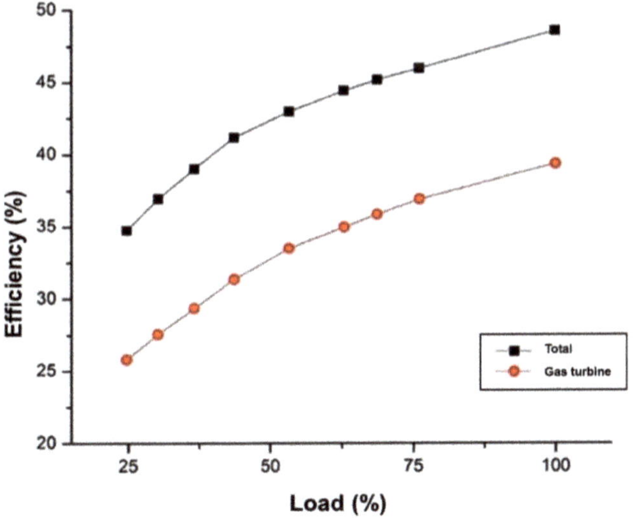

Figure 8. Efficiency of the system versus engine load.

Figure 9 explains the relationship between the load of the system and the speed that can be achieved by the ship. This simulation was carried out on the conditions of an empty ship or ballast condition as well as the ship at loaded condition. Both conditions are at maximum propulsion power loading. The loaded condition ship which is containing the cargo can reach a maximum speed of 20.67 knots, whereas for the ballast condition ship which is not containing a cargo can reach a maximum speed of 21.7 knots. Both at maximum propulsion power load.

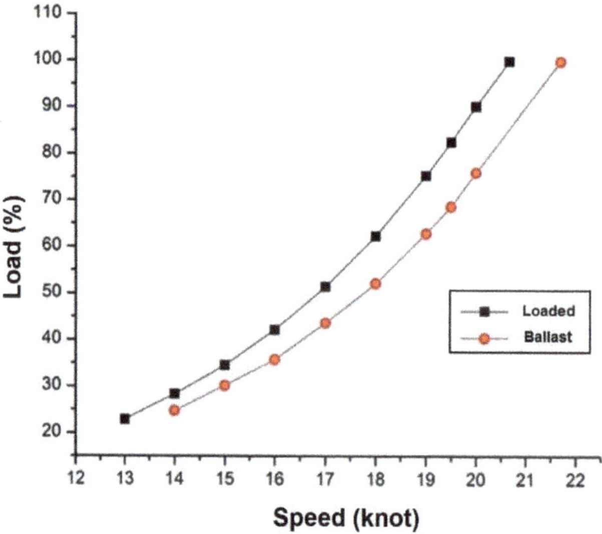

Figure 9. Graph of loading of the system and ship speed.

In Figure 10 a and b, it can be seen that the relationship between the required fuel consumption and the boil-off gas availability in LNG vessels, at the time of fully loaded cargo delivery (a) and when returning to port for loading (b). Based on the simulation results, there is no problem on the fuel consumption at the cargo delivery conditions ship. However if the ship sails on low speed, some considerations must be taken to anticipate problems from the increasing of the boil-off gas producing in the cargo space due to the lower fuel consumption with longer journey time. For loaded conditions ship to the destination port, the fuel availability from the boil-off gas produced in shipping will be 725,788.8 kg at maximum speed, whereas at lower speed of 13 knots, will be 1,154,004.3 kg. Then there will be differences of residual boil-off gas that was not consumed as engine fuel. It will be 235,896.2 kg after maximum speed sailing and 877,524.32 kg after sailing speed of 13 knots.

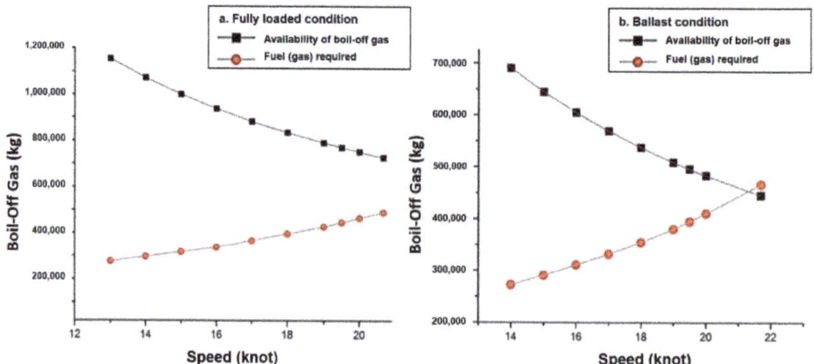

Figure 10. Availability of boil-off gas and fuel requirements of ships at the time of fully loaded condition (**a**) and ballast condition (**b**).

For empty ship conditions or ballast conditions with maximum sailing speed, the fuel needed during the cruise is 466,639.6 kg, while the available boil-off gas will be 445,401.7 kg. The fuel consumption will exceed the producing of availability of boil-off gas in the cargo. So, therefore a sailing speed selection is important for maintaining the fuel availability for ship. At a speed of 20 knots, for fully loaded condition ship the propulsion system fuel requirements is 488,320.86 kg, whereas for ballast condition ship the needed fuel consumption is 410,832 kg. This shows that the choice of speed and load of the engine greatly affects the producing availability of boil-off gas for engine fuel.

4. Conclusions

In this study, the design of the combined-cycle propulsion system was carried out on an LNG tanker with analytical calculation and simulation approaches using the Cycle-Tempo software. Combined cycle was used in the LNG tanker propulsion system with COGES (combined gas–electric steam) turbine configuration with the main components: gas turbine, steam turbine, heat recovery steam generator (HRSG), condenser, pump, deaerator, and generator. In accordance with the limits of gas turbine performance on the compressor performance characteristic map, gas turbines have a minimum loading limit of 22.8% due to the limits of the surge line. The maximum temperature of the gas turbine cycle at maximum loading is 1,215.87 °C with a flue gas temperature of 530 °C, and the inlet pressure on the steam turbine is 25 bar. From the simulation results, the maximum power from the resulting system was 28,139.25 kW. With the maximum power, the ship can cruise with a maximum speed of 20.67 knots at fully loaded conditions and 21.7 knots at ballast ship conditions. There is a maximum speed difference of 1.03 knots between fully loaded conditions and ballast conditions. In addition, the availability of fuel from boil-off gas in shipping to the port of destination is calculated to be 725,788.8 kg at maximum speed, while at a speed of 13 knots is 1,154,004.3 kg. There are residual boil-off gas differences that are not utilized as engine fuels of 235,896.2 kg at maximum speed and 877,524.32 kg at a speed of 13 knots. From the results, it can be concluded that the combined-cycle propulsion system using boil-off gas is feasible for LNG vessels. From the designed system, at fully loaded conditions, a maximum power of the system of about 28,122.23 kW is obtained with fuel consumption of about 1.173 kg/s, system efficiency of about 48.49% and the vessel speed can be reached up to 20.67 knots as well.

Author Contributions: Conceptualization, W.N., M.A.B. and R.M.; methodology, W.N., M.A.B. and R.M.; software, W.N., M.A.B. and R.M.; validation, W.N., M.A.B. and R.M.; formal analysis, W.N., M.A.B. and R.M.; investigation, W.N., M.A.B. and R.M.; resources, W.N.; data curation, W.N., M.A.B. and R.M.; writing—original draft preparation, W.N., R.M.; writing—review and editing, W.N., M.A.B.; visualization, W.N., M.A.B.; supervision, W.N.; project administration, W.N., M.A.B.; funding acquisition, W.N., M.A.B. All authors have read and agreed to the published version of the manuscript.

Funding: The research received only internal funding from the University of Indonesia as a research grant UI/PUTI Q2/2020 program number NKB-1683/UN2.RST/HKP.05.00/2020.

Acknowledgments: The authors would like to express their gratitude to the Directorate Research and Development of Universitas Indonesia (DRPM UI) for the grant support. This paper and its publication are supported by the UI/PUTI Q2/2020 program number NKB-1683/UN2.RST/HKP.05.00/2020. The authors also thank the Department of Mechanical Engineering at Universitas Indonesia for providing the supporting facilities during the research.

Conflicts of Interest: The authors are declaring that there are no conflicts of interests in this research. The funders, i.e., the University of Indonesia, had no role in the design of the study; in the collection, analyses, or interpretation of data; in the writing of the manuscript, or in the decision to publish the results.

References

1. Javanmardi, J.; Nasrifar, K.; Najibi, S.; Moshfeghian, M. Economic evaluation of natural gas hydrate as an alternative for natural gas transportation. *Appl. Therm. Eng.* **2005**, *25*, 1708–1723. [CrossRef]
2. Brito, A.; de Almeida, A.; de Almeida, A.T. Multi-attribute risk assessment for risk ranking of natural gas pipelines. *Reliab. Eng. Syst. Saf.* **2009**, *94*, 187–198. [CrossRef]
3. Folga, S.M. *Natural Gas Pipeline Technology Overview*; Argonne National Lab: Argonne, IL, USA, 2007. [CrossRef]
4. Montiel, H.; Vílchez, J.A.; Arnaldos, J.; Casal, J. Historical analysis of accidents in the transportation of natural gas. *J. Hazard. Mater.* **1996**, *51*, 77–92. [CrossRef]
5. Budiyanto, M.A.; Riadi, A.; Buana, I.S.; Kurnia, G. Study on the LNG distribution to mobile power plants utilizing small-scale LNG carriers. *Heliyon* **2020**, *6*, e04538. [CrossRef]

6. Bennett, C.P. *Marine Transportation of LNG at Intermediate Temperature*; Springer: Boston, MA, USA, 1980; pp. 751–756. [CrossRef]

7. Wagner, J.V.; Van Wagensveld, S. Marine Transportation of Compressed Natural Gas a Viable Alternative to Pipeline or LNG. In Proceedings of the SPE Asia Pacific Oil & Gas Conference and Exhibition, Society of Petroleum Engineers, Melbourne, Australia, 8–10 October 2002. [CrossRef]

8. Stenning, D.G.; Cran, J.A. Ship Based System for Compressed Natural Gas Transport. 1997. Available online: https://patents.google.com/patent/US5803005A/en (accessed on 20 January 2020).

9. Feng, L.; Shuanshi, F. Transmission and economy of shipping NGH. *Tsinghua Tongfang Knowl. Netw.* **2005**, 7. Available online: http://en.cnki.com.cn/Article_en/CJFDTOTAL-TRQG200507043.htm (accessed on 20 January 2020).

10. Sinor, J.; Consultants, J. *Comparison of CNG and LNG Technologies for Transportation Applications*; Department of Energy: Golden, CO, USA, 1992. [CrossRef]

11. Maxwell, D.; Zhu, Z. Natural gas prices, LNG transport costs, and the dynamics of LNG imports. *Energy Econ.* **2011**, *33*, 217–226. [CrossRef]

12. International Gas Union. *2018 World LNG Report*; International Gas Union: Barcelona, Spain, 2018. Available online: https://www.igu.org/sites/default/files/node-document-field_file/IGU_LNG_2018_0.pdf (accessed on 20 January 2020).

13. The Oxford Institute for Energy Studies. *The LNG Shipping Forecast: Costs Rebounding, Outlook Uncertain*; The Oxford Institute for Energy Studies: Oxford, UK, 2018. Available online: https://www.oxfordenergy.org/wpcms/wp-content/uploads/2018/02/The-LNG-Shipping-Forecast-costs-rebounding-outlook-uncertain-Insight-27.pdf (accessed on 20 January 2020).

14. Mrzljak, V.; Poljak, I.; Mrakovčić, T. Energy and exergy analysis of the turbo-generators and steam turbine for the main feed water pump drive on LNG carrier. *Energy Convers. Manag.* **2017**, *140*, 307–323. [CrossRef]

15. Pamitran, A.S.; Budiyanto, M.A.; Maynardi, R.D.Y. Analysis of ISO-Tank Wall Physical Exergy Characteristic: Case Study of LNG Boil-off Rate from Retrofitted Dual Fuel Engine Conversion. *Evergreen* **2019**, *6*, 134–142. [CrossRef]

16. Haglind, F. A review on the use of gas and steam turbine combined cycles as prime movers for large ships. Part II: Previous work and implications. *Energy Convers. Manag.* **2008**, *49*, 3468–3475. [CrossRef]

17. Senary, K.; Tawfik, A.; Hegazy, E.; Ali, A. Development of a waste heat recovery system onboard LNG carrier to meet IMO regulations. *Alex. Eng. J.* **2016**, *55*, 1951–1960. [CrossRef]

18. Attah, E.E.; Bucknall, R. An analysis of the energy efficiency of LNG ships powering options using the EEDI. *Ocean Eng.* **2015**, *110*, 62–74. [CrossRef]

19. Budiyanto, M.A.; Pamitran, A.S.; Wibowo, H.T.; Murtado, F.N. Study on the Performance Analysis of Dual Fuel Engines on the Medium Speed Diesel Engine. *J. Adv. Res. Fluid Mech. Therm. Sci.* **2020**, *68*, 163–174. [CrossRef]

20. Fernández, I.A.; Gómez, M.R.; Gómez, J.R.; Insua, Á.B. Review of propulsion systems on LNG carriers, Renew. *Sustain. Energy Rev.* **2017**, *67*, 1395–1411. [CrossRef]

21. Chang, D.; Rhee, T.; Nam, K.; Chang, K.; Lee, D.; Jeong, S. A study on availability and safety of new propulsion systems for LNG carriers. *Reliab. Eng. Syst. Saf.* **2008**, *93*, 1877–1885. [CrossRef]

22. Sinha, R.P.; Nik, W.M.N.W. Investigation of propulsion system for large LNG ships. In Proceedings of the 1st International Conference on Mechanical Engineering Research 2011 (ICMER2011), IOP Conference Series: Materials Science and Engineering, Kuantan, Pahang, Malaysia, 5–7 December 2011; pp. 1–16. [CrossRef]

23. Haglind, F. A review on the use of gas and steam turbine combined cycles as prime movers for large ships. Part III: Fuels and emissions. *Energy Convers. Manag.* **2008**, *49*, 3476–3482. [CrossRef]

24. Lee, D.-M.; Kim, S.-Y.; Moon, B.-Y.; Kang, G.-J. Layout design optimization of pipe system in ship engine room for space efficiency. *J. Korean Soc. Mar. Eng.* **2013**, *37*, 784–791. [CrossRef]

25. Budiyanto, M.A.; Nasruddin; Nawara, R. The optimization of exergoenvironmental factors in the combined gas turbine cycle and carbon dioxide cascade to generate power in LNG tanker ship. *Energy Convers. Manag.* **2020**, *205*, 112468. [CrossRef]

26. Hansen, J.F.; Lysebo, R. Comparison of Electric Power and Propulsion Plants for LNG Carriers with Different Propulsion Systems. 2007. Available online: http://www.dieselduck.info/machine/02propulsion/2007ComparisonLNGcarrierpropulsion.pdf (accessed on 20 January 2020).

27. Wiggins, E.G. COGAS propulsion for LNG ships. *J. Mar. Sci. Appl.* **2011**, *10*, 175–183. [CrossRef]

28. Dimopoulos, G.; Frangopoulos, C. Thermoeconomic Simulation of Marine Energy Systems for a Liquefied Natural Gas Carrier. *Int. J. Thermodyn.* **2008**, *11*, 195–201. Available online: https://dergipark.org.tr/en/pub/ijot/issue/5770/76768 (accessed on 20 January 2020).
29. Jefferson, M.; Zhou, P.L.; Hindmarch, G. Analysis by computer simulation of a combined gas turbine and steam turbine (COGAS) system for marine propulsion. *J. Mar. Eng. Technol.* **2003**, *2*, 43–53. [CrossRef]
30. Oka, M.; Kazuyoshi, H.; Kenji, T. Development of Next-Generation LNGC Propulsion Plant and HYBRID System. *MHI Tech. Rev.* **2004**, *41*. Available online: http://www.mhi.co.jp/technology/review/pdf/e416/e416322.pdf (accessed on 20 January 2020).
31. Hanochem Shipping, LNG AQUARIUS. Available online: http://www.gts-internasional.com/lng-aquarius (accessed on 20 January 2020).
32. Cuneo, J. Service Experience with 125.000 m^3 LNG Vessel of Spherical-Tank Design. *Soc. Nav. Arch. Mar. Eng. Trans.* **1980**, *88*. Available online: https://www.sname.org/pubs/journals1 (accessed on 20 January 2020).
33. BMKG. Meteorology, Climatology, Data Online Center Database. Available online: http://dataonline.bmkg.go.id/home (accessed on 16 July 2019).
34. Bentley Systems, Maxsurf 20.00 V8i Release Notes—MAXSURF|MOSES|SACS—Wiki—MAXSURF|MOSES|SACS—Bentley Communities. Available online: https://communities.bentley.com/products/offshore/w/wiki/14169/maxsurf-20-00-v8i-release-notes (accessed on 21 January 2020).
35. Garjito, A.; Sumarno, A. *Indonesia LNG and the Badak Plant*; Indonesian Petroleum Association: South Jakarta, Indonesia, 1981; pp. 459–474. Available online: http://archives.datapages.com/data/ipa/data/010/010001/459_ipa0100459.htm (accessed on 20 January 2020).
36. Aviation, G.E. Marine Gas Turbine. 2014. Available online: https://www.geaviation.com/sites/default/files/datasheet-25mw.pdf (accessed on 15 January 2020).
37. Asimptote, B.V. Cycle-Tempo Manual Technical Notes. Available online: http://www.asimptote.nl/assets/media/7d155f62-ffe2-4a9e-9f33-bb003c80bd2b.pdf (accessed on 20 January 2020).
38. Følgesvold, E.R.; Skjefstad, H.S.; Riboldi, L.; Nord, L.O. Combined Heat and Power Plant on Offshore Oil and Gas Installations. 2011. Available online: http://papers.itc.pw.edu.pl/index.php/JPT/article/view/842 (accessed on 20 January 2020).
39. Saravanamuttoo, H.I.; Rogers, G.F.; Cohen, H. *Gas Turbine Theory*; Addison-Wesley Longman: Boston, MA, USA, 1996.

Journal of
*Marine Science
and Engineering*

Article

Analysis of the Impact of Split Injection on Fuel Consumption and NO$_x$ Emissions of Marine Medium-Speed Diesel Engine

Vladimir Pelić [1], Tomislav Mrakovčić [2,*], Radoslav Radonja [1] and Marko Valčić [2]

[1] Faculty of Maritime Studies, University of Rijeka, Studentska ulica 2, 51000 Rijeka, Croatia; vpelic@pfri.hr (V.P.); radonja@pfri.hr (R.R.)

[2] Faculty of Engineering, University of Rijeka, Vukovarska 58, 51000 Rijeka, Croatia; mvalcic@riteh.hr

* Correspondence: tomislav.mrakovcic@riteh.hr; Tel.: +385-51-651-520

Received: 29 September 2020; Accepted: 16 October 2020; Published: 20 October 2020

Abstract: The medium-speed diesel engine in diesel-electric propulsion systems is increasingly used as the propulsion engine for liquefied natural gas (LNG) ships and passenger ships. The main advantage of such systems is high reliability, better maneuverability, greater ability to optimize and significant decreasing of the engine room volume. Marine propulsion systems are required to be as energy efficient as possible and to meet environmental protection standards. This paper analyzes the impact of split injection on fuel consumption and NO$_x$ emissions of marine medium-speed diesel engines. For the needs of the research, a zero-dimensional, two-zone numerical model of a diesel engine was developed. Model based on the extended Zeldovich mechanism was applied to predict NO$_x$ emissions. The validation of the numerical model was performed by comparing operating parameters of the basic engine with data from engine manufacturers and data from sea trials of a ship with diesel-electric propulsion. The applicability of the numerical model was confirmed by comparing the obtained values for pressure, temperature and fuel consumption. The operation of the engine that drives synchronous generator was simulated under stationary conditions for three operating points and nine injection schemes. The values obtained for fuel consumption and NO$_x$ emissions for different fuel injection schemes indicate the possibility of a significant reduction in NO$_x$ emissions but with a reduction in efficiency. The results showed that split injection with a smaller amount of pilot fuel injected and a smaller angle between the two injection allow a moderate reduction in NO$_x$ emissions without a significant reduction in efficiency. The application of split injection schemes that allow significant reductions in NO$_x$ emissions lead to a reduction in engine efficiency.

Keywords: marine diesel engine; split injection; fuel consumption; NO$_x$ emissions

1. Introduction

Energy efficiency and environmental friendliness are the basic criteria when choosing the optimal technology in any industry, so this is also the case with the transport of goods. It is known that the transport of goods by sea is the most efficient mode of transport. Nevertheless, maritime transport is facing increasing demands on energy efficiency and the lowest possible environmental impact. The requirements for reducing air pollution with pollutants from marine power plants are defined in MARPOL 73/78 (International Convention for the Prevention of Pollution from Ships), Annex VI (Prevention of Air Pollution from Ships, enforced since 19 May 2005). For marine diesel engines with a rated power of more than 130 kW the NOx emission limits are divided into Tier I, Tier II and Tier III according to the IMO (International Maritime Organization). The limit values are applied depending on engine power and speed, the date of construction and the area of navigation, as shown in Figure 1.

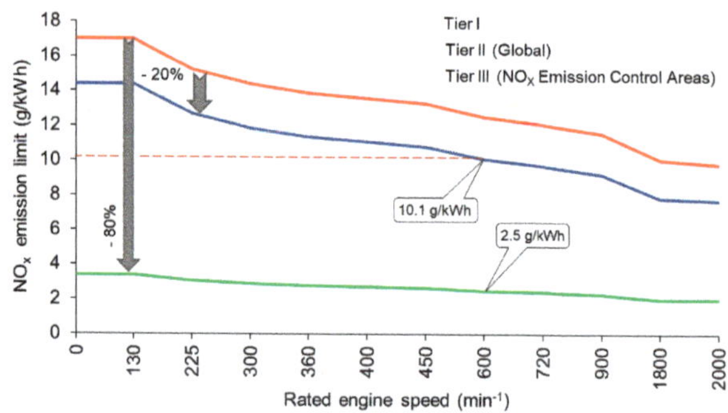

Figure 1. NO_x emission limits for marine engines [1].

Tier I refers to all ships built since 2000. Tier II is enforced since 2011. Due to the Tier II, the NO_x emission limits are reduced approximately 20% compared to Tier I. The Tier III requires approximately 80% reduction in emissions for ships operating in ECA (Emission Control Areas). Depending on their operating area of navigation, many ships are affected by Tiers II and III. It is therefore necessary to optimize the emissions of marine diesel engines. An example of determining the permissible NO_x emission for a marine diesel engine with a speed of 600 rpm is shown in Figure 1 by dashed red lines.

Most merchant ships are powered by a two-stroke low-speed diesel engine whose overall efficiency exceeds 50% under certain operating conditions. Marine medium-speed four-stroke diesel engines have approximately 3–5% lower efficiency than marine low-speed engines. Medium-speed diesel engines are half the size at the same rated power and NO_x emissions are considerably lower. The advantages of marine medium-speed diesel engines are especially pronounced if they are used in diesel-electric and hybrid propulsion systems. Slightly lower energy efficiency of four-stroke engines can be compensated by utilizing waste heat of the engine. This is supported by the fact that due to significantly higher exhaust gas temperatures of four-stroke engines, their exergy is significantly higher than that of two-stroke engines.

In marine diesel engines, various technologies are used to reduce emissions of harmful substances and in particular NO_x emissions, to the level required by regulations. These technologies usually are divided into primary and secondary measures. Primary measures involve modifying the process in the engine cylinder. Secondary measures include exhaust after treatment. Fuel type and quality also have a significant influence on emissions. Technologies for reducing NO_x emissions are listed in Table 1.

Table 1. NO_x emission reduction technologies [2].

No	NO_x Emission Reduction Technology	Expected Reduction
1	Two-stage turbocharging and Miller process	~40%
2	Combustion process adjustment	~10%
3	EGR—exhaust gas recirculation	~60%
4	Higher humidity of the scavenging air	~40%
5	Adding water to the fuel before injecting	~25%
6	Direct injection of water into the cylinder	~50%
7	SCR—selective catalytic reduction	~80%
8	Replacing liquid fuel with gaseous fuel	~85%

NO_x emission reduction technologies, which are marked 1, 2, 7 and 8 in Table 1, have the most favourable impact on energy efficiency and specific fuel oil consumption (SFOC). The implementation of other listed technologies leads to an increase in specific fuel consumption.

The adjustment of the combustion process in the engine cylinder by increasing the compression ratio while simultaneously reducing the amount of fuel injected per crankshaft revolution theoretically enables the approximately constant pressure of the combustion process. This leads to a lower maximum pressure and a lower maximum temperature, which is beneficial because NO_x emissions are largely temperature-dependent. By using modern electronically controlled fuel injection systems, this technology does not lead to a significant increase in specific fuel consumption.

Numerical modelling of internal combustion engines is today an indispensable tool that speeds up the development of the engine while reducing development costs. The available literature offers different approaches to numerical modelling internal combustion engines. The target area of research, the required accuracy and the time available for calculations are the basic parameters for model selection [3–5]. Zero-dimensional single-zone models are an efficient tool for predicting motor performance in stationary and dynamic operating conditions using modest computing resources and fast performance of simulations [6–10]. Multi-zone combustion models [11–16] allow the prediction of emissions of NO_x and other pollutants such as soot. In addition to the mentioned advantage of multi-zone models, a longer calculation time is associated. These models are usually not suitable for determining the overall performance of engines and their energy balances but are mostly adapted to predict emissions. Rakopoulus et al. [17] described in detail the development and verification of a numerical model of a direct fuel injection diesel engine. The described model implies the division of the combustion space inside the cylinder into two zones. The chemical equilibrium method was used to calculate the concentration of individual pollutants in the exhaust gas. The development and application of a complex multi-zone model to simulate the operation of turbocharged diesel engines is described in Reference [18]. This model divides the fuel jet injected into the engine cylinder into a number of zones. The interaction of the jet with the cylinder walls, the influence of the injection angle and the conditions of fuel evaporation for each zone are taken into account. Scappin et al. [19] have successfully applied a zero-dimensional model with two zones for predicting NO_x emissions in electronically controlled low speed two-stroke marine diesel engine. A study of the impact of split fuel injection on diesel engines using the FIRE computer program is presented in Reference [20]. The paper investigates the influence of split fuel injection on the emission of solid particles and nitrogen oxides, using three different injection schemes. In Reference [21], the development and application of a zero-dimensional model with three zones for the analysis of the operating parameters of a high-speed diesel engine are presented. In Reference [22], a three-zone model is described that is applicable in real-time applications. Compared to other similar models, this study uses a procedure that does not require iterative resolution thus significantly shortening the computational time. Baldi et al. [23] presented a numerical model of a marine medium-speed diesel engine in which a zero-dimensional model is used to model the high-pressure part of the process while the mean value model is applied for the rest of process in the engine cylinder. More recently [24], the impact of multiple fuel injections on NO_x emissions has been investigated. Simulations show that by applying split injection it is possible to achieve a reduction in NO_x emissions without a significant increase in fuel consumption. The paper [25] describes the development of a semi-empirical multi-zone model for predicting nitrogen oxide emissions in high-speed diesel engines with direct fuel injection. As in most other papers, the extended Zeldovich mechanism of NO_x formation is applied here as well. The development of another semi-empirical model that allows good prediction of NO_x emissions under stationary operating conditions and engine loads is presented in Reference [26]. Model testing was performed on several diesel engines under different operating conditions and with simultaneous application of different methods to reduce NO_x emissions. While the research in these papers describes in detail different models of internal combustion engines to simulate nitrogen oxide emissions, few or no investigate the impact of split fuel injection with application to marine medium-speed diesel engines of 5000 kW and more.

The aim of this paper is to examine the impact of different split fuel injection schemes on the specific fuel consumption and nitrogen oxide emissions of a marine medium-speed diesel engine using a two zone combustion numerical model.

2. Numerical Model of a Four-Stroke Diesel Engine

The numerical model is based on the laws of conservation of energy and mass and solving the resulting differential equations described in References [6,27]. A one-zone, zero-dimensional model of the four-stroke diesel engine presented in Reference [28] was upgraded to two-zone model with possibility to predict NO_x emissions. The model has additional features such as variable integration step selection, variable inlet valve closing angle, adjustment of the turbocharger air mass flow and graphic display of the results.

The main advantages of the applied model compared to multidimensional and multi-zone models are lower complexity, higher execution speed, adaptability and satisfactory accuracy of the obtained results, which are comparable to more complex models.

An four-stroke diesel engine consists of the following interconnected subsystems (Figure 2): engine cylinder, inlet manifold, exhaust manifold, turbocharger, intercooler, fuel injection subsystem, piston mechanism and valve timing mechanism.

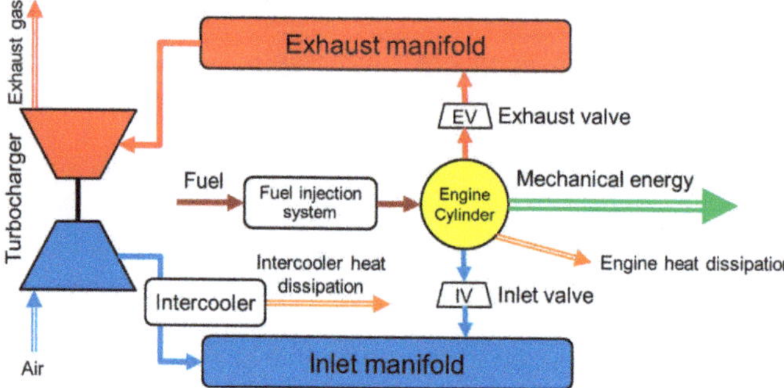

Figure 2. Diesel engine subsystems within implemented zero-dimensional numerical model.

The control volumes are interconnected by appropriate connections, which allow the exchange of the working medium. In the cylinder control volume, the heat is exchanged through the walls between the working medium and the cooling water. The heat is also exchanged with the ambient air through the walls of the inlet and outlet manifold. The heat generated by friction in the bearings is taken into account via the mechanical losses mean pressure, while the heat dissipated by radiation is neglected as it does not exceed 1% of the total heat input. Pressure and temperature in the control volumes are determined by solving differential equations derived from the laws of conservation of energy and mass. The properties of the working medium are determined according to References [29,30].

The software was developed in the C programming language. The model has been validated using data provided by the engine manufacturer and sea trial data. The data obtained from the operation of Wärtsilä 12V50DF engine on LNG ships with diesel-electric propulsion power systems. Rated power of one engine is 11.7 MW.

2.1. Mass Conservation Law

The mass change dm in the engine cylinder, inlet and exhaust manifold during the angle of rotation of the crankshaft $d\varphi$ is caused by the flow of the working medium through the inlet and exhaust valves, the mass of the injected fuel and the mass loss due to leakage can be expressed as:

$$\frac{dm_i}{d\varphi} = \frac{dm_{i,in}}{d\varphi} + \frac{dm_{i,ex}}{d\varphi} + \frac{dm_{i,f}}{d\varphi} + \frac{dm_{i,leak}}{d\varphi} \tag{1}$$

where m_{in} is the mass of the medium entering the control volume and m_{ex} is the mass of the medium exiting the volume, m_f is the mass of fuel supplied, m_{leak} is the mass of the medium exiting the volume and subscript "i" denote control volume. If the engine is properly maintained the leaked mass from cylinder may be neglected.

2.2. Energy Conservation Law

Energy balance of the medium in the control volume is given by:

$$\frac{dQ_i}{d\varphi} = \frac{dQ_{i,f}}{d\varphi} + \frac{dQ_{i,w}}{d\varphi} + h_{in}\frac{dm_{i,in}}{d\varphi} + h_{ex}\frac{dm_{i,ex}}{d\varphi} + h_f\frac{dm_{i,f}}{d\varphi} - p \cdot dV \tag{2}$$

where $dQ_{i,f}$ denotes heat released through fuel combustion and $dQ_{i,w}$ heat exchanged through the walls. The variables h_{in} and h_{ex} represents the enthalpy of the medium entering or leaving the control volume and h_f is enthalpy of the fuel.

Assuming that the internal energy of the gas depends solely of temperature, the equation of the temperature change is given by

$$\frac{dT_i}{d\varphi} = \frac{1}{m_i\left(\frac{\partial u}{\partial T}\right)_i}\left[-p_i\frac{dV_i}{d\varphi} + \sum_j \frac{dQ_{i,j}}{d\varphi} + \sum_k h_{i,k}\frac{dm_{i,k}}{d\varphi} - u_i\frac{dm_i}{d\varphi} - m_i\left(\frac{\partial u}{\partial \lambda}\right)_i\frac{d\lambda_i}{d\varphi}\right] \tag{3}$$

where u_i denotes internal energy and λ is the equivalent ratio of air to fuel.

In previous equations, all variables containing the mass or enthalpy of the fuel refer only to the control volume in which the fuel burns or to the cylinder. The same applies to variables describing a change in volume. When the fuel burns in the cylinder, the chemical energy of the fuel is converted into heat, which increases pressure and temperature. The increased pressure acts on the piston, where the thermal energy is converted into mechanical work.

2.3. Indicated Work

Indicated mechanical work is determined by:

$$\frac{dW_c}{d\varphi} = p_c\frac{dV_c}{d\varphi} \tag{4}$$

The pressure p_c in the cylinder is determined using the equation of state for a gas:

$$p_c = \frac{m_c \cdot R_c \cdot T_c}{V_c} \tag{5}$$

Current cylinder volume V_c is derived from the crankshaft mechanism geometry:

$$V_c(\varphi) = \frac{V_s}{2}\left[(1 - \cos\varphi) + \frac{1}{\lambda_m}\left(1 - \sqrt{1 - \lambda_m^2\sin^2\varphi}\right) + \frac{2}{\varepsilon - 1}\right] \tag{6}$$

where V_s is the cylinder swept volume, ε is compression ratio and λ_m denotes ratio between crank radius and piston stroke.

2.4. Heat Exchange

Heat transfer through the cylinder walls can be expressed as:

$$\frac{dQ_{w,c}}{d\varphi} = \sum_i \alpha_c \cdot A_{w,c,i}(T_{w,i} - T_c)\frac{dt}{d\varphi} \tag{7}$$

According to References [31,32], there are no significant temperature changes under stationary operating conditions, therefore a mean cylinder wall temperature is assumed. Furthermore, relatively small deviations in the heat transfer coefficient can be neglected, so that the mean heat transfer coefficient can be applied in the calculations. For the calculation of the heat transfer coefficient an empirical expression [33] is used in this paper:

$$\alpha_c = C_1 \cdot V_c^{-0.06} \cdot p_c^{0.8} \cdot T_c^{-0.4} \cdot \left(c_{\text{mps}} + C_2\right)^{0.8}, \tag{8}$$

where C_1 and C_2 are the empirical coefficients and c_{mps} is the mean piston speed.

2.5. Heat Release

Numerical models that describe the complex process of fuel combustion inside the cylinder is divided according to References [34,35] into zero-dimensional, quasi-dimensional and multidimensional models.

Vibe [36] provided the heat release rate by the following expression:

$$\frac{x_f}{d\varphi} = C\,(m+1)\left(\frac{\varphi - \varphi_{\text{IS}}}{\varphi_{\text{CD}}}\right)^m \exp\left(-C\left(\frac{\varphi - \varphi_{\text{IS}}}{\varphi_{\text{CD}}}\right)\right)^{m+1} \tag{9}$$

where x_f is the relative portion of fuel burned, C is the constant that depends on the efficiency of fuel combustion. The subscript IS refers to the crankshaft angle at which ignition starts, while the subscript CD represents the duration of combustion. The exponent m is determined according to Reference [37] and the change in combustion duration $\Delta\varphi_{\text{CD}}$ is determined according to Reference [38].

The expressions for determining the ignition delay for diesel fuel are given in Reference [39]. Adjusted expression for heavy fuel are given in Reference [40].

It is assumed that the rate of injected fuel mass follows the heat release rate and that the combustion products are immediately mixed with the medium in the cylinder to form a homogeneous mixture. The total mass within the cylinder increases during combustion due to the injected fuel. The excess air in the engine cylinder is calculated from the mass of the gases in the engine cylinder and the mass of the injected fuel.

In the numerical sub-model of split fuel injection, a double Vibe function was used for pilot injection and a single Vibe function for main injection.

2.6. Change in Mass And Excess Air in The Cylinder

The change in mass in the engine cylinder due to fuel injection is expressed by:

$$\frac{dm_c}{d\varphi} = \frac{dm_{f,c}}{d\varphi} = \frac{dx_f}{d\varphi}m_{f,\text{proc}} = \frac{1}{\eta_{\text{comb}}LHV}\frac{dQ_f}{d\varphi} \tag{10}$$

Also, fuel injection affects the change in excess air ratio which is calculated as follows:

$$\frac{d\lambda_c}{d\varphi} = -\frac{\lambda_c}{m_{f,c}}\frac{dm_{f,c}}{d\varphi} \tag{11}$$

When the working medium flows out of the control volume, there is no change in the excess air ratio and there is no change in the gas composition. If gases flowing into the control volume have a different composition, there is also a change in the excess air ratio. The change in excess ratio as a function of the crankshaft angle is determined by the expression:

$$\frac{d\lambda_c}{d\varphi} == \frac{\frac{dm_{c,i}}{d\varphi}\left(1 - \frac{\lambda_c}{\lambda_i}\frac{S_{\text{AFR}}+1}{S_{\text{AFR}}+1}\right)}{S_{\text{AFR}}\,m_{g,c}} \tag{12}$$

where S_{AFR} is stoichiometric mass of air in mixture with fuel.

2.7. Working Medium Exchange in A 4-Stroke Engine Cylinder

The working fluid flows between the cylinder and the inlet and exhaust manifolds. The flow of the working fluid from one control volume to the other is determined by valves timing, the effective flow area and the pressure difference:

$$\frac{dm}{d\varphi} = \alpha_p \, A_{p,geo} \, \psi \, p_1 \sqrt{\frac{2}{R_1 T_1}} \frac{dt}{d\varphi} \tag{13}$$

In the previous equation, the geometric flow areas $A_{p,geo}$ of the inlet and exhaust valves are determined according to the camshaft cam geometry. The flow coefficient α_p is determined according to Reference [41]. The flow function ψ for the subcritical pressure ratio is determined according to Reference [42]

$$\psi = \sqrt{\frac{\kappa}{\kappa-1}\left[\left(\frac{p_2}{p_1}\right)^{\frac{2}{\kappa}} - \left(\frac{p_2}{p_1}\right)^{\frac{\kappa+1}{\kappa}}\right]}, \text{ if } 1 \geq \frac{p_2}{p_1} \geq \left(\frac{2}{\kappa+1}\right)^{\frac{\kappa}{\kappa+1}} \tag{14}$$

and flow function ψ for supercritical pressure is:

$$\psi = \left(\frac{2}{\kappa+1}\right)^{\frac{1}{\kappa-1}} \sqrt{\frac{\kappa}{\kappa+1}}, \text{ if } \frac{p_2}{p_1} < \left(\frac{2}{\kappa+1}\right)^{\frac{\kappa+1}{\kappa}} \tag{15}$$

Subscript 1 refers to the state in the upstream control volume, while subscript 2 refers to the state in the downstream control volume.

2.8. Turbocharger

For modelling the operation of a diesel engine under stationary operating conditions, the numerical model of the turbocharger does not require the use of suitable compressor map data. Instead, it is acceptable to assume that the air mass flow is known for a given engine load. Engine manufacturers typically provide inlet manifold pressure and air mass flow data for different engine loads in range between the 50% to 100% of engine MCR (Maximum Continuous Rating). The exhaust gases mass flow through the turbine is determined by the following expression:

$$\frac{dm_T}{d\varphi} = \alpha_T A_{T,geo} \, \psi \, p_{EM} \sqrt{\frac{2}{R_{EM} \, T_{EM}}} \frac{dt}{d\varphi} \tag{16}$$

where α_T is the flow coefficient, $A_{T,geo}$ denotes the cross-sectional area of the turbine, ψ is the flow function and p_{EM} is exhaust manifold pressure.

The temperature of the exhaust gases after the turbine is calculated according to:

$$T_{AT} = T_{EM} - \frac{|\Delta h_{T,is}|}{\eta_{T,is} \cdot c_{p,EG}} \tag{17}$$

2.9. Effective Engine Power

The indicated engine power is determined by integration of the total work of all cylinders during one duty cycle:

$$P_{ind} = \frac{n_M}{30 \, \tau} \sum_{i=1}^{z} \int \frac{dW_{C,i}}{d\varphi} \, d\varphi \tag{18}$$

where z denotes number of cylinders and n_M is crankshaft speed in rpm.

Effective engine power is calculated by the following equation:

$$P_{ef} = \frac{z\, n_M}{30\, \tau} V_S\, p_{mep} = P_{ind} \frac{p_{mep}}{p_{mip}} \tag{19}$$

where P_{mep} is the mean effective pressure and P_{ind} is the mean indicated pressure. The mean effective pressure is determined by subtracting the mean pressure of the mechanical losses from the mean indicated pressure. The mean pressure of mechanical losses takes into account losses caused by friction and operation of oil and water pumps. In the developed numerical model, the mean pressure of mechanical losses is calculated using empirical expressions according to Reference [43].

For easier understanding and tracking of the interconnections between individual equations and submodels, the block diagram of the engine numerical model is shown in Figure 3.

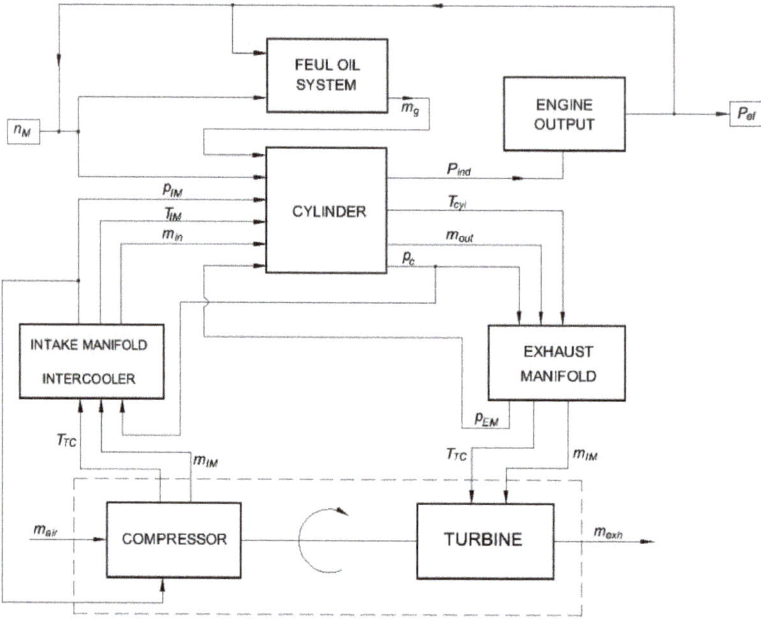

Figure 3. Block diagram of the engine numerical model.

3. Two-Zone Numerical Sub-Model

The formation of nitrogen oxides in the engine cylinder is exponentially dependent on the temperature at the boundary between the flame and the fresh medium in the cylinder. The formation of NO_x is exactly proportional to the available time in which the chemical reactions of formation and decomposition take place. Since the time in which the chemical reactions on which the formation of NO_x depends takes place is relatively short (depending on the engine speed), the process does not take place in conditions of chemical equilibrium.

The single-zone model only allows monitoring of the mean medium temperature in the engine cylinder. To predict the rate of NO_x formation with satisfactory accuracy, it is necessary to know the temperature at the boundary between the flame and the fresh medium in the engine cylinder. The following is a model in which during the part of the process in which the combustion and expansion of the working medium takes place, the control volume of the cylinder is divided into two zones. Such a model is also known in the literature as a quasi-dimensional combustion model and a detailed description of the two-zone model is presented in References [16,44,45].

In models with two or more zones, the formation of zones begins with the start of fuel combustion. After opening the exhaust valve, the process is observed as in models with single zone. Typically, a two-zone model divides the combustion space into a fresh medium zone and a zone made up of combustion products. The simplified model of the combustion process applied in this paper implies the division of the control volume of the engine cylinder into two zones:

- Zone 1—fresh mixture consisting of air, residual gases from the previous process and recirculated exhaust gases (only in case EGR is used), and
- Zone 2—combustion gases consisting of gaseous products of fuel combustion during stoichiometric combustion.

The model of the combustion process with two zones implies the following simplifications and assumptions:

- division of the working medium in the combustion space into two zones: the zone of fresh medium and the zone of combustion gases,
- the actual geometric shape of the zones is neglected and only their volume is taken into account,
- at the observed position of the crankshaft, the pressure in all zones is the same and does not depend on the position within the zone,
- at the observed position of the crankshaft, the temperature does not depend on the position within the zone and the same applies to the excess air,
- the working medium in each of the zones is a homogeneous mixture whose chemical composition and mass fractions of individual participants within the zone do not depend on the position within the zone,
- the formation of zones begins with the injection and combustion of fuel and until then there is only one zone,
- combustion in the combustion gas zone or in the edge layer ("flame front") takes place in the conditions of a slightly "poor" mixture,
- there is no heat exchange between zones,
- heat exchange takes place only between Zone 2 (combustion gas zone) and the environment,
- at the moment of opening the exhaust valve, both zones are instantly mixed into a homogeneous mixture.

A schematic representation of the formation of zones and changes in the mass of the medium depending on the crankshaft angle φ for the process in a four-stroke diesel engine is given in Figure 4. The process begins with the suction stroke, whereby the mass of the medium in the cylinder increases. After closing the inlet valve, the mass of the medium in the cylinder does not change during the compression stroke until the moment of fuel injection into the engine cylinder.

The applied numerical model assumes that the pressure in individual zones is equal to the pressure in the cylinder and that it forms a homogeneous pressure field. Therefore, the values for the pressure are obtained by the calculation using the single-zone model.

$$p_C = p_1 = p_2 \tag{20}$$

Subscript "1" refers to Zone 1 (fresh medium zone) and subscript "2" to Zone 2 (combustion gas zone).

The mass of the medium in the cylinder is calculated according to:

$$m_C = m_1 + m_2 \tag{21}$$

After closing the inlet valve, the mass of the medium in the cylinder does not change until the fuel injection begins. The total mass of fresh medium in the cylinder is the sum of the masses: clean air, residual combustion gases and recirculated combustion gases.

$$m_C = m_1 = m_A + m_{RG} + m_{EGR} \qquad (22)$$

Assuming that the combustion of fuel in the marginal layer (boundary between the zones) takes place with the prior mixing of the medium from Zone 1 with the injected fuel in a stoichiometric ratio, according to the expression:

$$\alpha_{st}^1 = \left(\frac{m_1}{m_f}\right)_{st} = \left(\frac{m_A + m_{RG} + m_{EGR}}{m_f}\right)_{st} \qquad (23)$$

The mass of the medium in Zone 2 is determined from the known mass of burned fuel (data obtained from the single-zone model) and the stoichiometric ratio for Zone 1, according to the expression:

$$m_2 = m_f\left(1 + \alpha_{st}^1\right) \qquad (24)$$

and the mass of the media in Zone 1 is calculated according to:

$$m_1 = (m_C + m_f) - \left[m_f\left(1 + \alpha_{st}^1\right)\right] = (m_A + m_{RG} + m_{EGR} + m_f) - m_2 \qquad (25)$$

At any time or position of the crankshaft, the sum of the volumes of both zones is equal to the volume of the cylinder.

$$V_C = V_1 + V_2 \qquad (26)$$

The volume of the zones is calculated using the equation of state of the ideal gas according to the expression:

$$V_i = \frac{m_i R_i T_i}{p_C} \qquad (27)$$

Subscript "i" refers to the zone.

The temperature of the gases in Zone 1 is calculated according to the expression for the adiabatic change of state:

$$T_{1,k} = T_{1,k-1}\left(\frac{p_{C,k}}{p_{C,k-1}}\right)^{\frac{\kappa-1}{\kappa}} \qquad (28)$$

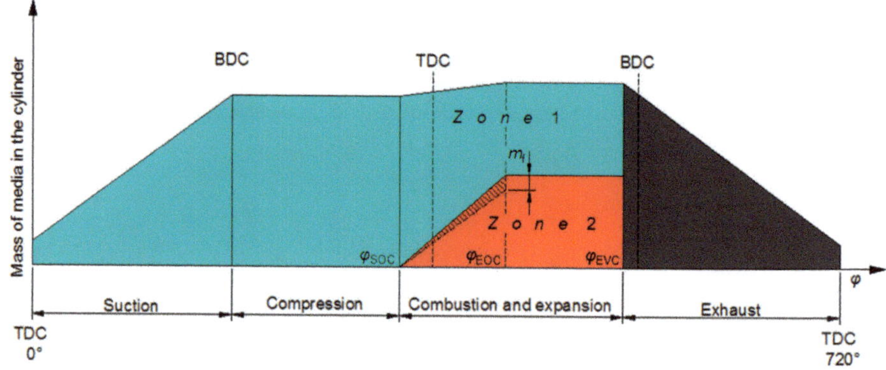

Figure 4. Formation of zones in the cylinder of a four-stroke diesel engine.

The subscript "k − 1" indicates the values of pressure and temperature from the previous calculation step.

The change in temperature in Zone 2 depending on the angle of the crankshaft is determined by applying the expression:

$$\frac{dT_2}{d\varphi} = \frac{1}{m_2\left(\frac{\partial u}{\partial T}\right)_2}\left[\frac{dQ_f}{d\varphi} + \frac{dQ_w}{d\varphi} - p_c\frac{dV_2}{d\varphi} - u_2\frac{dm_2}{d\varphi} + h_1\frac{dm_2}{d\varphi} - m_2\left(\frac{\partial u}{\partial \lambda}\right)_2\frac{d\lambda_2}{d\varphi}\right] \tag{29}$$

where subscript 1 correspond to Zone 1 and subscript 2 to Zone 2, f represents fuel and w denote cylinder walls.

The changes in the mass and temperature of the medium in the zones depending on the position of the crankshaft obtained by applying the described numerical model are shown in Figures 5 and 6.

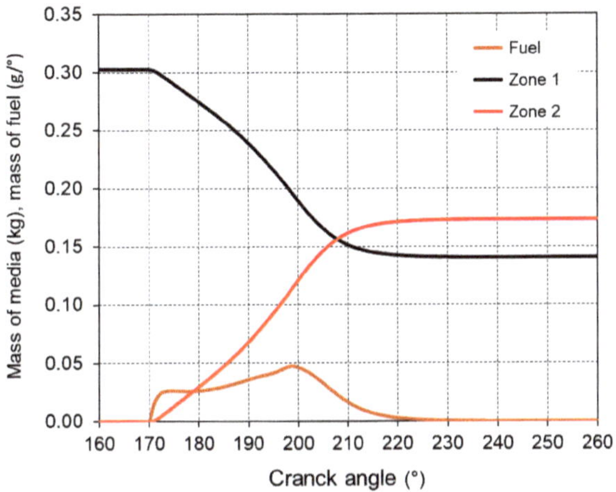

Figure 5. Change in mass of media in zones.

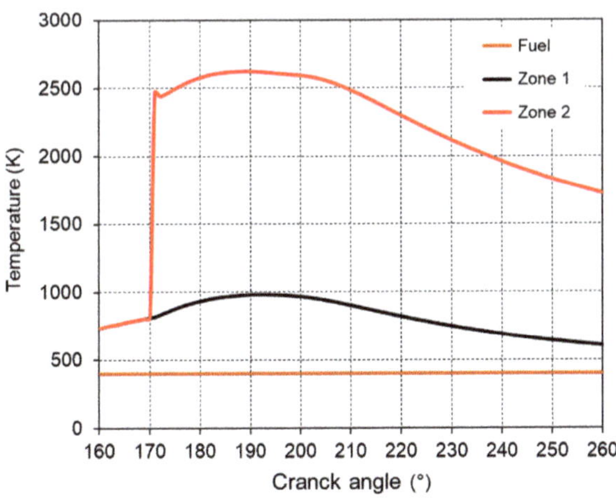

Figure 6. Change in temperature of media in zones.

4. Nitrogen Oxide Formation Submodel

The applied model of "thermal" nitrogen oxide formation is based on the extended Zeldovic mechanism. Models in which the formation of NO is described with three chemical reactions as in papers [19,22,46] are most often used in the literature. Chemical reactions of formation and decomposition of nitrogen monoxide are:

$$N_2 + O \leftrightarrow NO + N \tag{R1}$$

$$O_2 + N \leftrightarrow NO + O \tag{R2}$$

$$OH + N \leftrightarrow NO + H \tag{R3}$$

In the conditions prevailing during combustion and due to the short time due to the relatively high speed of the combustion process, equilibrium concentrations of NO do not occur. All other chemical reactions that take place in the combustion space are assumed to take place at high speed and that the concentration of chemical elements and compounds (O_2, H, H_2, OH, N, N_2, CO, CO_2 and H_2O) is in a state of chemical equilibrium.

Changes in the concentration of NO in the chamber in which the combustion process takes place are calculated according to the expression:

$$\frac{d[NO]}{dt} = k_{1,f}[O]_e[N_2]_e - k_{1,d}[NO][N]_e + k_{2,f}[N]_e[O_2]_e - k_{2,d}[NO][O]_e + k_{3,f}[N]_e[OH]_e - k_{3,d}[NO][H]_e \tag{30}$$

The concentrations of all elements in square brackets marked with the subscript "e" are calculated from the chemical equilibrium conditions. The coefficients of formation rate $k_{i,f}$ and decomposition rates $k_{i,d}$ are calculated using the expressions from Reference [47].

Concentrations of individual components under the condition of chemical equilibrium, depending on pressure, temperature and equivalent ratio of fuel and air Φ are calculated using the model described in Reference [48]. In this model, diesel fuel was replaced by a hydrocarbon $C_{12}H_{26}$.

5. Validation of Numerical Model of the Engine

The developed numerical model of the engine was validated using the engine manufacturer data and measurements acquired during sea trial of LNG ships with diesel-electric propulsion.

Validation of numerical model of the engine was based on the Wärtsilä 12V50DF engine data (Table 2). All presented data refers to engine performance running on heavy fuel oil (HFO).

Table 2. Wärtsilä 12V50DF engine data [49].

Engine General Technical Data	Value/Type
Bore, mm	500
Stroke, mm	580
Valves per cylinder (inlet/exhaust)	2/2
Inlet/outlet valve diameter, mm	165/160
Number of cylinders and configuration	12 cylinders, V/45°
Maximum continuous rating (MCR), kW	11,700
Engine speed, rpm	514
Mean piston speed, m s^{-1}	9.9
Number of turbochargers	2
Turbocharger type	ABB TPL71-C

Manufacturer's records at different loads (Table 3) and sea trial records LNG carrier (Table 4) were compared with the results obtained by numerical simulations.

Table 3. Manufacturer's data for Wärtsilä W 12V50 DF engine [49].

Engine Load	50%	75%	100%
Engine power, kW	5850	8775	11,700
Specific fuel oil consumption, g/kWh	196	187	189
Exhaust gas temperature after turbocharger, °C	337	336	352
Exhaust gas mass flow, kg/s	13.9	18.4	23.0

Table 4. Sea trial records for Wärtsilä 12V50 DF engine (sea trial).

Engine Load	40%	50%	71%
Engine power, kW	4680	5850	8307
Specific fuel oil consumption, g/kWh	199	197	190
Maximum cylinder pressure, bar	70	83	107
Exhaust gas temperature after turbocharger, °C	412	392	359

The simulation of engine operation under steady-state conditions was performed for five operating points in the range of 40% to 100% of the engine rated power. Model validation was performed by comparing data from Tables 3 and 4 with data obtained by simulating specific fuel consumption, maximum cylinder pressure, exhaust gas temperature and exhaust gas mass flow. Figure 7 shows closed indicating diagrams for five engine operating regimes obtained by developed engine operation simulation software.

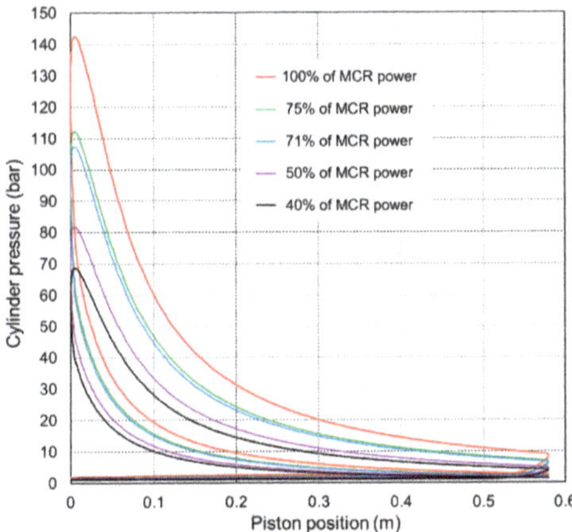

Figure 7. Closed indicated pressure diagrams at 40% to 100% of engine rated power.

Figure 8 shows the comparison of the specific fuel oil consumption measured on the test bed and during the sea trial with the values obtained by the engine simulation model. The largest deviation occurs at 40% of the engine load, which is approximately 3.5%, that is, 7 g/kWh. The smallest deviation from the measured data occurs between 50% and 71% of the maximum engine load and is less than 1%, that is, 2 g/kWh.

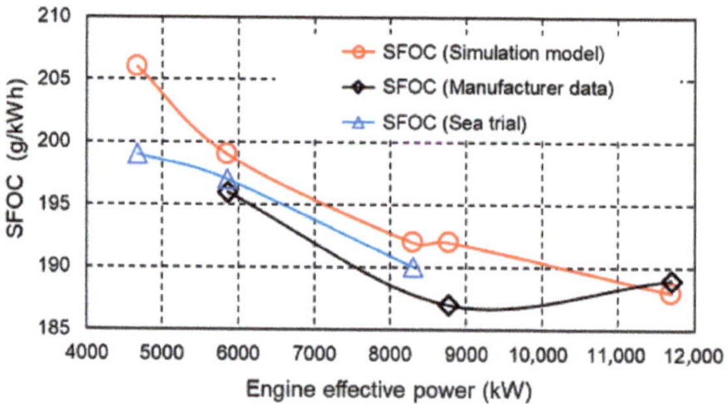

Figure 8. Comparison of the specific fuel oil consumption.

Figure 9 shows a comparison of the maximum pressures in the engine cylinder. The absolute pressure deviations are less than 1 bar at all observed operating regimes. The maximum pressure in the engine cylinder measured during the sea trial is presented as an average value of all 12 cylinders.

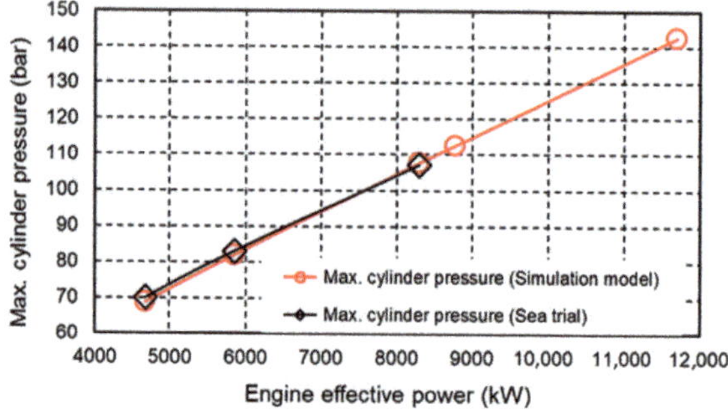

Figure 9. Comparison of maximum pressures in the engine cylinder.

Figure 10 shows the comparison of exhaust gas temperatures after the turbocharger. The biggest difference occurs at 50% of the engine load and its value is 10.4%, that is, 35 °C. The deviation at the same operating point compared to the sea trial data is 6.8%, that is, 20 °C. The difference between the exhaust gas temperatures after the turbocharger is only 1.7%, that is, 6 °C at full engine load.

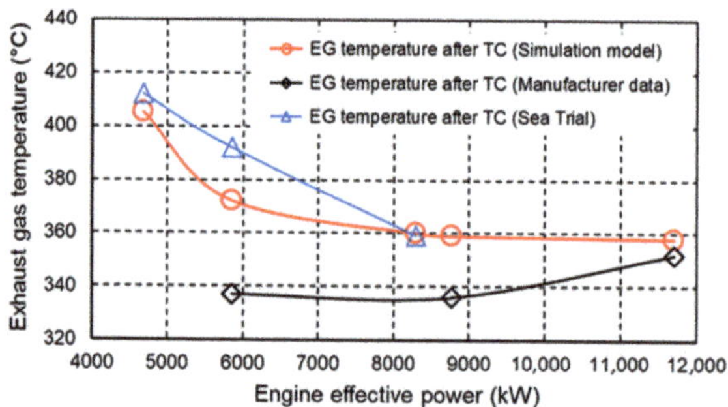

Figure 10. Comparison of the exhaust gas temperatures after turbocharger.

Comparison of the exhaust gas mass flows shown in Figure 11 indicate on very small deviations between manufacturer's data and results obtained by numerical model of the engine.

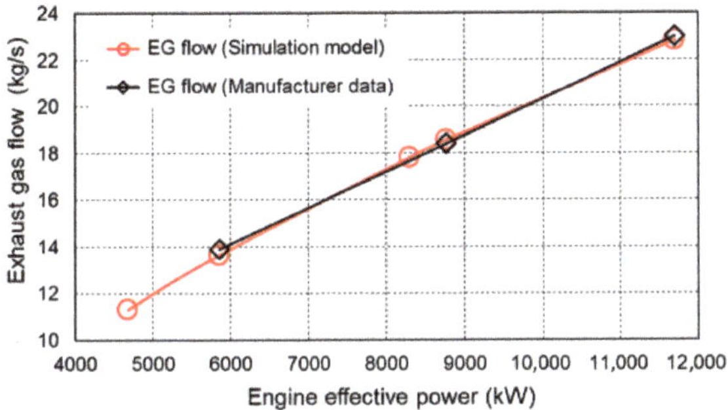

Figure 11. Comparison of exhaust gas mass flows.

6. Split Fuel Injection Impact on SFOC and NO$_x$ Emission

Split fuel injection can be used as a method to reduce NO$_x$ emissions during combustion due to decreasing of temperature and pressure in the cylinder. The process in the diesel engine cylinder from the beginning of the fuel injection to the end of the fuel combustion takes place in four phases, as shown in Figure 12. The first phase (1) is called the ignition delay and it involves the evaporation and mixing of the fuel until the conditions for ignition of the resulting fuel mixture are met. The second phase (2) is characterized by the relatively intensive combustion of the fuel mixture formed during the ignition delay period. The second phase is called the combustion of the previously formed mixture (premixed burning) and there is an intense release of heat which causes a sudden rise in temperature and pressure in the cylinder. After the initial sharp increase in the heat release rate, a third phase (3) follows in which the combustion rate is controlled by the rate of mixing of the remaining fuel with the fresh medium. In the third phase, called mixing controlled combustion, the heat release rate is lower than in the previous phase. During the fourth phase (4) of combustion, the remaining fuel burns out and this phase is called late combustion.

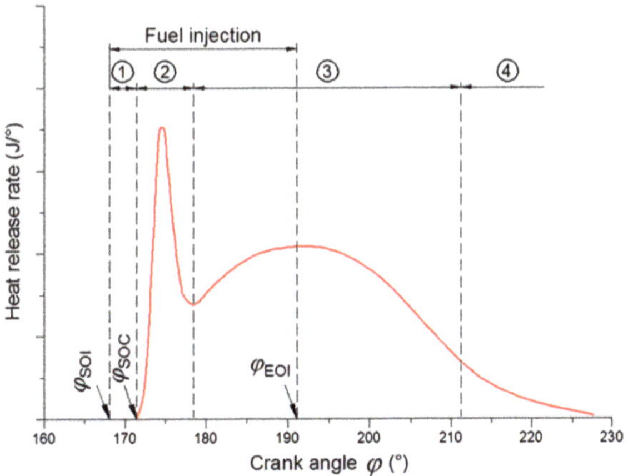

Figure 12. Heat release rate and phase of combustion process.

The use of split injection can significantly affect the course of the combustion process or the amount of harmful substances in the exhaust gases of diesel engines. Pilot injection has a significant impact on the reduction of NO$_x$ emissions as well as noise generated during the combustion. While subsequent fuel injection can achieve a reduction in soot emissions, as well as an increase in the exhaust gas temperature required when applying secondary exhaust aftertreatment measures. The basic principle of split fuel injection into the engine cylinder is shown in Figure 13.

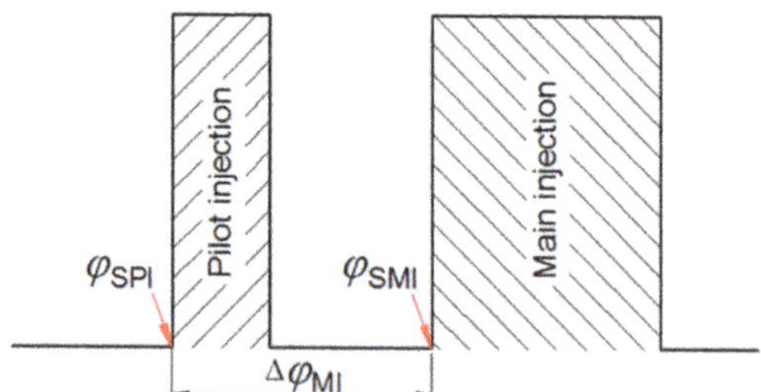

Figure 13. The basic principle of split fuel injection.

In Figure 13, the following labels were used:

φ_{SPI}—start of pilot injection,

φ_{SMI}—start of the main injection,

$\Delta\varphi_{MI} = \varphi_{SMI} - \varphi_{SPU}$.

If applying an appropriate split injection scheme, it is possible to effectively control emissions. However, the attention has to be paid to the specific fuel consumption. It is necessary to choose an injection scheme in which a compromise will be reached between reducing emissions and increasing specific fuel consumption.

The results of experimental research [50–52] have shown that increasing the amount of pilot injected fuel increases NO_x emissions, while increasing the difference $\Delta\varphi_{MI}$ between the pilot and the main injection leads to an increase in specific fuel consumption.

In this study, the analysis of split injection was performed for stationary engine operating conditions at 50%, 75% and 100% of the rated engine power. For research purpose, nine schemes of fuel injection were selected.

For all three load cases, engine operation was simulated with 10%, 20% and 30% of pilot injection for total amount of fuel injected. Difference $\Delta\varphi_{MI}$ was varied as 3, 6 and 9 degrees of crankshaft angle for each pilot injection.

The corresponding injection schemes (Schemes 1–9) are marked as xx(y)zz, with "xx" and "zz" respectively giving the amount of fuel injected in the first pilot or second main injection phase. While "y" represents the angle of rotation of the crankshaft between the pilot and the main injection. Fuel quantities are expressed as percentages of the total amount of fuel injected into the cylinder per process.

The results obtained by computer simulation are presented and compared with the "basic" motor in the form of a diagram.

Figure 14 shows the heat release rate curves for the nine fuel injection schemes (Schemes 1–9) while Figure 15 shows the effect of split injection on the pressure in the cylinder during the high-pressure part of the process. Both figures are showing curves for 75% of MCR power and the shape of curves obtained for 50% and 100% are very similar.

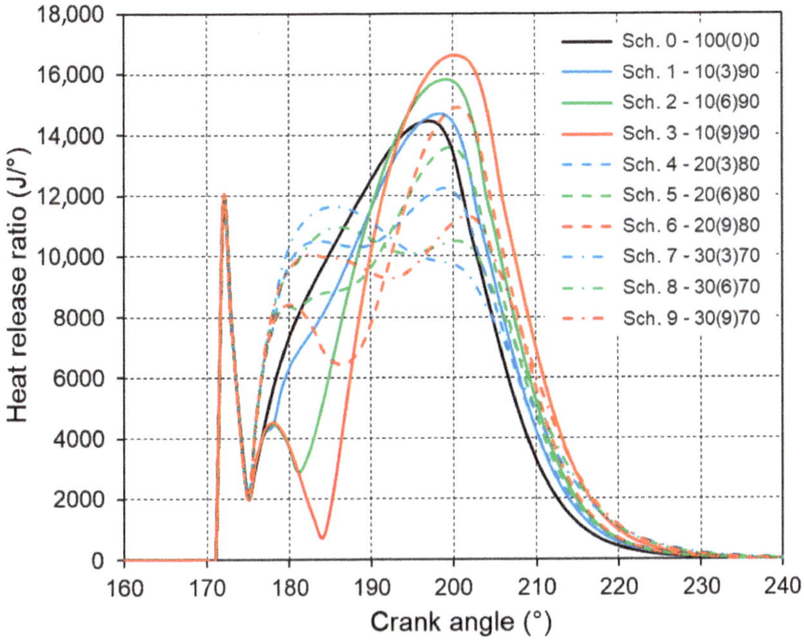

Figure 14. Influence of split injection on heat release at 75% of MCR power.

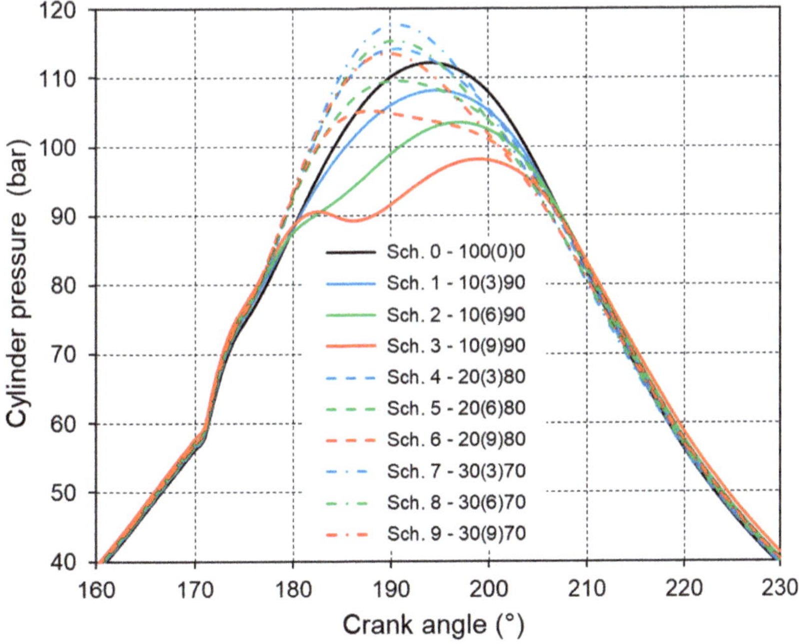

Figure 15. Influence of split injection on cylinder pressure at 75% of maximum continuous rating (MCR) power.

The diagrams shown in Figure 14 and the corresponding heat release rate curves show that split combustion consisting of "pilot" and "main" injection results in a reduction in heat release rate in the first part of the mixing controlled combustion phase whose intensity is determined by the rate of fuel mixture formation. In this case, a greater reduction in the rate of heat release in the first part of the mixing controlled combustion phase occurs with an increase in the difference $\Delta\varphi_{MI}$ between injections. As the proportion of injected "pilot" fuel increases, the effect on reducing the heat release rate decreases in the first part of the mixing controlled combustion phase but increases in the second part. The impact of split fuel injection on cylinder pressure is shown in Figure 15. There is a noticeable trend of decreasing cylinder pressure with increasing the difference $\Delta\varphi_{MI}$ between "pilot" and "main" injection. While increasing the amount of "pilot" fuel leads to a decrease in this effect. Further increase in the amount of "pilot" fuel also leads to an increase in the maximum pressure in the cylinder.

The effects of different nine split fuel injection schemes (Sch. 1 to 9) on SFOC, NO_x emission and maximum cylinder pressure at different engine loads compared to the base engine are shown in Figures 16–18.

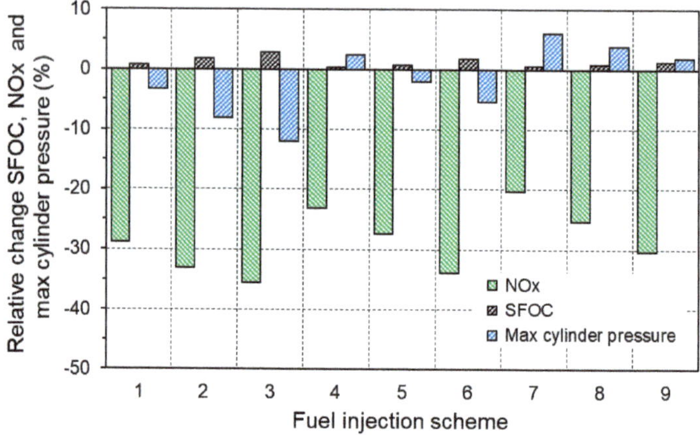

Figure 16. Influence of injection scheme on NO_x, specific fuel oil consumption (SFOC) and max cylinder pressure at 50% of MCR.

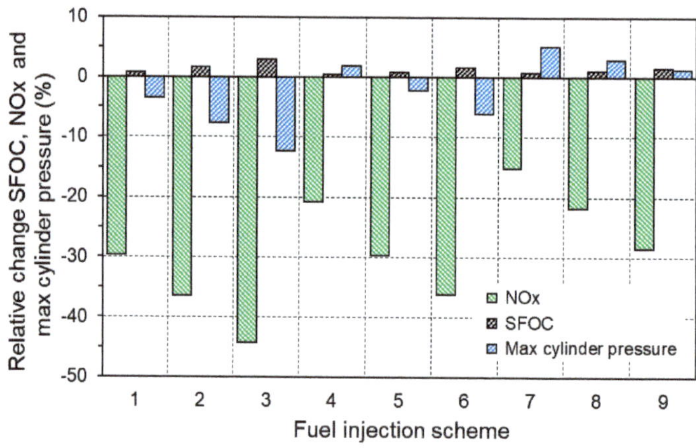

Figure 17. Influence of injection scheme on NO_x, SFOC and max cylinder pressure at 75% of MCR.

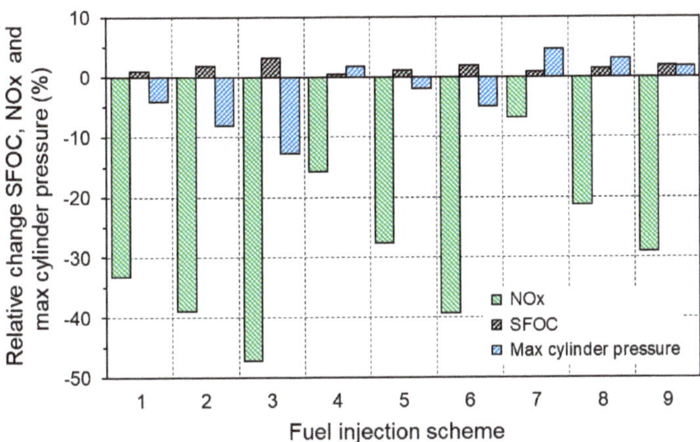

Figure 18. Influence of injection scheme on NO_x, SFOC and max cylinder pressure at 100% of MCR.

7. Conclusions

Marine propulsion systems are required to be as energy efficient as possible and to meet environmental protection standards. This paper analyzes the impact of split injection on fuel consumption and NO_x emissions of marine medium-speed diesel engines.

For the needs of the research, a zero-dimensional, two-zone numerical model of a diesel engine was developed. Comparison of the results obtained by the simulation with the available data of the engine manufacturer and from sea trials showed relatively small deviations of the numerical model in relation to the real engine. Sub-model based on the extended Zeldovich mechanism was applied to predict NO_x emissions.

The operation of the motor that drives synchronous generator was simulated under stationary conditions for three operating points and nine injection schemes. The results obtained by numerical simulations of engine operation indicate that by using split injection it is possible to achieve a relatively large reduction in NO_x emissions. However, all analyzed split injection schemes lead to increases in SFOC. Depending on the engine load, the NO_x emission is reduced from approximately 29% to 33% and the increase in fuel consumption specificity does not exceed 1%. It is possible to achieve greater reductions in NO_x emissions but with significant reductions in engine efficiency. The results are showing that NO_x emission reduction of up to 46% is achievable. However, this increases SFOC by approximately 3%. When increasing the angle between injection the maximum pressure decreases, so the amount of pilot injection is increased to compensate the difference.

Based on the results obtained by numerical simulations of engine operation, it is to be concluded that properly applied split fuel injection is an effective method for reducing NO_x emissions. If the appropriate scheme is applied it can be done without significant reduction in engine efficiency.

To continue research in this field an appropriate algorithm is to be developed since there are unlimited number of schemes that can be simulated and designed. Such an algorithm would allow faster and more accurate determination of the optimal injection scheme depending on the operating conditions of the engine.

Author Contributions: Formal analysis, R.R. and M.V.; Funding acquisition, M.V.; Investigation, V.P., T.M. and R.R.; Resources, M.V.; Software, V.P.; Validation, V.P. and T.M.; Writing—original draft, V.P.; Writing—review & editing, T.M. and R.R. All authors have read and agreed to the published version of the manuscript.

Funding: This work was partially supported by the Croatian Science Foundation under the project IP-2018-01-3739. This work was also supported by the University of Rijeka (project no. uniri-tehnic-18-18 1146 and uniri-tehnic-18-266 6469).

Conflicts of Interest: The authors declare no conflict of interest.

References

1. IMO-International Maritime Organization. Nitrogen Oxides (NOx)-Regulation 13. Available online: http://www.imo.org/en/OurWork/Environment/PollutionPrevention/AirPollution/Pages/Nitrogen-oxides-(NOx))-%E2%80%93-Regulation-13.aspx (accessed on 1 September 2020).
2. Wik, C. Reducing medium-speed engine emissions. *J. Mar. Eng. Technol.* **2010**, *9*, 37–44. [CrossRef]
3. Grimmelius, H.; Mesbahi, E.; Schulten, P.; Stapersma, D. The use of diesel engine simulation models in ship propulsion plant design and operation. In Proceedings of the 25th CIMAC World Congress, Wien, Austria, 21–24 May 2007; p. 227.
4. Heywood, J.B.; Sher, E. *The Two-Stroke Cycle Engine*; Taylor & Francis: London, UK, 1999.
5. Sorenson, S.C. *Engine Principles and Vehicles*; Technical University of Denmark: Lyngby, Denmark, 2008.
6. Mrakovčić, T. Osnivanje i Vođenje Brodskog Pogonskog Postrojenja Primjenom Numeričke Simulacije. Ph.D. Thesis, Faculty of Engineering, University of Rijeka, Rijeka, Croatia, 2003.
7. Račić, N. Simulacija Rada Brodskog Propulzijskog Sustava sa Sporohodnim Dizelskim Motorom u Otežanim Uvjetima. Ph.D. Thesis, Faculty of Engineering, University of Rijeka, Rijeka, Croatia, 2008.
8. Asay, R.J.; Svensson, K.I.; Tree, D.R. *An Empirical, Mixing-Limited, Zero-Dimensional Model for Diesel Combustion*; 2004-01-0924; SAE Technical Paper: Warrendale, PA, USA, 2004. [CrossRef]
9. Tauzia, X.; Maiboom, A.; Chesse, P.; Thouvenel, N. A new phenomenological heat release model for thermodynamical simulation of modern turbocharged heavy duty. *Diesel Engines* **2006**, *26*, 1851–1857. [CrossRef]
10. Descieux, D.; Feidt, M. One zone thermodynamic model simulation of an ignition compression engine. *Appl. Therm. Eng.* **2007**, *27*, 1457–1466. [CrossRef]
11. Jumg, D.; Assanis, D.A. *Multi-Zone DI Diesel Spray Combustion Model for Cycle Simulation Studies of Engine Performance and Emissions*; 2001-01-1246; SAE Technical Paper: Warrendale, PA, USA, 2001. [CrossRef]
12. Kulkarni, A.M.; Shaver, G.M.; Popuri, S.S.; Frazier, T.R.; Stanton, D.W. Computationally efficient whole-engine model of a Cummins 2007 turbocharged diesel engine. *J. Eng. Gas Turbines Power* **2010**, *132*. [CrossRef]
13. Andersson, M.; Johansson, B.; Hultquist, A.; Nöhre, C. *A Real Time NOx Model for Conventional and Partially Premixed Diesel Combustion*; 2006-01-0195; SAE Technical Paper: Warrendale, PA, USA, 2006. [CrossRef]
14. Maiboom, A.; Tauzia, X.; Shah, S.R.; Hetet, J.F. New phenomenological six-zone combustion model for direct injection diesel engines. *Energy Fuels* **2009**, *23*, 690–703. [CrossRef]
15. Durgun, O.; Sahin, Z. Theoretical investigation on heat balance in DI diesel engines for neat diesel fuel and gasoline fumigation. *Energy Convers. Manag.* **2009**, *50*, 43–51. [CrossRef]
16. Škifić, N. Analiza Utjecajnih Parametara Opreme na Značajke Dizelskog Motora. Ph.D. Thesis, University of Rijeka, Rijeka, Croatia, 2003.
17. Rakopoulos, C.D.; Rakopoulos, D.C.; Kyritsis, D.C. Development and validation of a comprehensive two-zone model for combustion and emissions formation in a DI diesel engine. *Int. J. Energy Res.* **2003**, *27*, 1221–1249. [CrossRef]
18. Kuleshov, A.S. *Use of Multi-Zone DI Diesel Spray Combustion Model for Simulation and Optimization of Performance and Emissions of Engines with Multiple Injection*; 2006-01-1385; SAE Technical Paper: Warrendale, PA, USA, 2006. [CrossRef]
19. Scappin, F.; Stefansson, H.S.; Haglind, F.; Andreasen, A.; Larsen, U. Validation of a zero-dimensional model for prediction of NOx and engine performance for electronically controlled marine two-stroke diesel engines. *Appl. Therm. Eng.* **2012**, *37*, 344–352. [CrossRef]
20. Jafarmadar, S. The Effect of Split Injection on the Combustion and Emissions in DI and IDI Diesel Engines. In *Diesel Engine-Combustion, Emissions and Condition Monitoring*; Bari, S., Ed.; InTech: Rijeka, Croatia, 2013.
21. Juntarakod, P.; Soontornchainacksaeng, T. A Quasi-dimensional Three-zone Combustion Model of the Diesel Engine to Calculate Performances and Emission Using the Diesel-Ethanol Dual Fuel. *Contemp. Eng. Sci.* **2014**, *7*, 19–37. [CrossRef]
22. Finesso, R.; Spessa, E. A real time zero-dimensional diagnostic model for the calculation of in-cylinder temperatures, HRR and nitrogen oxides in diesel engines. *Energy Convers. Manag.* **2014**, *79*, 498–510. [CrossRef]

23. Baldi, F.; Theotokatos, G.; Andersson, K. Development of a combined mean value-zero dimensional model and application for a large marine four-stroke Diesel engine simulation. *Appl. Energy* **2015**, *154*, 402–415. [CrossRef]

24. Sindhu, R.; Amba Prasad Rao, G.; Madhu, M.K. Effective reduction of NOx emissions from diesel engine using split injections. *Alex. Eng. J.* **2018**, *57*, 1379–1392. [CrossRef]

25. Savva, N.S.; Hountalas, D.T. Evaluation of a Semiempirical, Zero-Dimensional, Multizone Model to Predict Nitric Oxide Emissions in DI Diesel Engines'Combustion Chamber. *J. Combust.* **2016**, *2016*, 6202438. [CrossRef]

26. Sharma, S.; Sun, Y.; Vernham, B. Predictive Semi-Empirical NOx Model for Diesel Engine. *Int. J. Energy Power Eng.* **2019**, *13*, 365–376. [CrossRef]

27. Medica, V. Simulation of Turbocharged Diesel Engine Driving Electical Generator under Dynamic Working Conditions. Ph.D. Thesis, Faculty of Engineering, University of Rijeka, Rijeka, Croatia, 1988.

28. Pelić, V.; Mrakovčić, T.; Bukovac, O.; Valčić, M. Development and validation of 4 stroke diesel engine numerical model. *Pomor. Zb.* **2020**, *Special Edition*, 359–372.

29. Woschni, G. Die Berechnung der Wandwärmeverluste und der thermischen Belastung der Bauteile von Dieselmotoren. *MTZ-Mot. Z.* **1970**, *31*, 491–499.

30. Jankov, R. *Matematičko Modeliranje Strujno-Termodinamičkih Procesa i Pogonskih Karakteristika Dizel-Motora, I i II Dio*, 1st ed.; Naučna Knjiga: Beograd, Serbia, 1984.

31. Pflaum, W.; Mollenhauer, K. *Wärmeübergang in der Verbrennungskraftmaschine*, 1st ed.; Springer: Vienna, Austria, 1977; Volume 3. [CrossRef]

32. Löhner, K.; Döhring, E.; Chore, G. Temperaturschwingungen an der Innenwand von Verbrennungskraftmaschinen. *MTZ-Mot. Z.* **1956**, *12*, 413–418.

33. Hohenberg, G.F. *Advanced Approaches for Heat Transfer Calculations*; SAE Technical Papers: Warrendale, PA, USA, 1979. [CrossRef]

34. Heywood, J.B. Engine Combustion Modelling-An Overview. In *Combustion Modelling in Reciprocating Engines*; Plenum Press: New York, NY, USA, 1980; pp. 1–35.

35. Boulochos, K.; Papadopulos, S. Zur Modellbildung des motorischen Verbrennungsablaufes. *MTZ-Mot. Z.* **1984**, *45*, 21–26.

36. Vibe, I.I. *Brennverlauf und Kreisprozess von Verbrennungsmotoren*; Verlag Technik: Berlin, Germany, 1970.

37. Woschni, G.; Anisits, F. Eine Methode zur Vorausberechnung der Änderung des Brennverlaufs mittelschnellaufender Dieselmotoren bei geänderten Betriebsbedingungen. *MTZ-Mot. Z.* **1973**, *34*, 106–115.

38. Betz, A.; Woschni, G. Umsetzungsgrad und Brennverlauf aufgeladener Dieselmotoren im instationären Betrieb. *MTZ-Mot. Z.* **1986**, *47*, 263–267.

39. Sitkei, G. Über den dieselmotorischen Zündverzug. *MTZ-Mot. Z.* **1963**, *26*, 190–194.

40. Boy, P. Beitrag zur Berechnung des Instationären Betriebsverhaltens von Mittelschnellaufenden Schiffsdieselmotoren. Ph.D. Thesis, Universität Hannover, Hannover, Germany, 1980.

41. Chapman, K. *Engine Airflow Algorithm Prediction, Introduction to Internal Combustion Engines*; Kansas State University: Manhattan, KS, USA, 2001. [CrossRef]

42. Bošnjaković, F. *Nauka o Toplini II.*; Tehnička Knjiga: Zagreb, Croatia, 1976.

43. Maass, H.; Klier, H. *Kräfte, Momente und Deren Ausgleich in der Verbrennungskraftmaschine*; Springer: Vienna, Austria, 1981. [CrossRef]

44. Hohlbaum, B. Beitrag zur Rechnerischen Unrersuchung der Stickstoffoxid-Bildung Schellaufender Hohleistungsdieselmotoren. Ph.D. Thesis, Universitat Fridricijana Karlsruche, Karlsruche, Germay, 1992.

45. Heider, G.; Woschni, G.; Zeilinger, K. 2-Zonen Rechenmodell zur Vorausrech nung der NO-Emission von Dieselmotoren. *MTZ-Mot. Z.* **1998**, *59*, 770–775. [CrossRef]

46. Provataris, S.A.; Savva, N.S.; Chountalas, T.D.; Hountalas, D.T. Prediction of NO_x emissions for high speed DI Diesel engines using a semi-empirical, two-zone model. *Energy Convers. Manag.* **2017**, *153*, 659–670. [CrossRef]

47. Weisser, G.A. Modelling of Combustion and Nitric Oxide Formation for Medium-Speed DI Diesel Engines: A Comparative Evaluation of Zero- and Thre-Dimensional Approaches. Ph.D. Thesis, Universitat Fridricijana Karlsruche, Karlsruhe, Germay, 2001.

48. Rakopoulos, C.D.; Hountalas, D.T.; Tzanos, E.I.; Taklis, G.N. A fast algorithm for calculating the composition of diesel combustion products using 11 species chemical equilibrium scheme. *Adv. Eng. Softw.* **1994**, *19*, 109–119. [CrossRef]

49. *Wärtsilä 50DF-Product Guide*; Wärtsilä, Marine Solutions: Vaasa, Finland, 2019.

50. Nehmer, D.A.; Reitz, R.D. Measurement of the Effect of Injection Rate and Split Injections on Diesel Engine Soot and NO_x Emissions. In Proceedings of the SAE International Conference, Detroit, MI, USA, 28 February–3 March 1994.

51. Pierpont, D.A.; Montgomery, D.T.; Reitz, R.D. Reducing Particulate and NO_x Using Multiple Injections and EGR in a D.I. Diesel. In Proceedings of the SAE International Conference, Detroit, MI, USA, 27 February–2 March 1995.

52. Han, Z.; Uludogan, A.; Hampson, G.J.; Reitz, R.D. Mechanism of Soot and NO_x Emission Reduction Using Multiple-injection in a Diesel Engine. In Proceedings of the SAE International Conference, Detroit, MI, USA, 26–29 February 1996.

Publisher's Note: MDPI stays neutral with regard to jurisdictional claims in published maps and institutional affiliations.

Journal of
Marine Science and Engineering

Article

Improvement of Marine Steam Turbine Conventional Exergy Analysis by Neural Network Application

Sandi Baressi Šegota [1], Ivan Lorencin [1], Nikola Anđelić [1], Vedran Mrzljak [2,*] and Zlatan Car [1]

[1] Department of Automation and Electronics, Faculty of Engineering, University of Rijeka, Vukovarska 58, 51000 Rijeka, Croatia; sbaressisegota@riteh.hr (S.B.Š); ilorencin@riteh.hr (I.L.); nandelic@riteh.hr (N.A.); car@riteh.hr (Z.C.)

[2] Department of Thermodynamics and Energy Engineering, Faculty of Engineering, University of Rijeka, Vukovarska 58, 51000 Rijeka, Croatia

* Correspondence: vmrzljak@riteh.hr; Tel.: +385-51-651-551

Received: 13 October 2020; Accepted: 3 November 2020; Published: 5 November 2020

Abstract: This article presented an improvement of marine steam turbine conventional exergy analysis by application of neural networks. The conventional exergy analysis requires numerous measurements in seven different turbine operating points at each load, while the intention of MLP (Multilayer Perceptron) neural network-based analysis was to investigate the possibilities for measurements reducing. At the same time, the accuracy and precision of the obtained results should be maintained. In MLP analysis, six separate models are trained. Due to a low number of instances within the data set, a 10-fold cross-validation algorithm is performed. The stated goal is achieved and the best solution suggests that MLP application enables reducing of measurements to only three turbine operating points. In the best solution, MLP model errors falling within the desired error ranges (Mean Relative Error) MRE < 2.0% and (Coefficient of Correlation) $R^2 > 0.95$ for the whole turbine and each of its cylinders.

Keywords: exergy destruction; exergy efficiency; marine steam turbine; MLP neural network; turbine cylinders

---

1. Introduction

The dominant usage of steam turbines worldwide is related to electrical generator drive and electricity production [1,2]. Steam power plants, with steam turbines as essential components, can be assembled by following various methodologies. Along with conventional steam power plants [3] and nuclear power plants [4,5], the novel approach in steam power plant design is the usage of various renewable energy sources which can notably improve steam power plant operation and its efficiency, and which are very beneficial to the environment [6,7]. The reduction of harmful emissions from such plants is today one of the most important research and scientific topic and multiple researchers are developing various techniques and processes with a goal of emissions reduction [8–10].

In addition to being used as stand-alone systems, steam power plants can be integrated into more complex systems, such as combined cycle power plants [11]. In combined cycle power plants, waste heat from the gas turbine is used for superheated steam production—in such a way, the environment is protected from huge waste heat amount and the reduction in harmful emissions (in comparison to pure gas or pure steam power plants) is also notable [12,13]. Such an operation of combined cycle power plants results in high efficiency, much higher in comparison to the conventional or nuclear steam power plants [14,15].

In marine power systems, internal combustion engines take a dominant share in the entire world fleet [16]. Due to internal combustion engine dominancy, various researchers are involved in investigating improvements as well as in minimizing the overall harmful impact on the environment [17–19]. Steam power plants are generally rarely used in marine power systems. However, there are several marine engineering fields in which steam power plants are still dominant. Additionally, steam power plants can be found as a part of new, complex marine systems that are currently under development [20,21]. All aforementioned benefits of combined cycle power plants are also present in the marine power systems [22,23]. The complexity of combined cycle power plants, especially for marine usage, requires adequate control and regulation systems for its proper (or if possible, optimal) operation.

One of the marine fields in which steam power plants are still predominantly used is the propulsion of LNG (Liquefied Natural Gas) carriers, but it should be noted that the utilization share of internal combustion engines, especially dual-fuel engines, is also increasing [24,25]. There are several steam turbines in the steam propulsion plant of any steam-powered LNG carrier. Along with the main turbine used for the propulsion propeller (or several propellers) drive, in such plants two or more steam turbines are mounted for the electrical generators drive (turbo-generators) [26,27] and low power steam turbine for the main feed water pump drive [28].

Analysis of any component from the steam power plant can be performed by using various approaches and techniques presented in the literature [29,30]. For the analysis of the main marine steam turbine observed in this research, exergy analysis is used. Exergy analysis of any component or the entire system is a technique that offers many benefits in comparison to other analysis methods. Exergy analysis does not take into consideration processes that occur inside any component—for the exergy analysis, only fluid flows and heat transfer (to and from the analyzed component) as well as used or produced mechanical power are necessary. Therefore, for the exergy analysis, details about the analyzed component's inner structure are not required, simplifying all necessary measurements [31,32]. On the other hand, a lack of information about the inner structure of the observed component does not allow research and analysis of many details and processes inside the component. All the exergy analysis benefits can be seen in a variety of scientific papers that take this analysis as the baseline.

If considering the analyses of the entire power plants, Ahmadi and Toghraie [33] applied exergy analysis for the investigation of Montazeri steam power plant in Iran, while Si et al. [34] used the same analysis for the investigation of a 1000 MW double reheat ultra-supercritical power plant. Ibrahim et al. [35] analyzed the thermal performance of the gas turbine power plant, while Aghbashlo et al. [36] observed the performance assessment of a wind power plant by using exergy analysis. AlZahrani and Dincer, I. [37] observed parabolic trough solar power plant and Abuelnuor et al. [38] investigated Garri "2" combined cycle power plant also by using exergy analysis.

Exergy analysis is successfully applied in the performance analysis of many components and processes from various plants. Zhao et al. [39] used exergy analysis for the investigation of the turbine system in a 1000 MW double reheat ultra-supercritical power plant. Medica-Viola et al. [40] used exergy analysis for the performance analysis of low-power steam turbine with one extraction used in marine applications. Presciutti et al. [41] applied exergy analysis for the investigation of glycerol combustion in an innovative flameless power plant. Szablowski et al. [42] used exergy analysis for the investigation of an adiabatic compressed air energy storage system. Arshad et al. [43] performed a review of the exergy analysis usage in the investigation of fuel cells. Lorencin et al. [44] applied exergy analysis for the analysis of steam mass flow rate leakage through steam turbine labyrinth (gland) seals.

Exergy analysis can also be a baseline for the economic analysis of various power plants or its components [45–47]. From the literature, it can be found that exergy analysis is used in the investigation and observation of many other plants, processes and components.

Along with exergy analysis, an extensive literature review also shows that many scientists and researchers used various artificial intelligence methods and processes in the analysis of power plants or its components.

One of the most used artificial intelligence methods in the energy sector is MLP (Multilayer Perceptron) neural network. Sun et al. [48] developed a new MLP-based soft sensor for SO_2 power plant emissions detection. Several researchers [49–52] used MLP for predicting electrical power output from various complex power plants. Wahid et al. [53] applied MLP for the prediction of energy consumption in the buildings. Tahan et al. [54] used MLP for condition-based maintenance of gas turbine, while Lorencin et al. [55] also used MLP for condition-based maintenance, but not for the gas turbine only, then for the entire marine CODLAG (Combined Diesel and Gas) propulsion system. Many other authors also used MLP for the condition-based maintenance problems of various plants and components [56,57].

MLP can also be used for predicting ship speed by using some of the ship propulsion system parameters [58]. Detecting and diagnosing faults by applying MLP in a steam turbine that operates in a thermal power plant was presented Dhini et al. [59], while Tian et al. [60] and Ayo-Imoru and Cilliers [61] used MPL for detecting various losses and prevention of accidents in the nuclear power plants. Various other neural network applications can also be found in the literature in many energy sectors and processes [62–64].

An extensive literature review shows that MLP is not used currently for tracking operating parameters or performances of the main marine steam turbine and all its cylinders. Additionally, the possibility that measurements are reduced by MLP neural network application is not investigated. During the possible reduction of measurements, the dominant goal for MLP must be high accuracy and precision in the prediction of any required operating parameter. The intention of this paper is not only to fill the literature gap but also to show possibilities that neural network applications offer in marine systems (or its components) and to be a guideline for other researchers interested in this field.

In the presented paper, exergy analysis of the main marine steam turbine (as well as both of its cylinders) is performed. The analysis is based on the measurement data obtained during steam turbine exploitation at 24 different loads. At the beginning, the conventional exergy analysis is performed, which requires many measurements at each turbine load. After conventional analysis, an exergy analysis is performed by the MLP neural network application. The application of MLP can significantly reduce the amount of performed measurements, while the accuracy and precision of the obtained exergy analysis parameters remain high, regardless of the observed load. This analysis can be a guideline for reducing control and measurement equipment inside the marine steam power plant, especially on new ships.

2. Description and Operation Principle of the Analyzed Main Marine Steam Turbine

The main marine steam turbine analyzed in this paper operates at the 100,450 tons (gross tonnage) commercial LNG carrier. A steam turbine is used for the LNG carrier propulsion. The maximum mechanical power that can be produced by the observed turbine is 29,420 kW, according to the manufacturer specifications [65]. The general scheme of the analyzed turbine, along with operating points required for the exergy analysis, is presented in Figure 1.

The main marine steam turbine is composed of two cylinders and these are High Pressure Cylinder (HPC) and Low Pressure Cylinder (LPC). Each marine steam propulsion system consists of two parallel operating steam generators which produce superheated steam and delivers the majority of cumulatively produced steam mass flow rate (it depends on the system load) to the HPC inlet [66]. HPC has one steam extraction used for steam delivery to various auxiliary steam plant processes—HFO (Heavy Fuel Oil) heater, BOG (Boil Off Gas) heater, water heater for the crew requirements, etc. Both steam generators in this marine steam plant simultaneously use HFO and BOG during operation. After extraction, the remaining steam mass flow rate expands through HPC until the cylinder outlet. HPC consists of one Curtis and seven Rateau stages.

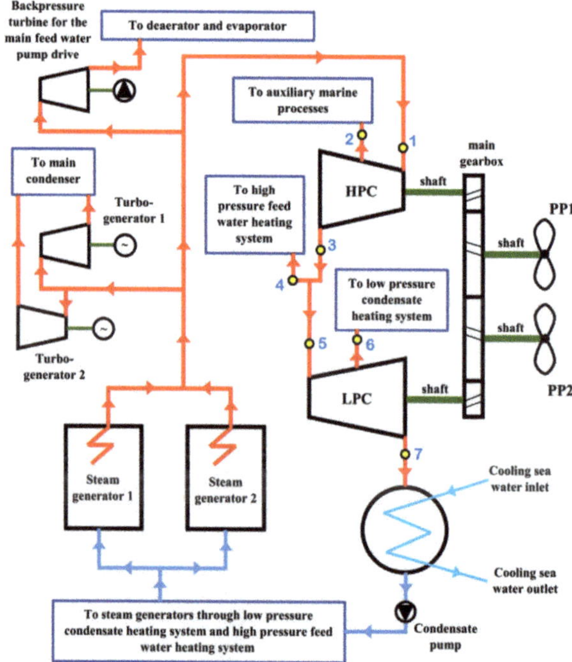

Figure 1. Scheme of the main marine steam turbine along with operating points required for the exergy analysis. HPC: High Pressure Cylinder; LPC: Low Pressure Cylinder; PP1: The first Propulsion Propeller; PP2: The second Propulsion Propeller.

The analyzed main steam turbine is an older variant of marine propulsion steam turbines and it is designed without steam reheating. Newer variants of marine propulsion steam turbines have one additional cylinder, Intermediate Pressure Cylinder (IPC), and steam reheating (steam reheaters are mounted inside steam generators) [67,68]. Such an upgrade increases plant overall efficiency, but simultaneously increases plant complexity and requires more stringent maintenance in comparison to older marine propulsion steam turbines.

Between the HPC and LPC of the analyzed turbine, one additional extraction is mounted for steam delivery to a high pressure feed water heating system. In the observed commercial LNG carrier, the high pressure feed water heating system consists of one high pressure feed water heater and deaerator [69]. In some operating regimes, when required, a part of steam extracted in this extraction (operating point 4, Figure 1), is delivered to air heaters used for heating of air at the steam generators entrance.

The remaining steam mass flow rate (operating point 5, Figure 1) expands through LPC. LPC, similar to HPC, has one steam extraction which is used for steam delivery to low pressure condensate heating system, which in the observed plant consists of one low pressure condensate heater and evaporator. An evaporator is a component used for the freshwater production (from sea water) and simultaneously for condensate heating [70]. After expansion in LPC, the remaining steam mass flow rate is delivered to the main marine steam condenser for condensation [71]. LPC consists of eight Rateau stages.

All three steam extractions from the observed turbine are not open all the time during the main turbine operation. Regulation valves opened and closed each of these extractions (and regulate extracted steam mass flow rate in each extraction) according to the predefined regulation procedure (following operation dynamic of the whole plant).

Both cylinders of the observed turbine are connected to the main marine gearbox through which one or two propulsion propellers are driven (in Figure 1 are shown two propulsion propellers, PP1 and PP2) [72].

It should be noted that in this exergy analysis (both conventional and with MLP application) several additional losses that occur in the plant, related to the main steam turbine, are neglected. For example, these losses are steam mass flow rate leakage through gland seals of each cylinder [73], heat losses in the pipelines and through the housing of each cylinder, mechanical losses [74], etc. Although important, all of these losses have a minor impact on the exergy analysis results of the observed steam turbine and each turbine cylinder.

3. Conventional Exergy Analysis of Main Marine Steam Turbine and Each of its Cylinders

Conventional exergy analysis of the main marine steam turbine and its cylinders is characterized by the fact that three steam operating parameters (steam temperature, pressure and mass flow rate) must be measured in each operating point presented in Figure 1 at each turbine load. Thus, each turbine load requires 21 measured data in order to be able to perform conventional exergy analysis at that particular load for the whole turbine and each cylinder. Change in turbine load requires new three measured data in each operating point. Therefore, conventional exergy analysis of observed main marine steam turbine and its cylinders require extensive measurements at each turbine load. Without these measurements (if any steam operating parameter, in any operating point from Figure 1 is missing), proper conventional exergy analysis cannot be performed or certain approximations must be used.

3.1. Overall Exergy Analysis Balances and Equations

In comparison to the energy analysis, of which results are not dependable on the ambient conditions [75,76], exergy analysis of any control volume or a system is dependable on the ambient conditions (ambient temperature and pressure) [77,78]. For the proper conventional exergy analysis of any control volume or a system, an overall exergy balance, mass flow rate balance and the most important variables should be defined. These overall equations and balances are valid in any exergy analysis, as well as in exergy analysis of the observed main marine steam turbine and both of its cylinders [79].

The overall steady-state exergy balance equation is defined as recommended in [80] by using an Equation (1):

$$\dot{Q}_{EX} + P_{INLET} + \sum \dot{Ex}_{INLET} = P_{OUTLET} + \sum \dot{Ex}_{OUTLET} + \dot{Ex}_{DES}. \tag{1}$$

where P is the mechanical power (used or produced) and $\dot{Ex}_{DES}$ is exergy destruction (exergy loss). It should be highlighted that in the overall exergy balance equation, potential and kinetic energies are disregarded due to its low influence on the overall balance. For the analyzed main marine steam turbine and its cylinders potential and kinetic energies in above balance are also low, and their inclusion will not bring meaningful change in the obtained results [81]. $\dot{Q}_{EX}$ is the exergy transfer by heat at the temperature T, of which the definition can be found in the literature [82] through the following Equation (2):

$$\dot{Q}_{EX} = \sum \left(1 - \frac{T_0}{T}\right) \cdot \dot{Q}, \tag{2}$$

where $\dot{Q}$ is an energy transfer by heat, T is temperature and index 0 corresponds to the state of the ambient. The last undefined variable from the overall exergy balance equation is a total exergy power of operating medium flow ($\dot{Ex}$), of which definition can be found in [83], Equation (3):

$$\dot{Ex} = \dot{m} \cdot \varepsilon. \tag{3}$$

where $\dot{m}$ is operating medium mass flow rate and ε is specific flow exergy of operating medium. Operating medium specific flow exergy is calculated according to the Equation (4), [84]:

$$\varepsilon = (h - h_0) - T_0 \cdot (s - s_0), \tag{4}$$

where h is operating medium specific enthalpy and s is operating medium specific entropy. During control volume or a system standard operation, mass flow rate leakage did not occur. By taking into account the fact that for the analyzed main marine steam turbine and its cylinders all the standard small steam leakages (as for example, leakage through gland seals of each cylinder) are neglected as described above, the valid mass flow rate balance is [85], Equation (5):

$$\sum \dot{m}_{\text{INLET}} = \sum \dot{m}_{\text{OUTLET}}. \tag{5}$$

The overall definition of the exergy efficiency can be presented as proposed in [86], Equation (6):

$$\eta_{\text{EX}} = \frac{\text{CUMULATIVE EXERGY OUTLET}}{\text{CUMULATIVE EXERGY INLET}}, \tag{6}$$

with the note that exergy efficiency of any observed control volume or a system can significantly differ from the overall definition, which depends on operating characteristics and operation principles of each control volume or a system.

3.2. Equations for the Exergy Aanalysis of Main Marine Steam Turbine and Its Cylinders

For each cylinder and the whole turbine, the first step is the calculation of the developed mechanical power. This step represents the essential element in exergy analysis equations. After developing the mechanical power equations, equations for the calculation of exergy destruction and exergy efficiency will be presented for each cylinder and the whole turbine. All the equations are defined according to recommendations from the literature [87,88] and are related to operating points presented in Figure 1.

3.2.1. High Pressure Cylinder (HPC)

Developed mechanical power, Equation (7):

$$P_{\text{HPC}} = \dot{m}_1 \cdot (h_1 - h_2) + \left(\dot{m}_1 - \dot{m}_2\right) \cdot (h_2 - h_3). \tag{7}$$

Exergy destruction (exergy loss), Equation (8):

$$\dot{Ex}_{\text{DES,HPC}} = \dot{Ex}_1 - \dot{Ex}_2 - \dot{Ex}_3 - P_{\text{HPC}}. \tag{8}$$

Exergy efficiency, Equation (9):

$$\eta_{\text{EX,HPC}} = \frac{P_{\text{HPC}}}{\dot{Ex}_1 - \dot{Ex}_2 - \dot{Ex}_3}. \tag{9}$$

3.2.2. Low Pressure Cylinder (LPC)

Developed mechanical power, Equation (10):

$$P_{\text{LPC}} = \dot{m}_5 \cdot (h_5 - h_6) + \left(\dot{m}_5 - \dot{m}_6\right) \cdot (h_6 - h_7). \tag{10}$$

Exergy destruction (exergy loss), Equation (11):

$$\dot{Ex}_{\text{DES,LPC}} = \dot{Ex}_5 - \dot{Ex}_6 - \dot{Ex}_7 - P_{\text{LPC}}. \tag{11}$$

Exergy efficiency, Equation (12):

$$\eta_{\text{EX,LPC}} = \frac{P_{\text{LPC}}}{\dot{E}x_5 - \dot{E}x_6 - \dot{E}x_7}. \tag{12}$$

3.2.3. Whole Turbine (WT)

Developed mechanical power, Equation (13):

$$P_{\text{WT}} = P_{\text{HPC}} + P_{\text{LPC}}. \tag{13}$$

Exergy destruction (exergy loss), Equation (14):

$$\dot{E}x_{\text{DES,WT}} = \dot{E}x_1 - \dot{E}x_2 - \dot{E}x_4 - \dot{E}x_6 - \dot{E}x_7 - P_{\text{WT}}. \tag{14}$$

Exergy efficiency, Equation (15):

$$\eta_{\text{EX,WT}} = \frac{P_{\text{WT}}}{\dot{E}x_1 - \dot{E}x_2 - \dot{E}x_4 - \dot{E}x_6 - \dot{E}x_7}. \tag{15}$$

In the equations for the exergy destruction and exergy efficiency of the whole turbine and each cylinder, total exergy power of steam flow ($\dot{E}x$) and steam specific flow exergy (ε) should be calculated using Equations (3) and (4) in each operating point from Figure 1. Additionally, the overall exergy balance equation, Equation (1), and steam mass flow rate balance, Equation (5), should always be satisfied for each cylinder and the whole turbine at each load.

4. Exergy Analysis of Main Marine Steam Turbine and Each of its Cylinders by MLP Neural Network Application

MLP is a neural network that consists of artificial neurons arranged into multiple layers. MLP consists of at least three layers—an input layer, output layer and one or more hidden layers [89,90]. Neurons in input layers are used to set inputs for the MLP model. The number of neurons in that layer is equal to the number of inputs of the data set, and their values are set to the number of input values contained within the data set [89,91]. The subsequent layers consist of neurons whose values are calculated depending on inputs and connection weights—with each artificial neuron in the subsequent layer being connected to all the artificial neurons in the preceding neurons with weighted connections [92,93]. If it is assumed that the value of the neuron is y_i^k, where k represents the layer number and i the neuron number within the layer, then the value of that particular artificial neuron is calculated as the activated weighted sum of artificial neurons of the previous layer [89], Equation (16):

$$y_i^k = F\left(\sum_{j=0}^{n_k} \theta_{j,i}^{k-1} \cdot y_j^{k-1} \right), \tag{16}$$

where $\theta_{j,i}^{k-1}$ represents the weight of the connection between the artificial neuron j in layer $k-1$ (y_j^{k-1}) and the artificial neuron i in the layer k (y_i^k). F represents the activation function—the function used to map the value of neuron into the desired range of values. Commonly used activation functions may [94,95]:

- Eliminate the unwanted values such as Rectified linear unit—ReLU ($y = \max(0,x)$)—used to eliminate negative values [95],
- Map the input files to a certain range such as sigmoid (logistic) function which maps the values to a range of $[0,1]$ ($y = \frac{1}{1+e^{-x}}$) or hyperbolic tangent function which maps them to the range of $[-1,1]$ ($y = tanh(x)$),

- Simply map the input directly to output as is the case with the identity activation function $(y = x)$ [96,97].

As MLP belongs to the family of machine learning algorithms, it has the ability to adjust itself to the data used for training it. This is done through the process of training, divided into forward and backward propagation [89,98]. In the forward propagation part of the training process, a single set of the input data values (steam temperature, pressure and mass flow rate in used operating points, Figure 1, along with the ambient pressure and temperature) are used as input neuron values (where the number of input neurons equals the total number of inputs). Then, weights of inter-neuron connections are set randomly and the values of neurons in the hidden layer. Finally, the output layer values are calculated using Equation (16) [89,92]. The output value is compared to the value of one of the outputs—either exergy destruction (exergy loss) or exergy efficiency—and the general output contained within the dataset is marked with y. It can be expected that the MLP output, marked $\hat{y}$, will have a certain error $\epsilon = \sqrt{(y - \hat{y})^2}$. This error is then used in the backward propagation process in order to adjust the weights based on the gradient of the error (with a higher error values causing a larger adjustment being made to the weights). If the vector of the weights in a layer k is marked as $\Theta^k = \left[\theta_1^k \theta_2^k \cdots \theta_{n_k}^k\right]$, and learning rate with α, this can be written as [58,89], Equation (17):

$$\Theta_{new}^k = \Theta_{old}^k - \alpha \frac{\partial \epsilon}{\partial \Theta_{old}^k}. \tag{17}$$

By repeating the described training process for multiple sets of input and output values, the MLP weights can be finely adjusted and provide a very low error when used as a trained model. The dataset consists of 125 points, each of which has data entries for the ambient temperature and pressure, as well as steam mass flow rate, temperature and pressure in each of the seven operating points (Figure 1), as well as values of exergy destruction and exergy efficiency for high pressure cylinder, low pressure cylinder and the whole turbine. This means that each data point has 23 input values and 6 output values. It should be noted that by its nature, MLP can only regress a single value within a model, and the number of inputs need to be fixed. Due to this fact, each output and each separate input set need to have a separate model trained.

Each of the listed input combinations will have two models trained—one for each of the possible outputs—exergy destruction and exergy efficiency. Due to this, the total number of final models is 72. Each of the models trained will need to have hyperparameters adjusted to achieve a quality regression performance. Hyperparameters are values which describe the general architecture of the neural network used to train the model. Hyperparameters of the MLP need to be varied to achieve the best regression models for each case. One of the varied hyperparameters is the earlier described activation function of the hidden layer neurons [96]. Further hyperparameters include the number of hidden layers and the number of neurons per hidden layer expressed as $(k_1, k_2, \cdots k_n)$ in which the total number of layers is n and k_i represents the number of neurons in layer i. The algorithm used for calculating the weight values during the training process, called a solver, is also one of the varied hyperparameters [99]. Additional varied hyperparameters are the learning rate, which adjusts the rate of the weight adjustment during the backpropagation process, as well as the type of the learning rate—whether its value will remain constant or scale depending on the number of iterations [100]. Finally, hyperparameter is the L2 regularization parameter, which, if high, penalizes the inputs which have a high individual influence on the MLP output value—which can result in underfitted models [101].

In order to find the optimal set of hyperparameters, the Grid Search (GS) algorithm can be used. GS works in a way that it calculates all possible hyperparameter combinations. Then, a neural network is trained with each of this hyperparameter combinations [102]. In this manner, a wide range of hyperparameters can be tested. While the algorithm might not find the best possible combination of hyperparameters, with enough hyperparameters, it can find a hyperparameter combination which is close to the best one [103]. If needed, for example, if none of the yielded models provide satisfactory performance, possible hyperparameter values can be expanded or further refined, around the hyperparameter combination that provides the best results [58,104].

Finally, the metrics that will define the quality of the model solution need to be defined. In machine learning algorithms, models are evaluated by splitting the data set into training and testing portions [89,98]. The training portion is used during the previously described training process, and the trained model is evaluated on the testing portion. This is done by performing solely the forward propagation of the training process in order to obtain pairs of predicted values ($\hat{y}_i$) which can then be compared to the real values (y) from the data set. This will provide the value that describes the performance of the model. Two such metrics are used in this paper and those are coefficient of correlation (R^2) and mean absolute error (MAE).

R^2 defines the ratio of variance which exists inside the data set with the amount of variance contained in the results of the trained models [105]. Less unexplained variance means that the model is tracking the real data better, with higher values of R^2, which is defined in the range from 0 to 1 [106]. R^2 is defined by the Equation (18), [105]:

$$R^2 = 1 - \frac{S_{RESIDUAL}}{S_{TOTAL}} = 1 - \frac{\sum_{i=1}^{n}(y_i - \hat{y})^2}{\sum_{i=1}^{n}\left(y_i - \frac{1}{n}\sum_{i=1}^{n} y_i\right)^2}. \tag{18}$$

MAE provides a clearer, direct value of the error which using the MLP model introduces when used for regression [107]. MAE is defined as [107,108], Equation (19):

$$MAE = \frac{1}{n}\sum_{i=1}^{n}\left|y_i - \hat{y}_i\right|. \tag{19}$$

As the data set used in this research is relatively small, the need for cross-validation arises. Cross-validation is a technique which allows a larger amount of data to be used for testing. As the splits of training and testing data are randomized, a situation can happen in which a bad model performs well on a randomized testing set—while its real performance on the entire dataset is comparatively low [50,109,110]. A K-fold cross-validation, with 10 folds, is performed. This technique is applied in the following manner: first, the data set is split into K splits. Then, the model with the architecture provided from the grid search is trained on the training set consisting of the mix of $K - 1$ subsets, with the remaining 1 being used as the testing set [92,111]. This process is performed K times, with no repetitions of testing splits—in other words, until all the splits have been used as the testing split exactly once. In this manner, the entire dataset is used as the testing set, providing more detailed information on the model performance. In the case of the 10-fold K-fold cross-validation, each split is trained with 90% of the dataset (9 folds) used as the training set, and 10% used as the testing set (1 fold). The metrics defined with the Equations (18) and (19) are applied on all K training/testing set combinations, and the final scoring is expressed as the average score over the K fold, along with the standard error of that value. To summarize, for each of the 72 input/output combinations, 6144 different MLP model architectures are trained for 10 cross-validation folds, and evaluated using R^2 and MAE metrics across all folds. This provides the average and standard error values which allow the models for each parameter combination to be compared and the best achieved model hyperparameter values to be determined. This process is illustrated in Figure 2.

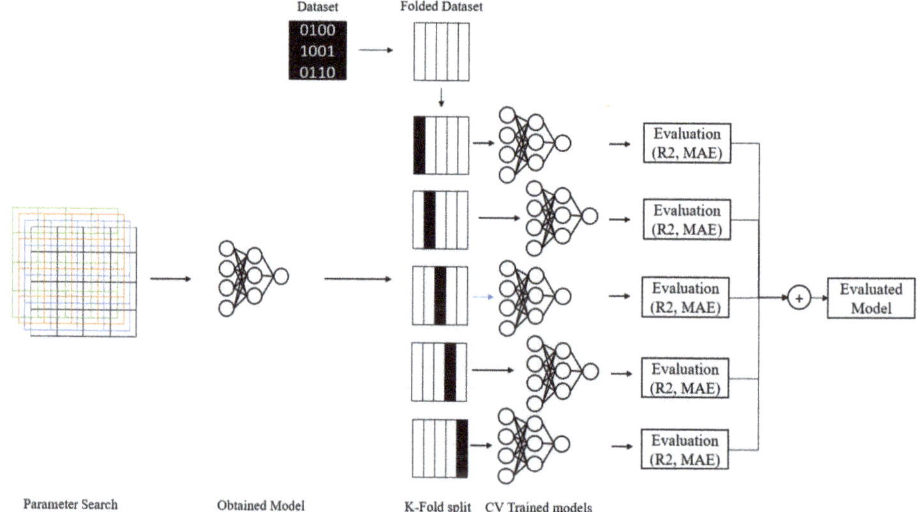

Figure 2. The illustration of the Multilayer Perceptron (MLP) process used in the research, starting with the parameter search performed using a grid search (GS), yielding the MLP model, which is then applied on separate K-Fold splits and evaluated using R^2 and *MAE* metrics.

The implementation of the described algorithm is done in the Python programming language, using Scikit-Learn machine learning library. Scikit-Learn was chosen for this research because it implements all the necessary algorithms for the presented research. Namely, the described Multilayer Perceptron, as used for a regression problem, is implemented within the MLPRegressor function, which takes the hyperparameters as the input [99]. Grid Search and K-Fold cross validation were implemented using the GridSearchCV function, which takes the number of folds (10), the selected algorithm (MLPRegressor), possible hyperparameter values stored within a dictionary data structure (with the names of hyperparameters being used as keys) and the list of desired metrics as inputs [99,100]. Helpfully, the metrics used are also implemented within Scikit-Learn, or to be more precise, within the metrics module as mean absolute error and R^2; both of which take the least actual dataset values and a list of predicted values as inputs [99,101].

The training of the models was performed using University of Rijeka's Bura supercomputer. Models for each output and input combination were trained on a single node, so a total of 72 nodes were used. Each node of the Bura supercomputer consists of an Intel Xeon E5 CPU, which provides 24 physical or 48 logical cores, and 64 GB of RAM. At the time of the research Bura supercomputer used Red Hat Enterprise Linux operating system, with kernel version 3.10.0-957 [112]. Use of Scikit-Learn was enabled through Anaconda Data Science platform, version 4.8.4. [113].

5. Steam Operating Parameters Required for the Exergy Analysis

For the purpose of conventional exergy analysis and exergy analysis by MLP neural network application of the main marine steam turbine and both its cylinders, in this research measurements are performed in each operating point from Figure 1 at 24 different turbine loads. Turbine load will be presented in relation to maximum turbine power (29,420 kW) specified by the turbine manufacturer. The maximum measured load equals to 84.31% of maximum power, which corresponds with minimum specific fuel consumption of the steam propulsion plant. Measurements on the LNG carrier were obtained at 24 steady-state conditions. It should be highlighted that the authors did not have any permission to get involved in the ship operation or to influence the ship crew.

The conventional exergy analysis the complete data will be presented in three turbine loads—low, medium, and high load, which correspond to 7.03%, 49.79% and 83.22% of the maximum turbine power, respectively.

For the turbine exergy analysis by MLP neural network application, all the collected data are used at all measured turbine loads. Those data will not be fully presented, for each measured steam operating parameter in each point from Figure 1 will be presented data range (from minimum to maximum) collected in all 24 turbine loads.

5.1. Conventional Exergy Analysis

Steam operating parameters in each operating point from Figure 1 at three different loads, required for the conventional exergy analysis of the main marine steam turbine and its cylinders are presented in Tables 1–3. In Table 1 steam data are presented for low turbine load, which corresponds to 7.03% of turbine maximum power, in Table 2 data are presented for medium turbine load (49.79% of maximum turbine power) and in Table 3 data are presented for high turbine load (83.22% of turbine maximum power).

Table 1. Steam operating parameters at low load (7.03% of maximum power).

Operating Point *	Temperature (°C)	Pressure (MPa)	Mass Flow Rate (kg/h)
1	487	6.2	9622
2	-	-	0
3	235	0.097	9622
4	-	-	0
5	235	0.097	9622
6	-	-	0
7	62.13	0.00511	9622

* Operating point numeration refers to Figure 1.

Table 2. Steam operating parameters at medium load (49.79% of maximum power).

Operating Point *	Temperature (°C)	Pressure (MPa)	Mass Flow Rate (kg/h)
1	511	6.065	51,419
2	-	-	0
3	259	0.401	51,419
4	-	-	0
5	259	0.401	51,419
6	158	0.085	2985
7	28.85	0.00397	48,434

* Operating point numeration refers to Figure 1.

Table 3. Steam operating parameters at high load (83.22% of maximum power).

Operating Point *	Temperature (°C)	Pressure (MPa)	Mass Flow Rate (kg/h)
1	500	5.795	95,570
2	354	1.558	3398
3	250	0.590	92,172
4	250	0.590	13,172
5	250	0.590	79,000
6	154	0.120	4636
7	34.80	0.00557	74,364

* Operating point numeration refers to Figure 1.

Steam-specific enthalpy and specific entropy in each operating point at all loads are calculated from the measured steam temperature and pressure by using NIST-REFPROP 9.0 software [114]. Steam-specific flow exergy in each operating point for all turbine loads is calculated by using

Equation (4). The steam specific flow exergy calculation requires definition of the ambient state in which analyzed steam turbine operates—in this research, the ambient state is defined as proposed in the literature [115] through the ambient temperature of 25 °C and the ambient pressure of 1 bar.

By observing data from Tables 1–3, operation dynamics of the analyzed steam turbine can be seen during its load variations. At low turbine load (Table 1) all three steam extractions are closed—an increase in turbine load results in steam extractions opening, due to an increase in the steam mass flow rate delivered to the main turbine. The third and last steam extraction (operating point 6, Figure 1) is the first which will be open, through which a certain steam mass flow rate will be delivered to the components of low pressure condensate heating system (Table 2). A further increase in the turbine load results with opening of second steam extraction (operating point 4, Figure 1), and at high load follows the opening of first extraction. At the highest measured turbine loads, all three steam extractions will be opened, and through all of them, steam mass flow rate will be delivered to all required system components. Therefore, the first steam extraction (operating point 2, Figure 1) from the HPC is the last extraction which will be open at high turbine loads, Table 3.

Steam expansion process in the Mollier *h-s* diagram for all three main turbine loads in conventional exergy analysis are presented in Figure 3 [114]. Operating points numeration is performed according to Figure 1, while markings a, b and c denote 7.03%, 49.79% and 83.22% of turbine maximum power, respectively. From Figure 3 it can be seen that at low turbine load, steam after expansion (operating point 7a) is still superheated. In marine steam power systems at low load, in the steam flow stream at the main condenser entrance is injected certain amount of water. Injected water cool superheated steam and transferred it to the saturated state (changing of steam aggregate state in condenser at any load can be performed only if the steam is saturated). At higher loads (medium and high load), steam after expansion through the main turbine cylinders is saturated, and it did not require additional cooling (operating points 7b and 7c, Figure 3). It is also important to observe that an increase in turbine load shifts the whole expansion process closer to the saturation line. Such occurrence resulted with a fact that steam at the LPC outlet has higher content of water droplets as turbine load increases.

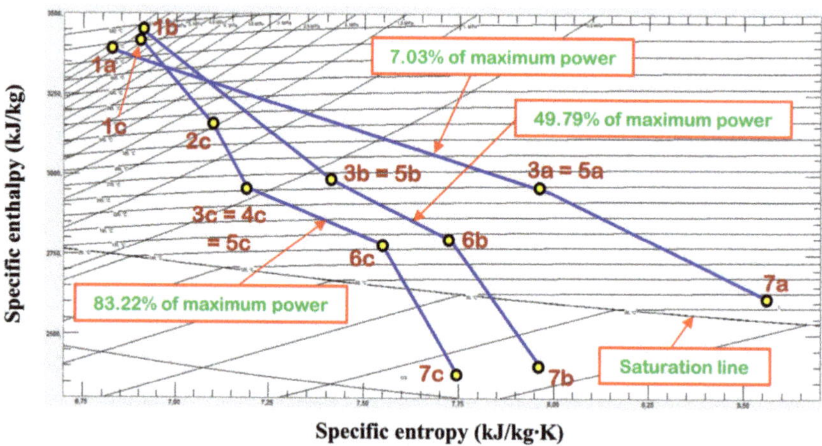

Figure 3. Steam expansion process in Mollier *h-s* diagram for three observed turbine loads.

5.2. Exergy Analysis by MLP Neural Network Application

The inputs (operating points) in Table 4 have been selected based on the physical relation to each output and part (HPC, LPC and WT) that was modeled. Performing the modeling of all possible combinations of input parameters (operating points) for each desired output would significantly increase the computational complexity. It should also be noted that each operating point consists of measurements for steam temperature, pressure and mass flow rate. In addition to the steam temperature, pressure and mass flow rate in each of the listed points, each input set also includes the ambient temperature and pressure, with the first combination in all cases only including those two values.

Table 4. Operating points used for regression models for each set of outputs. For each operating point, trained models use steam mass flow rate, temperature and pressure—in addition to ambient pressure and temperature.

Operating Points Combination	HPC * (Outputs: $\dot{Ex}_{DES,HPC}$, $\eta_{EX,HPC}$)	LPC * (Outputs: $\dot{Ex}_{DES,LPC}$, $\eta_{EX,LPC}$)	WT * (Outputs: $\dot{Ex}_{DES,WT}$, $\eta_{EX,WT}$)
1	-	-	-
2	1,2,3,4,5,6,7	1,2,3,4,5,6,7	1,2,3,4,5,6,7
3	1,2,3,4,5	3,4,5,6,7	1,2,3,5
4	1,2,3	4,6,7	2,6,7
5	1,2	5,6,7	1,2,3
6	1,3	6,7	5,6,7
7	2,3	5,6	4,6,7
8	3,4	4,6	3,6,7
9	1,4	5,7	1,2,3,4,5
10	2,4	4,7	2,4,6,7
11	-	-	1,3,7
12	-	-	1,4,7
13	-	-	1,5,7
14	-	-	2,4,6
15	-	-	2,6,7
16	-	-	1,3,5,7
Count	10	10	16

* Operating point numeration refers to Figure 1.

Exergy analysis of the observed steam turbine and both its cylinders by using the MLP neural network is performed as follows:

(1) By using all collected data, developed mechanical power, exergy destruction and exergy efficiency of each cylinder are calculated as well as the whole turbine at each of the 24 loads with the conventional exergy analysis.

(2) Results obtained by conventional exergy analysis are then used for MLP training and testing.

(3) MLP is trained for every hyperparameter combination given in Table 5, which results in a total of 442,368 models when the aforementioned cross-validation process is applied.

(4) The results of 442,368 models are compared across 72 input/output parameter combinations, given in Table 5, in order to determine the best possible model architecture for each of the aforementioned combinations.

Table 5. Possible values of hyperparameters used in grid search, with the number of the possible hyperparameter values being given in the column titled Total Count.

Hyperparameter	Possible Hyperparameter Values	Total Count
Hidden Layer Sizes	(84,84,84,84) (84,84,84) (84,84) (84) (42,42,42,42) (42,42,42) (42,42) (42) (21,21,21,21) (21,21,21) (21,21) (21) (84,42,42,21) (42,21,21) (84,42,21) (42,21)	16
Activation Function	'relu' 'identity' 'logistic' 'tanh'	4
Solver	'adam' 'lbfgs'	2
Learning Rate Type	'constant' 'adaptive' 'inverse scaling'	3
Initial Learning Rate Value	0.5 0.1 0.01 0.00001	4
L2 Regularization parameter	0.1 0.01 0.001 0.0001	4

The results achieved with each trained MLP model are evaluated using the MAE and R^2 metrics described in previous section. This allows the determination of the best models achieved by the MLP method. This also allows the comparison between various operating points in order to determine the best possible measurement combination, or in other words, such a combination which will allow us to obtain the output value with as few operating points, under the condition that the metrics are satisfactory.

Steam operating parameters range (minimum–maximum) in each operating point of the observed main steam turbine (Figure 1) for all 24 measured turbine loads are presented in Table 6. From this table, it can be seen that steam temperature and pressure at the HPC inlet (operating point 1) did not deviate significantly for the variety of turbine loads. This fact shows that both parallel operating steam generators delivers to the HPC steam with the highest temperature and pressure (specified by steam generators manufacturer) in the whole range of main turbine loads. Steam at the LPC outlet (operating point 7) can have high temperature at low turbine loads (which can reach up to 100 °C, Table 6).

Table 6. Steam operating parameters range (min–max) in each operating point for all 24 measured turbine loads.

Operating Point *	Temperature (°C)	Pressure (MPa)	Mass Flow Rate (kg/h)
1	485–513	5.795–6.2	3835–96,789
2	283–365	0.08–1.565	0–3398
3	229–279	0.048–0.593	3835–93,521
4	229–279	0.048–0.593	0–13,202
5	229–279	0.048–0.593	3835–80,319
6	121–169	0.009–0.121	0–4772
7	28.616–100.02	0.00392–0.00561	3835–75,547

* Operating point numeration refers to Figure 1.

Figures 4 and 5 present the ranges of input parameters used in MLP training. Figure 4 shows the values of parameters used as inputs, with subfigure (a) showing the ranges of temperature, (b) the ranges of pressure and (c) the ranges of mass flow rate. The range of values is shown for each operating point, as indicated on the labels, along with the environment value range for temperature and pressure.

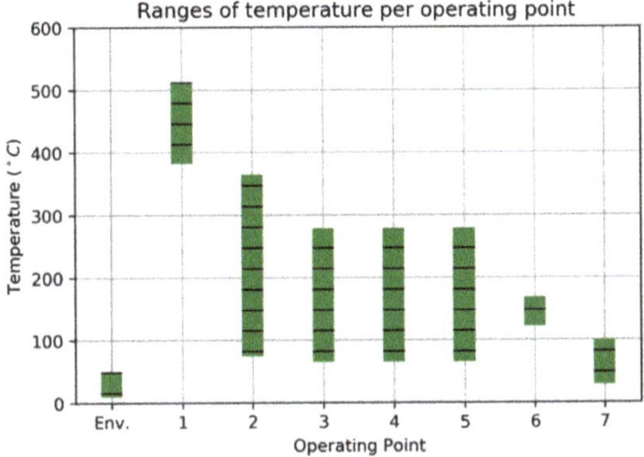

(a) Ranges of temperature in each operating point.

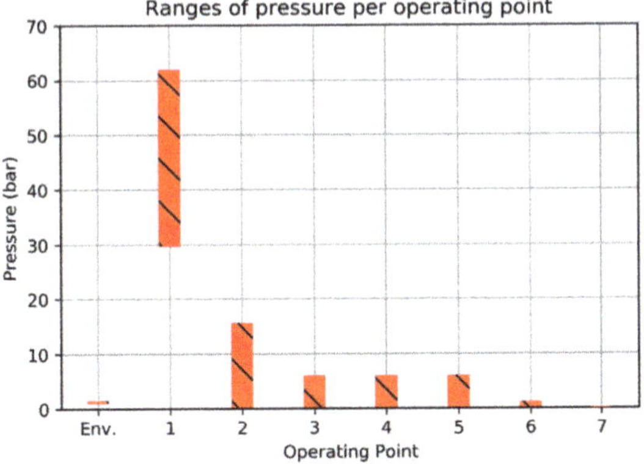

(b) Ranges of pressure in each operating point.

Figure 4. *Cont.*

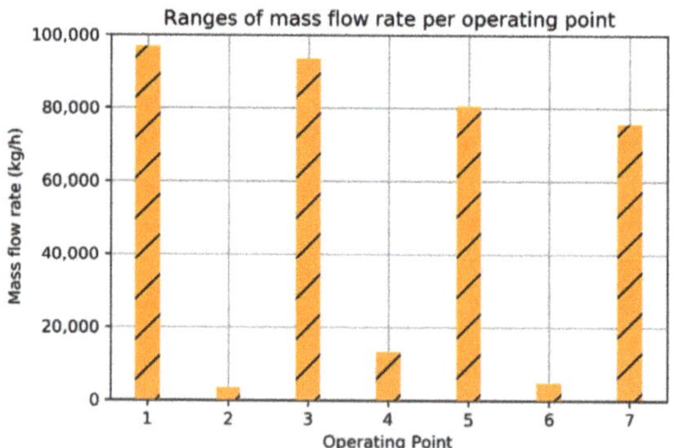

(c) Ranges of mass flow rate in each operating point.

Figure 4. The ranges of measured values used as inputs into the MLP for (**a**) temperature, (**b**) pressure and (**c**) mass flow rate in each operating point, along with the environmental values for temperature and pressure (labeled as Env.).

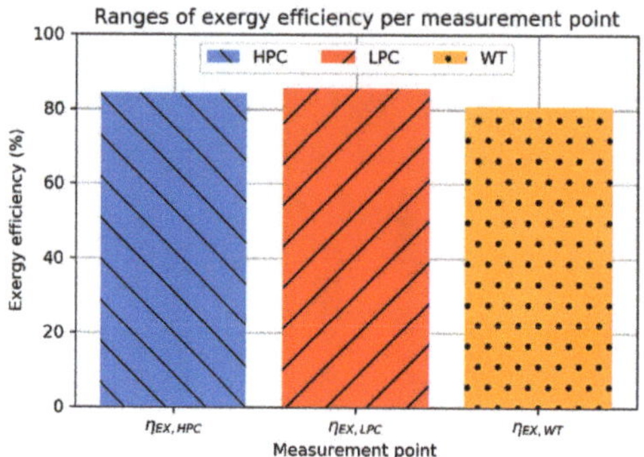

(a) Ranges of exergy efficiency in each output point.

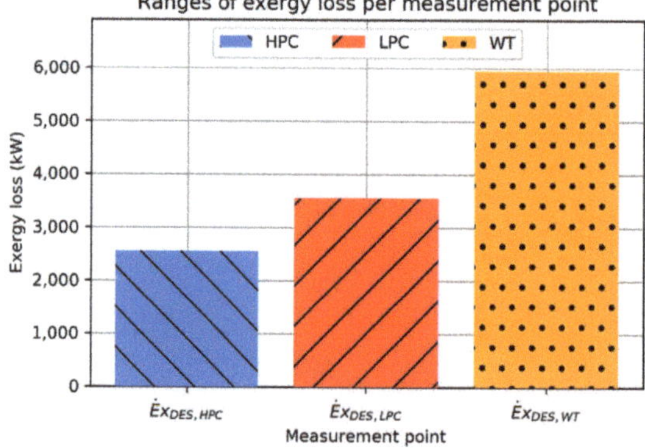

(b) Ranges of exergy loss in each output point.

Figure 5. The range of measured values used as outputs into the MLP for (**a**) exergy efficiency and (**b**) exergy loss measured for HPC, LPC and WT.

Figure 5 shows the range of output values, with subfigure (a) demonstrating the range of exergy efficiency values, while the subfigure (b) shows the exergy loss (exergy destruction) at each measurement point (HPC, LPC and WT).

It should be stated that exergy efficiency and exergy loss (exergy destruction) range starts from zero for the whole observed turbine and each of its cylinders. The reason of such occurrence is the first observed turbine load (of 24 overall loads). The first observed load is actually the heating of turbine (before start). In that load, the turbine and its cylinders did not produce useful power, so the exergy efficiencies and losses are equal to zero (losses are so small that it can be neglected).

5.3. Measuring Equipment

Measurements of steam temperature, pressure and mass flow rate in each operating point from Figure 1 are performed with a standard measuring equipment, calibrated and already mounted inside the power plant. That measuring equipment is used for the main steam turbine process control and regulation during exploitation. The list of used measuring equipment is presented in Table 7, while the detail specification of each measuring device can be found on the manufacturer website (provided in the list of references). Details related to measurement equipment accuracy and range of operation can be found in the Appendix A at the paper end. The measurement error did not have a significant influence on the obtained exergy analysis results.

Table 7. Used measuring equipment.

Operating Point *	Temperature (Immersion Probes) [116]	Pressure (Pressure Transmitters) [117]	Mass Flow Rate (Differential Pressure Transmitters) [118]
1	Greisinger GTF 601-Pt100	Yamatake JTG960A	Yamatake JTD960A
2			
3			
4		Yamatake JTG940A	Yamatake JTD930A
5	Greisinger GTF 401-Pt100		
6			Yamatake JTD920A
7			Yamatake JTD910A

* Operating point numeration refers to Figure 1.

6. Results and Discussion

This section will present the results obtained first by the conventional exergy analysis, followed by the results obtained by the application of described AI methodology.

6.1. The Results of the Conventional Exergy Analysis

In the conventional exergy analysis of any steam turbine, from measured steam operating parameters firstly should be calculated produced mechanical power. For the case of this particular marine steam turbine, produced mechanical power is calculated not only for the whole turbine, but also for both turbine cylinders (HPC and LPC), see Figure 6. An increase in turbine load is followed by the increase in developed mechanical power of the whole turbine. However, from Figure 6, the share of each turbine cylinder in the cumulative developed mechanical power is interesting and important to observe. At low load, the dominant mechanical power producer is HPC. An increase in turbine load results in a change in cylinder developed mechanical power share—at middle load the dominant mechanical power producer is LPC. At the highest measured loads, the share of both turbine cylinders in cumulative developed mechanical power is approximately the same; therefore, at the highest loads each cylinder develops approximately 50% of cumulative power. The same general conclusion about this type of marine steam turbines related to cylinder share in cumulative developed mechanical power at various loads can be found in the literature [66].

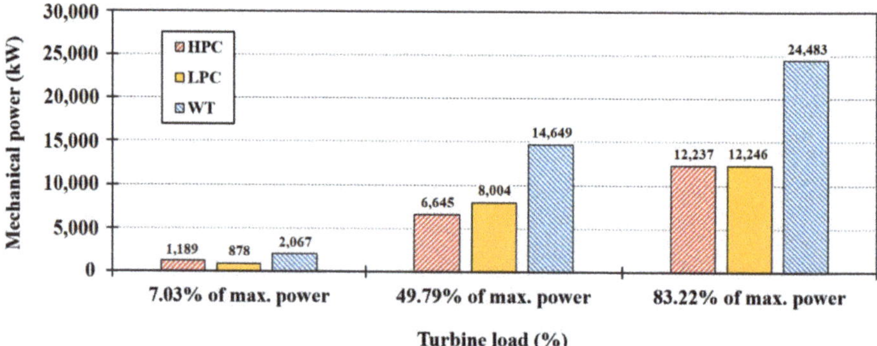

Figure 6. Mechanical power developed by the main turbine and each of its cylinders at three observed loads.

For three observed turbine loads in the conventional exergy analysis, developed mechanical power of HPC and LPC increases from 1189 kW and 878 kW at low load, to 6645 kW and 8004 kW at middle load and finally to 12,237 kW and 12,246 kW at high turbine load, respectively (Figure 6). It should be noted that developed mechanical power at each load presented in Figure 6 and calculated in each of 24 turbine loads is mechanical power calculated according to measured steam operating parameters and real (polytropic) steam expansion process throughout each cylinder. According to Figure 1, the mechanical power used for propulsion propellers drive in each turbine load is lower than the mechanical power calculated and presented in this analysis. The reason of such a difference is, as mentioned earlier, neglecting mechanical and other losses in the bearings, shafts and main gearbox.

Conventional exergy analysis at three observed turbine loads results with exergy destruction (exergy loss) and exergy efficiency of the whole main steam turbine and both of its cylinders, Figure 7. Comparison of Figures 6 and 7 shows that the developed mechanical power and exergy destruction of each cylinder and the whole turbine are directly proportional—higher developed mechanical power results in higher exergy destruction and vice versa. It is interesting to note that at a high load (83.22% of maximum power), regardless of the low difference between HPC and LPC developed mechanical power, LPC exergy destruction is significantly higher in comparison to HPC. Exergy destruction of the whole turbine increases during the load increase (proportional to developed mechanical power)—from 1380.09 kW at the low load, to 4437.86 kW at the middle load and finally to 5814.43 kW at the high load.

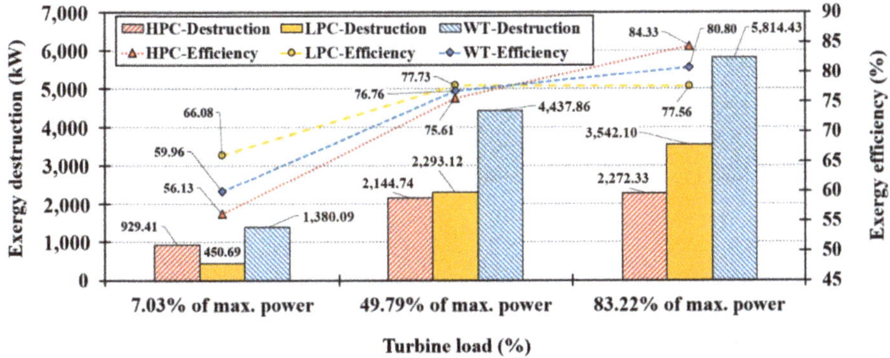

Figure 7. Exergy destruction and exergy efficiency of the main turbine and each of its cylinders at three observed loads.

Exergy efficiency results at three observed turbine loads in conventional exergy analysis show that an increase in turbine load increases the exergy efficiency of the whole turbine and each of its cylinders. The only deviation from this statement can be seen for the LPC, of which exergy efficiency slowly decreases from middle to high load (from 77.73% to 77.56%). Such a trend for LPC cannot be taken as relevant because the difference in exergy efficiency between middle and high load is so small that it can be the result of measurement equipment accuracy. By observing turbine cylinders, for low and high load a reverse proportionality between exergy destruction and exergy efficiency is valid—higher cylinder exergy destruction results in lower exergy efficiency. The mentioned reverse proportionality is not valid for the middle turbine load, where LPC, which has higher exergy destruction than HPC, also has higher exergy efficiency. An increase in the whole turbine load results in a simultaneous increase in its exergy efficiency—from 59.96% at a low load, followed by 76.76% at a middle load, to 80.80% at a high load (Figure 7).

In conclusion to previously described observations can be stated that the main marine steam turbine operation should be maintained at a high load where the whole turbine develops high mechanical power and has the highest exergy efficiency. Simultaneously, at a high load it should be taken into consideration that the whole turbine exergy destruction will be the highest, in comparison to lower loads. At a high load, both turbine cylinders will almost equally participate in cumulative developed mechanical power, but the exergy efficiency of HPC will be higher and its exergy destruction will be lower in comparison to LPC.

6.2. Exergy Analysis Results by MLP Neural Network Application

The results of MLP analysis are presented below. In order, the results for HPC, LPC and WT exergy destruction and efficiency are given. The graphs present the best results achieved using MLP per each given input combination, for each of the separate goals. In addition to *MAE*, Mean Relative Error (*MRE*) is used for presentation in order to demonstrate the error as a percentage of the range of the measured output. The results are not given for the case in which no input operating points have been used, using only ambient pressure and temperature, as the models for all possible hyperparameter combinations in all possible cases have failed to converge to a solution and as such have not provided any viable or meaningful results. The failure to converge in such a case suggests that such a regression, using only the ambient values, is insufficient for the desired outputs due to low or non-existent correlation between the ambient values and the desired outputs.

As for the acceptable error range, any *MRE* value that is smaller than 2% of the output range for a given output is considered to be acceptable, as this is precise enough estimation for the practical purposes in determining the exergy destruction and efficiency of a turbine in the marine environment. In the same manner, any R^2 value that is higher than 0.95 is considered within the acceptable range.

This means that all models obtained with a certain input combination that have achieved an R^2 value higher than 0.95 and an *MRE* lower than 2% are to be considered when the final model selection is performed.

For all figures in this subsection, numbers or ranges written on the abscissa (combination of input parameters) are related to operating points presented in Figure 1.

MAE and *MRE* scores in the case of HPC exergy destruction (exergy loss) estimation display low errors across all inputs (*MRE* < 2%), but there are some outliers. Namely, input combinations lacking the operating point 2 (3,4; 1,4; 1,3) display the highest error, suggesting that operating point 2 is necessary to achieve a low model error in this case. This is shown in Figure 8.

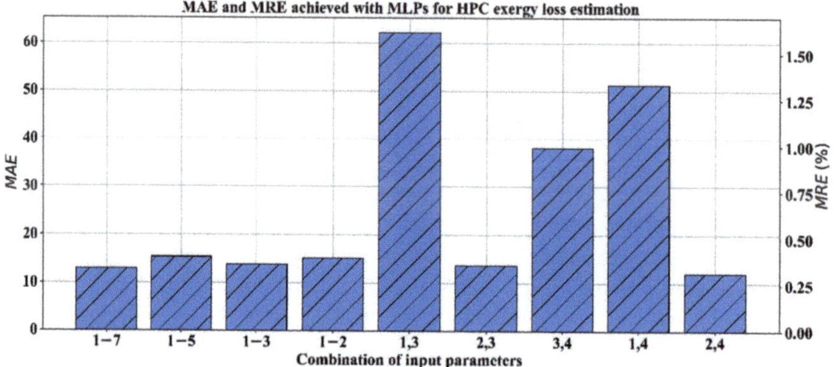

Figure 8. *MAE* and *MRE* values for HPC exergy destruction (exergy loss).

Figure 9 shows the R^2 scores achieved being high across all input combinations, with all of them achieving scores in excess of 0.99. These suggest that all the input combinations may be used for modeling, since all models track the outputs well, as long as a higher error of some models does not present an issue. Still, it is evident that those combinations of operating points which do not include the operating point 2 achieve a lower R^2 score, which corresponds with the larger errors provided by those models. This further shows the importance of the operating point 2 in the modeling of HPC exergy loss—if higher precision than the one sought in this document is needed.

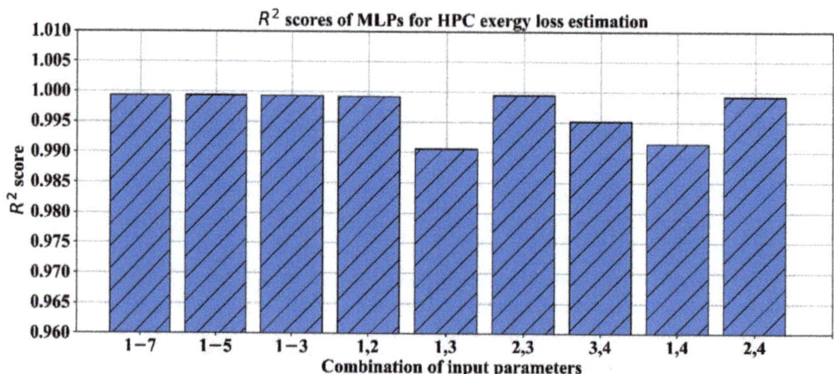

Figure 9. R^2 values for HPC exergy destruction (exergy loss).

While observing the *MAE* and *MRE* for HPC exergy efficiency estimation, given in Figure 10, it can be seen that all the input combinations achieve very low errors, namely below 1%. Interestingly, operating point 2, which seemed to have a large benefit, does not have as much importance. It seems that this role is replaced by the operating points 3 and 4 used in all combinations that achieve lower errors. As with the previous models, all the models here achieve *MAE* scores low enough to be considered for use.

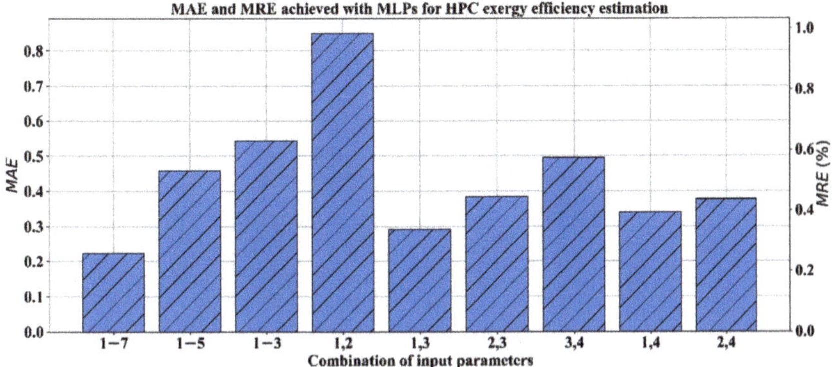

Figure 10. *MAE* and *MRE* values for HPC exergy efficiency.

Figure 11 shows that R^2 scores achieved for HPC exergy efficiency models are high, except in the case of input combination 1,2. Due to this combination being the only one which does not include operating points 3 or 4, it suggests the importance of these inputs in HPC exergy efficiency model as the previous graph. In line with the *MAE* scores, the R^2 scores achieved are high enough ($R^2 > 0.95$) to be considered usable in modeling HPC exergy efficiency.

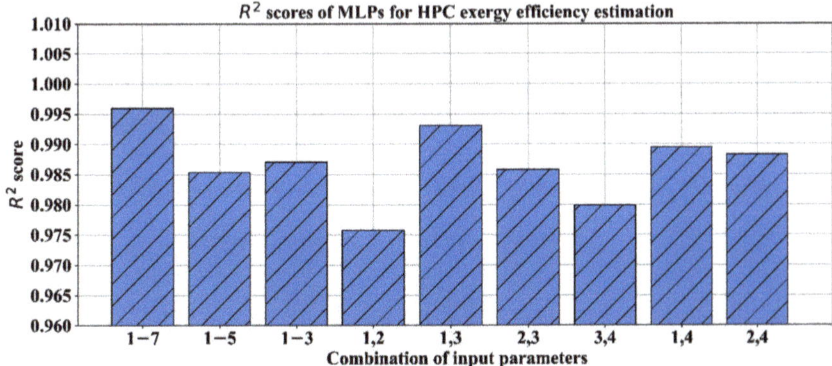

Figure 11. R^2 values for HPC exergy efficiency.

While all the input combinations provide satisfactory metric values in terms of *MRE* (*MRE* < 2%) for HPC exergy destruction and exergy efficiency estimators, it can be seen that there are some that do not achieve as high an R^2 value for exergy efficiency estimation ($R^2 > 0.99$). Namely, only input combinations which achieve R^2 scores 0.99 or higher are 1–7; 1,3 and 1,4. It can also be seen that R^2 scores of those operating point combinations which achieve a higher error are lower, which is to be expected.

By observing all the achieved scores for the output models of HPC it can be shown that the best scores are achieved, for both exergy efficiency and exergy destruction, when the input operating point combinations used are 1,4; 2,4; 2,3 and 1–7. These operating point combinations are obtained by considering both errors and R^2 value scores for both exergy efficiency and exergy loss, and while they may not present the best possible model for an individual output, a high enough performance is evidenced in both observed output cases for HPC.

Figure 12 demonstrates the best *MAE* and *MRE* achieved by the MLP for each input combination in the case of exergy destruction estimation for LPC. It can be seen that the best results are achieved when all the inputs are used (1–7), with comparable results being achieved for all the input combinations, which include operating points 5, 6 and 7. Still, all the operating point combinations provide a satisfactory error below 1.5%, signifying that they may all be used in modeling the LPC exergy loss.

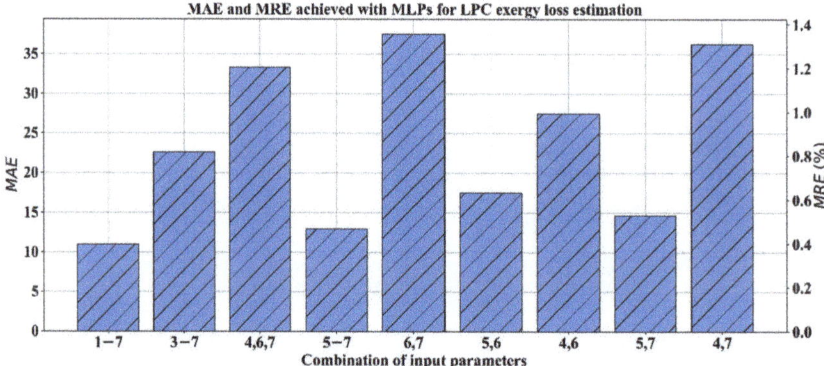

Figure 12. *MAE* and *MRE* values for LPC exergy destruction (exergy loss).

While observing the R^2 scores for LPC exergy destruction estimation presented in Figure 13, it can be seen that all input combinations achieve relatively high R^2 scores ($R^2 > 0.95$), meaning that all the models may be used in the LPC exergy destruction modeling. It should be noted that the input combination 5–7, despite a relatively low *MAE* in comparison to other results, has the lowest R^2 score. This points to the fact that, despite achieving a low error, this input combination does not explain all the variations in the test data. It can be concluded that this is due to the lack of information being contained within this input combination. Through comparison with the input combination of operating points 5 and 7, which has achieved a higher score, it can be concluded that the inclusion of operating point 6 as an input is actually detrimental to the model, lowering instead of increasing performance.

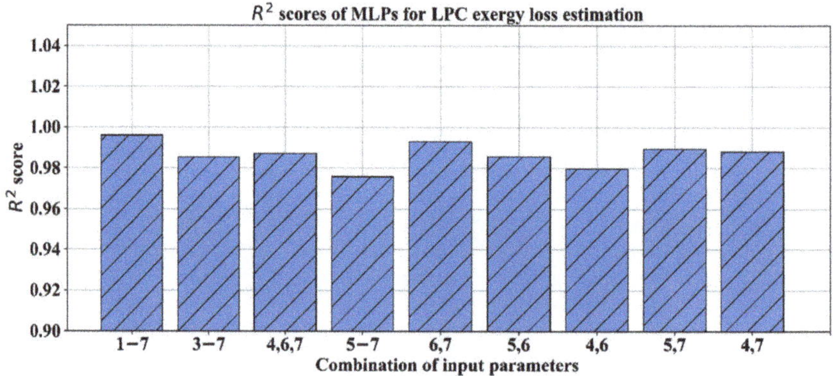

Figure 13. R^2 values for LPC exergy destruction (exergy loss).

For the model of LPC exergy efficiency, as in the previous case the best results are achieved when all inputs are used together (case 1–7), with comparable results being achieved for input combination 3–7. Despite this, all input combinations have achieved a low error (*MRE* < 1%), with the highest error occurring when input combination 6,7 is used. Interestingly, the input combination 5–7 shows a lower error than combination 5,7, meaning that the inclusion of operating point 6 is beneficial in this case, but the use of the measurements in operating point 5 is largely beneficial to it. Because of this, we can conclude that the combination of values in operating points 5 and 6 is important to achieve extremely low scores in the LPC exergy efficiency modeling. However, as with the previous models, all the error values are low enough to conclude that all models may be used. These results are shown in Figure 14.

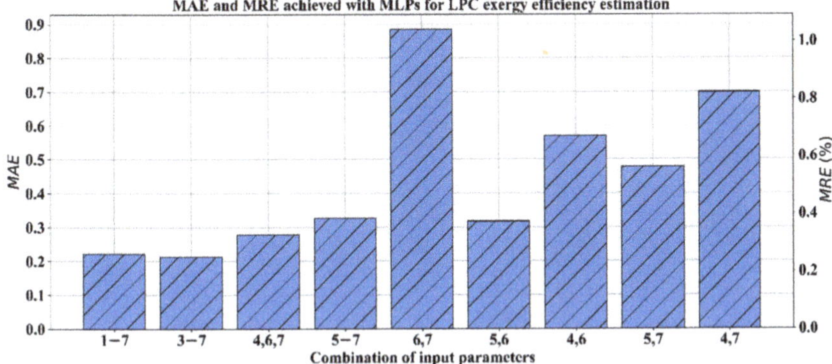

Figure 14. *MAE* and *MRE* values for LPC exergy efficiency.

As for the R^2 scores of the model for LPC exergy efficiency in Figure 15, it can be seen that the lowest score is achieved by the input combination 6,7, which has the highest error (as can be seen in Figure 14). Still, this combination, and all others, provide a quality enough model of LPC exergy efficiency. This points to the lack of information necessary for a higher quality model when these inputs are used. Higher quality models being achieved in all other cases, which points to the fact that LPC exergy efficiency models require the use of operating point 4 or 5, in order to achieve higher regression quality, and the scores achieved show that operating points 5 and 6—when used in unison—provide enough information for MLP to successfully converge to a high-quality solution.

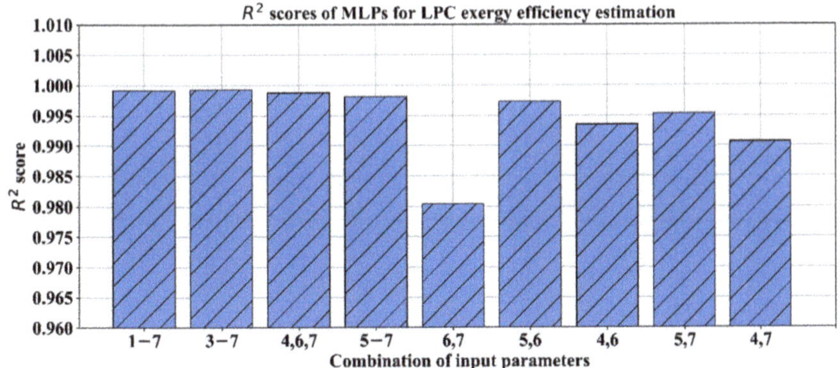

Figure 15. R^2 values for LPC exergy efficiency.

When observing the LPC exergy destruction and exergy efficiency estimation outputs (Figures 12–15), it can be seen that the best results are achieved when all the inputs are used, but satisfactory results are achieved for all input combinations with R^2 values in excess 0.97, and *MRE* below 1.5%. Observing all the scores points towards the fact that the best results for the LPC outputs, including both exergy destruction and exergy efficiency, are obtained when points used are 1–7; 5–7; 4,6,7 and 4,7.

Figure 16 demonstrates *MAE* and *MRE* values obtained by the models for WT exergy loss estimation. It can be seen that all the error values are extremely low. The only input combination which exceeds the error of 1% is the combination of operating points 1–3,5, which achieves an error below 1.2%. However, such an error is still within the acceptable error range, meaning that all the models of WT exergy loss achieve satisfactory performance in regards to the *MAE*.

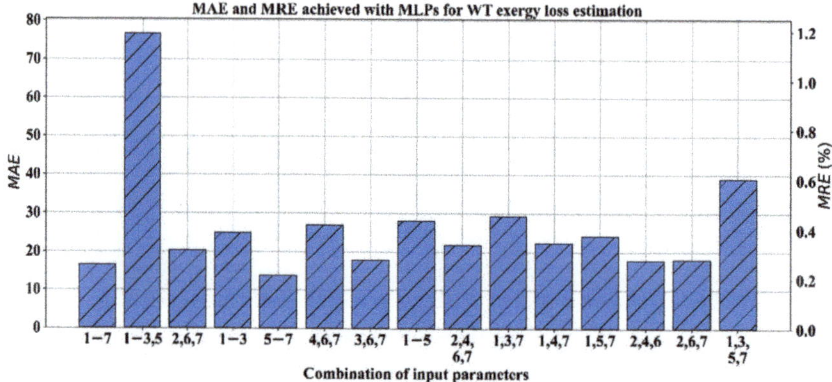

Figure 16. *MAE* and *MRE* values for WT exergy destruction (exergy loss).

Figure 17 demonstrates the R^2 scores achieved by the WT exergy loss estimation fall in line with the *MAE* and *MRE* results achieved, with all the inputs achieving R^2 scores in excess of 0.985; meaning all are high enough to be considered for modeling. The lowest score is achieved by the input combination 1–3,5 which, with the value of 0.99, is still inside the acceptable range.

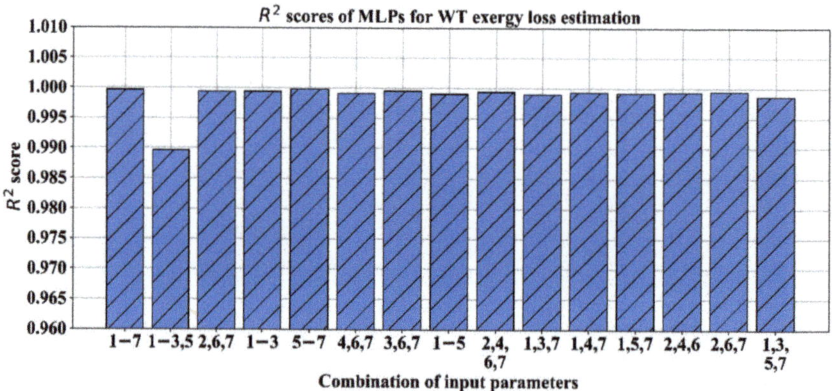

Figure 17. R^2 values for WT exergy destruction (exergy loss).

Observing both scores for WT exergy loss, it can be concluded that all models achieve similar high performance, with the exclusion of operating points 1,3,5,7 and 1–3,5, which achieve comparatively poorer scores. Still, all the model's scores fall well within the satisfactory ranges, indicating that it is possible to use them for the given task.

Through observing Figure 18, *MAE* and *MRE* of the WT exergy efficiency estimation can be seen. The output combination 1–5 achieves the lowest results, which is, interestingly, higher than the input combinations of 1–3,5 and 1,3,5,7, which do not include the operating point 4. This signifies that operating point 4 may be detrimental in this particular case, but not so much that its inclusion should be avoided, considering the error introduced by it is at its maximum below 0.3%. Due to all *MRE* values being below 0.5, this should not have a large influence on model selection.

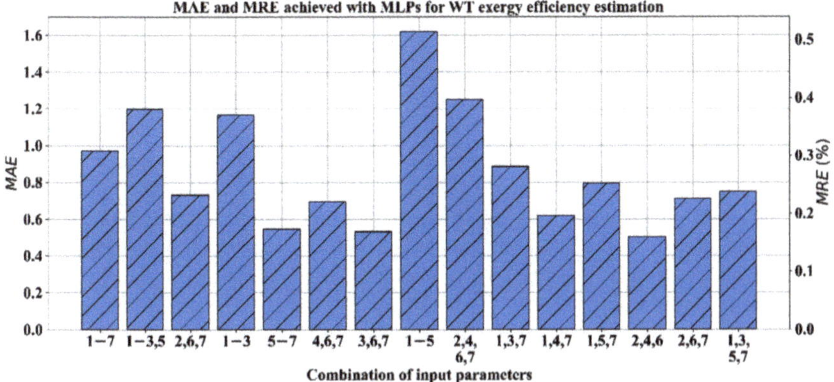

Figure 18. *MAE* and *MRE* values for WT exergy efficiency.

Figure 19 shows the R^2 values achieved by the models for WT exergy efficiency estimation. The figure shows that the R^2 scores achieved mostly fall into line with the *MAE* achieved, with those models that show a higher error, also showing a lower R^2 score, but with less drastic differences. It can be seen that all the scores are in the excess of 0.99, which confirms that all the models are of high quality and that models with all the input combinations may be used during the model selection.

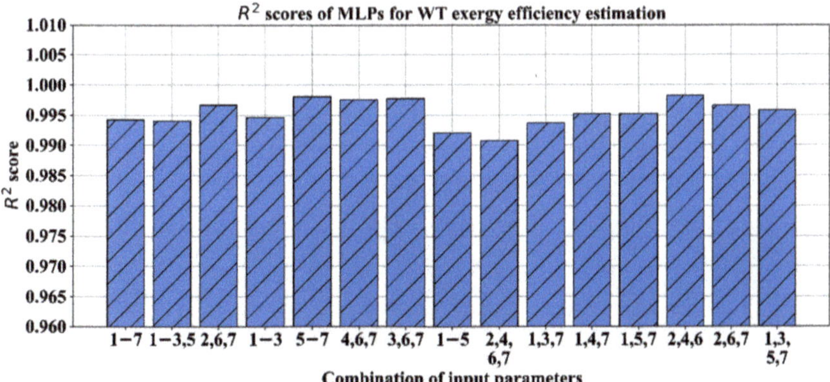

Figure 19. R^2 values for WT exergy efficiency.

In the case of the output analysis of the whole turbine, it can be seen that error values are extremely low, except for the model of exergy destruction when the input combination of operating points 1–3,5 is used. When the input of 1–3,5 operating points is used, the R^2 value drops below 0.99 and *MRE* grows above 1%. While these values are still within the limits of satisfactory results it can be concluded that they present the worst input combination in term of WT exergy destruction modeling, but may still be used if necessary. The best results achieved for both WT exergy destruction and efficiency are obtained when the following input parameters are used: 1–7; 1,4,7; 5–7; 2,4,6.

Considering that all the used operating point combinations achieve *MAE* of below 2% for the output value range and R^2 score higher than 0.95, all may be used for modeling the required outputs. For example, if a situation is observed in which the measuring equipment may already exist at the given points and installing it in different ones may pose a difficulty, using a model with slightly poorer scores (but still within the satisfactory range) can be a good choice. All the input operating point combinations which achieve quality models show the importance of including domain experts in the artificial intelligence-based research. In the presented research, this allowed for lowering the number of input combinations that were tested and consequentially for a lower computational complexity—while still generating useable models.

By observing the presented graphs, the best input combination can be determined. This can be done by cross referencing the best scores achieved in order to find the combination of inputs that provides the best scores. For example, while the best scores across all the observed cases are obtained when input combinations 1–7 are used, this does not lower the amount of operating points needed. Due to all of the output metrics falling well within the satisfactory error range, the selection can concentrate on finding such a combination of input operating points, which allows for an as low as possible number of operating points. If the goal is to achieve satisfactory measurements with as few operating points as possible, then it can be concluded that this combination is 1, 4 and 7—or namely by utilizing operating points 1 and 4 for HPC, 4 and 7 for LPC and 1,4,7 for WT. The respective values achieved for each cylinder and the whole turbine are given in the graphs below.

Selected operating points (1, 4 and 7) reduce the number of the required measurements for more than half (from 21 to 9 overall measurements of the steam mass flow rate, temperature and pressure). Selected operating points are the most dominant operating points related to the marine steam turbine operation because steam operating parameters at the turbine inlet and outlet also gives information about steam generators and main steam condenser operation. Additionally, knowing steam operating parameters in only one extraction, mounted between turbine cylinders, will be satisfactory for MLP estimation of turbine exergy analysis parameters. Therefore, MLP application can be very beneficial for reducing the costs of measurement and regulation equipment. The same idea, applied to the main marine steam turbine and its cylinders in this paper, can be applied to the whole marine steam propulsion plant.

Figure 20 demonstrates *MAE* and *MRE* of the selected input combinations. All the selected models achieve the errors of below 1.5%, with the models for HPC and LPC exergy destruction estimation achieving *MRE* of 1.2%, and the WT exergy destruction model achieving the *MRE* of 0.6%.

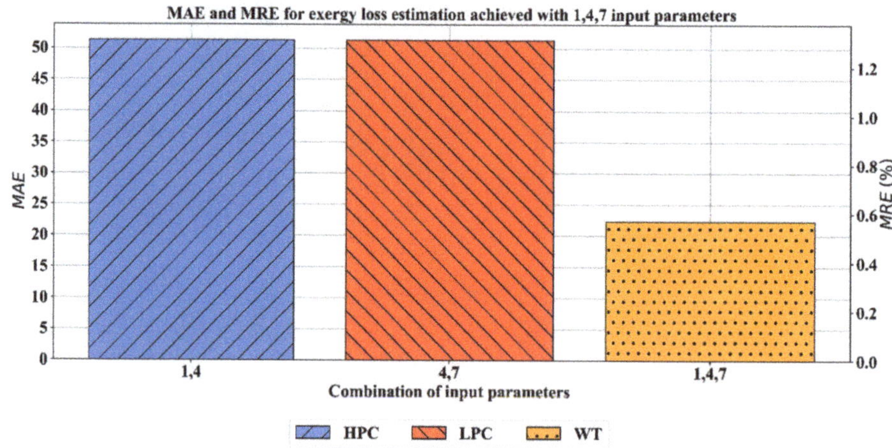

Figure 20. *MAE* and *MRE* values of the selected input combination (1,4,7) for exergy destruction (exergy loss).

R^2 scores for exergy loss of the whole turbine and each cylinder, presented in Figure 21, show that the models for HPC and WT achieve high fidelity with R^2 scores in excess of 0.99. The lowest R^2

score is achieved by the LPC model which reach the R^2 score of 0.97. While lower than some scores, when coupled with the low error seen in Figure 20, it can be concluded that this model is satisfactory. The importance of using multiple metrics when evaluating the artificial intelligence-based models is shown here, as models with a high R^2 value may achieve a critically low error or vice-versa. Similarly, a model that may seem to achieve a relatively poor score when a single metric is used may show good performance when evaluated with other metrics, leading to the conclusion that such a model is still usable (providing all the metrics are within the satisfactory ranges).

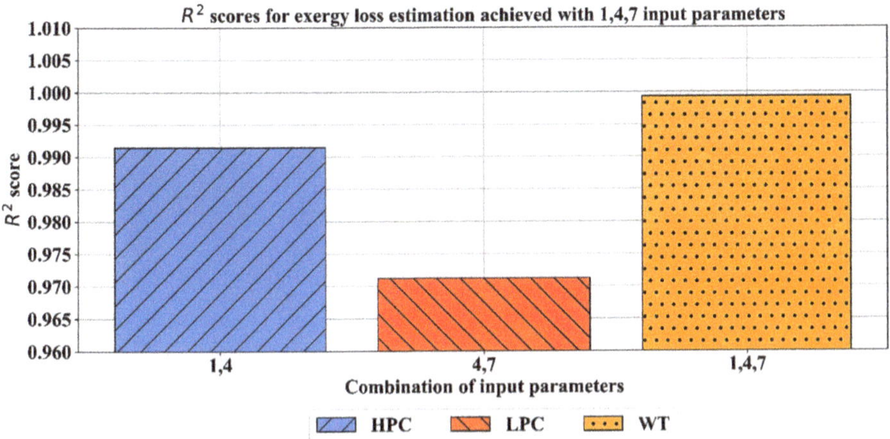

Figure 21. R^2 values of the selected input combination (1,4,7) for exergy destruction (exergy loss).

By observing Figure 22, *MAE* and *MRE* values of selected models for exergy efficiency can be seen. As shown, all the models achieve a low error (below 1%), with the error for the HPC exergy efficiency model being less than 0.5% and WT exergy efficiency model showing the error of below 0.3%. The highest error is achieved for the LPC exergy efficiency model, which achieves an error of 0.8%, what is well within the satisfactory error range.

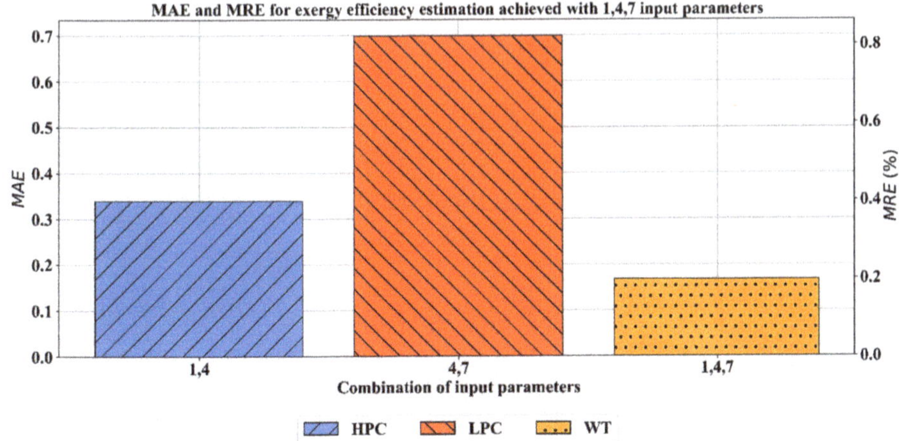

Figure 22. *MAE* and *MRE* values of the selected input combination (1,4,7) for exergy efficiency.

Finally, Figure 23 demonstrates the R^2 scores for the exergy efficiency estimation. With all the scores being in excess of 0.99, it can be concluded that all the selected models for exergy efficiency estimation of LPC, HPC and WT achieve satisfactory results. All the results are relatively close, with the lowest being the result for the HPC exergy efficiency model, achieving a score just slightly lower than 0.99.

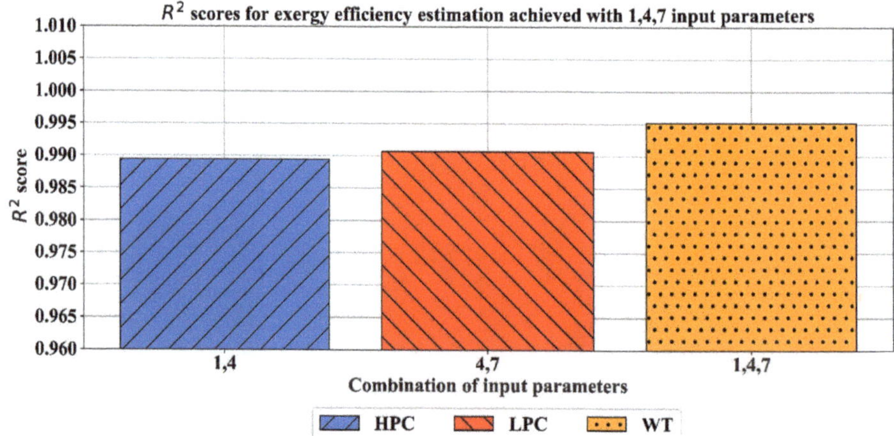

Figure 23. R^2 values of the selected input combination (1,4,7) for exergy efficiency.

While individually the input parameters may not have the best scores achieved, they all achieve relatively high scores. Considering that errors of models are below 1.5% when these input combinations are used, this is satisfactory for the measurement needs, especially since only three operating points are used. The hyperparameter values and numerical values of metrics for exergy destruction achieved are given in Table 8, and the values for the hyperparameters used in the best models are presented in Table 9. In the same way, values are given for exergy efficiency in Tables 10 and 11.

Table 8. Metrics achieved by selected operating points, for modeling of exergy destruction.

Operating Point	R^2	+/-	MAE	+/-
1,4 (HPC)	0.9914541639	0.02378781401	51.27713297	36.21271356
4,7 (LPC)	0.9712139368	0.09161658705	36.22307872	50.75836187
1,4,7 (WT)	0.9992643028	0.00172978046	20.44955009	12.32598431

Table 9. Hyperparameters used for best solutions achieved in selected operating points, for modeling of exergy destruction.

Operating Point	1,4 (HPC)	4,7 (LPC)	1,4,7 (WT)
Activation Function	ReLU	ReLU	ReLU
L2 Regularization	0.001	0.01	0.1
Hidden Layer Sizes	(84)	(84, 84, 84, 84)	(84, 84, 84, 84)
Learning Rate Type	Adaptive	Constant	Adaptive
Initial learning rate	0.1	0.01	1e-05
Solver	LBFGS	LBFGS	LBFGS

Table 10. Metrics achieved by selected operating points, for modeling of exergy efficiency.

Operating Point	R^2	+/-	MAE	+/-
1,4 (HPC)	0.9894154031	0.03141286439	0.3393632265	0.3058120431
4,7 (LPC)	0.9906770758	0.02055184947	0.6985228666	1.1429986910
1,4,7 (WT)	0.9951798924	0.01330619104	1.6066829045	0.0133061910

Table 11. Hyperparameters used for best solutions achieved in selected operating points, for modeling of exergy efficiency.

Operating Point	1,4 (HPC)	4,7 (LPC)	1,4,7 (WT)
Activation Function	ReLU	ReLU	ReLU
L2 Regularization	0.1	0.1	0.1
Hidden Layer Sizes	(84,42,21)	(84,84,84)	(84,84,84,84)
Learning Rate Type	Constant	Constant	Adaptive
Initial learning rate	0.1	0.01	1e-5
solver	LBFGS	LBFGS	LBFGS

Table 8 demonstrates the scores achieved for the exergy destruction with the selected input operating point combinations. It can be seen that the HPC and WT outputs modeled with operating points 1,4 and 1,4,7, respectively, achieve R^2 scores in excess of 0.99, while the model for the LPC output, modeled with operating points 4,7, achieves a lower, but sill satisfactory, R^2 score of 0.97. Maximal error ranges achieved during the cross-validation have also been provided next to the average of all 10 validation scores for both R^2 and MAE.

Table 9 provides the hyperparameters used to obtain models that have provided the solutions within the Table 8. The values of hyperparameters presented within Table 9 are a subset of hyperparameters provided within Table 5. Only the hyperparameters which have been used to obtain the best selected models are presented for brevity. It can be seen that a relatively large network, of 4 layers with 84 neurons each, has been used in the case of LPC and WT exergy destruction models, while the HPC exergy destruction model utilized a significantly smaller network—consisting of a single layer with 84 neurons. All the best models for exergy destruction used LBFGS solver and ReLU activation function.

Table 10 presents the scores for the exergy efficiency models, providing R^2 and MAE scores for LPC, HPC and WT models (input combinations being 1,4, 4,7 and 1,4,7 respectively). The table demonstrates that the R^2 scores achieved are in excess of 0.99 for LPC and WT exergy efficiency models, while the HPC model achieved the average R^2 score of 0.989 after the cross-validation process. MAE scores demonstrate an error of less than 1% of exergy efficiency for HPC and LPC models. MAE score grows to 1.6% for WT exergy efficiency, but the error is still within the previously stated satisfactory error range. As was the case in Table 8, the maximal error ranges obtained during the cross-validation are provided next to the obtained average R^2 and MAE scores in Table 10.

Hyperparameters used to obtain the solutions in the case of exergy efficiency are given in Table 11. As in Table 9, it can be seen that ReLU activation function and LBFGS solver have been used in all cases. Interestingly, the initial learning rates for each individual input operating point combination are the same as they were for exergy destruction models, indicating a similarity between the two regression problems. By observing the hidden layer size hyperparameter, it can be seen that the model for HPC exergy efficiency also required a lower number of neurons in comparison to the LPC and WT exergy efficiency models, as it did in the models of exergy destruction. Still, the number of neurons for HPC exergy efficiency model is much closer to the LPC and WT exergy efficiency models than it was the case for exergy destruction, indicating closer complexity in the problems.

From the hyperparameter tables (Tables 9 and 11), it can be seen that all solutions tended towards the larger number of hidden neurons. This, coupled with a high initial learning rate which was kept either constant or adaptive, signifies a hard to model problem [84]. It can also be seen that some models, namely the ones pertaining to the HPC exergy destruction and exergy efficiency, used smaller networks. The hidden layer sizes were (84) and (84, 42, 21) for exergy destruction and exergy efficiency, compared to the hidden layer sizes for other models, (84, 84, 84) and (84, 84, 84, 84), which are significantly higher. This may indicate that the HPC models were simpler to regress [84,93]. Still, it should be noted that the hidden layer values have tended towards the upper range of the possible hyperparameter values, indicating a relatively complex regression problem.

The regularization hyperparameter value is also the highest possible across all models, indicating that some of the inputs used initially had a larger influence on the output, which required lowering in order to achieve a precise model. It should be noted that this may not indicate an operating point by itself as an input, but one of the values (steam temperature, pressure or mass flow rate) that were measured in it. Additionally, it can be seen that across all models, regularization is kept high, the preferred solver is LBFGS and the activation function is ReLU in all cases. While these hyperparameters do not provide additional information about the problem complexity, they can be utilized in further research in order to lower the number of combinations in the GS algorithm and allow for faster training times.

7. Conclusions

From the presented results it can be concluded that the calculation of exergy destruction and exergy efficiency of a marine steam turbine and its cylinders can be performed using artificial intelligence methods, namely MLP. Through the analysis of the results it can be seen that, while the best results are still achieved when using all the operating points, good results can be achieved using only a subset with negligible loss in accuracy and precision when MLP is used for modeling.

By observing the hyperparameter values which defined the model architecture of the selected models it can be concluded that the best solutions used relatively complex architectures, pointing towards a high complexity of this problem. While it is possible that even more precise solutions could be found using even more complex architectures, as the results provided are within the satisfactory error range, a further increase of computational complexity is needless.

Furthermore, by selecting a subset of the inputs which achieves relatively high results in all six output measurements (exergy destruction and exergy efficiency for LPC, HPC and WT) it can be concluded that, through the application of an MLP model, the number of operating points which can be used to determine the outputs can be lowered. This means that the costs of measuring equipment can be significantly lowered when this method is utilized. This is especially apparent in the selected case of operating points used being 1, 4 and 7 considering that measuring points 1 and 7 are already equipped through the subsequent and prior equipment in the maritime environment that the turbine is placed in. While the selected points do not provide the best results within the tested operating point input combinations, they provide good results across the entire range of modeled outputs and allow for the lowest amount of measuring equipment to be utilized. It may be possible to find an input combination that provides an even better fit for the presented problem. However, it should be noted that this could only be done by performing an extensive search of all possible operating point combinations as inputs, which would significantly raise the computational complexity, when the presented methodology is applied. The increase in computational complexity, in combination with the fact that models obtained by selecting a subset of input combinations provide a high quality solution, means that such an approach is unnecessary for the presented problem.

Further research will be performed in the following direction:

(1) To determine optimal turbine operating points for the measurement of steam temperature, pressure and mass flow rate. The goal will be to find three or four operating points of which the measurement results, along with MLP application, can be used for exergy analysis parameters prediction of the whole turbine and each cylinder at any load, with the lowest possible errors.

(2) Extensive measurements during a long time period will allow determining performance degradation coefficients for the whole analyzed turbine and each of its cylinders. Implementation of such coefficients inside MPL structure will allow accurate and precise predicting of turbine exergy analysis parameters for the entire period of its operation.

(3) Investigate if the same technique can be applied for other main marine steam turbines (especially for newer variants, which consist of three cylinders and steam reheating).

The final goal will be to reduce the number of measurements (and to reduce the number of measuring equipment) for new ships with steam propulsion. The main idea is to perform MLP neural network training and testing on the manufacturer's test data at various loads along with the implementation of performance degradation coefficients inside the MLP structure. Such an approach will allow that onboard the ship, measurements of steam temperature, pressure and mass flow rate can be performed only in three or four optimal operating points (not in seven operating points as at the moment). By using the measurement results from a reduced number of operating points, the MLP neural network will be used for predicting the main steam turbine (and each cylinder) exergy destruction and exergy efficiency. In the end, the application of performance degradation coefficients will allow accurate and precise prediction of the whole turbine and its cylinders exergy analysis parameters for the entire steam power plant operation period.

If possible, the same idea presented in this paper for the main marine steam turbine and its cylinders will be applied to the entire marine steam propulsion plant.

Author Contributions: Conceptualization, S.B.Š., I.L., N.A., V.M. and Z.C.; methodology, V.M., I.L. and S.B.Š.; software, S.B.Š., N.A. and Z.C.; validation, I.L. and V.M.; formal analysis, N.A.; investigation, V.M., I.L., S.B.Š. and N.A.; resources, Z.C.; data curation, V.M.; writing—original draft preparation, V.M., I.L. and S.B.Š.; writing—review and editing, V.M, I.L., N.A. and S.B.Š.; visualization, I.L.; supervision, Z.C. and V.M.; project administration, Z.C.; funding acquisition, V.M. and Z.C. All authors have read and agreed to the published version of the manuscript.

Funding: This research has been supported by the Croatian Science Foundation under the project IP-2018-01-3739, CEEPUS network CIII-HR-0108, European Regional Development Fund under the grant KK.01.1.1.01.0009 (DATACROSS), project CEKOM under the grant KK.01.2.2.03.0004, CEI project "COVIDAi" (305.6019-20), University of Rijeka scientific grant uniri-tehnic-18-275-1447 and University of Rijeka scientific grant uniri-tehnic-18-18-1146.

Appendix A Specification of Used Measuring Equipment

Table A1. Temperature Measurements.

→ Greisinger GTF 601-Pt100	
Measuring range:	-200 **to** $+600\,^{\circ}$**C**
Response time:	approximate 10 s
Standard:	1/3 DIN class B
Error ranges:	$\pm\left(0.10 + 0.00167\,\lvert\text{Temp. in }^{\circ}\text{C}\rvert\right)$
→ Greisinger GTF 401-Pt100	
Measuring range:	-50 **to** $+400\,^{\circ}$**C**
Response time:	approximate 10 s
Standard:	DIN class B
Error ranges:	$\pm\left(0.30 + 0.00500\cdot\lvert\text{Temp. in }^{\circ}\text{C}\rvert\right)$

Table A2. Pressure Measurements.

→ Yamatake JTG960A	
Measuring span:	**0.7 to 14 MPa**
Setting range:	−0.1 to 14 MPa
Working pressure range:	2.0 kPa to 14 MPa
Accuracy:	±0.15% for $\psi \geq 2.1$ MPa
	$\pm\left(0.05 + 0.1\cdot\frac{2.1}{\psi}\right)\%$ for $\psi < 2.1$ MPa
→ Yamatake JTG940A	
Measuring span:	**35 to 3500 kPa**
Setting range:	−100 to 3500 kPa
Working pressure range:	2.0 kPa to 3500 kPa
Accuracy:	±0.1% for $\psi \geq 0.14$ MPa
	$\pm\left(0.025 + 0.75\cdot\frac{0.14}{\psi}\right)\%$ for $\psi < 0.14$ MPa

ψ = upper and lower limit of the calibration range (for both pressure measuring devices).

Table A3. Mass Flow Rate Measurements.

→ Yamatake JTG960A	
Measuring span:	**0.25 to 14 MPa**
Setting span:	−100 to 14 MPa
Working pressure range:	2.0 kPa to 14 MPa
Accuracy:	±0.15% for $\psi \geq 3.5$ MPa
	$\pm\left(0.1 + 0.05\cdot\frac{3.5}{\psi}\right)\%$ for $\psi < 3.5$ MPa
→ Yamatake JTD930A	
Measuring span:	**35 to 700 kPa**
Setting span:	−100 to 700 kPa
Working pressure range:	2.0 kPa to 14 MPa
Accuracy:	±0.1% for $\psi \geq 140$ kPa
	$\pm\left(0.025 + 0.075\cdot\frac{140}{\psi}\right)\%$ for $\psi < 140$ kPa
→ Yamatake JTD920A	
Measuring span:	**0.75 to 100 kPa**
Setting span:	−100 to 100 kPa
Working pressure range:	2.0 kPa to 14 MPa
Accuracy:	±0.1% for $\psi \geq 5.0$ kPa
	$\pm\left(0.025 + 0.075\cdot\frac{5.0}{\psi}\right)\%$ for $\psi < 5.0$ kPa
→ Yamatake JTD910A	
Measuring span:	**0.1 to 2 kPa**
Setting span:	−1 to 1 kPa
Working pressure range:	up to 210 kPa
Accuracy:	$\pm\left(0.15 + 0.15\cdot\frac{1.0}{\psi}\right)\%$

ψ = upper and lower limit of the calibration range (for all mass flow rate measuring devices).

References

1. Tontu, M.; Sahin, B.; Bilgili, M. An exergoeconomic–environmental analysis of an organic Rankine cycle system integrated with a 660 MW steam power plant in terms of waste heat power generation. *Energy Sources Part A Recovery Util. Environ. Eff.* **2020**, *2020*, 1–22. [CrossRef]
2. Elhelw, M.; Al Dahma, K.S. Utilizing exergy analysis in studying the performance of steam power plant at two different operation mode. *Appl. Therm. Eng.* **2019**, *150*, 285–293. [CrossRef]

3. Uysal, C.; Kurt, H.; Kwak, H.Y. Exergetic and thermoeconomic analyses of a coal-fired power plant. *Int. J. Ther. Sci.* **2017**, *117*, 106–120. [CrossRef]

4. Naserbegi, A.; Aghaie, M.; Minuchehr, A.; Alahyarizadeh, G. A novel exergy optimization of Bushehr nuclear power plant by gravitational search algorithm (GSA). *Energy* **2018**, *148*, 373–385. [CrossRef]

5. Wilding, P.R.; Murray, N.R.; Memmott, M.J. The use of multi-objective optimization to improve the design process of nuclear power plant systems. *Ann. Nucl. Energy* **2020**, *137*, 107079. [CrossRef]

6. Adibhatla, S.; Kaushik, S.C. Exergy and thermoeconomic analyses of 500 MWe sub critical thermal power plant with solar aided feed water heating. *Appl. Therm. Eng.* **2017**, *123*, 340–352. [CrossRef]

7. Mehrpooya, M.; Ghorbani, B.; Sadeghzadeh, M. Hybrid solar parabolic dish power plant and high-temperature phase change material energy storage system. *Int. J. Energy Res.* **2019**, *43*, 5405–5420. [CrossRef]

8. Idris, M.N.M.; Hashim, H.; Razak, N.H. Spatial optimisation of oil palm biomass co-firing for emissions reduction in coal-fired power plant. *J. Clean. Prod.* **2018**, *172*, 3428–3447. [CrossRef]

9. Kim, D.; Kim, K.T.; Park, Y.K. A Comparative Study on the Reduction Effect in Greenhouse Gas Emissions between the Combined Heat and Power Plant and Boiler. *Sustainability* **2020**, *12*, 5144. [CrossRef]

10. Li, X.; Teng, Y.; Zhang, K.; Peng, H.; Cheng, F.; Yoshikawa, K. Mercury Migration Behavior from Flue Gas to Fly Ashes in a Commercial Coal-Fired CFB Power Plant. *Energies* **2020**, *13*, 1040. [CrossRef]

11. Nazir, S.M.; Bolland, O.; Amini, S. Analysis of combined cycle power plants with chemical looping reforming of natural gas and pre-combustion CO_2 capture. *Energies* **2018**, *11*, 147. [CrossRef]

12. Javadi, M.A.; Hoseinzadeh, S.; Ghasemiasl, R.; Heyns, P.S.; Chamkha, A.J. Sensitivity analysis of combined cycle parameters on exergy, economic, and environmental of a power plant. *J. Ther. Anal. Calor.* **2020**, *139*, 519–525. [CrossRef]

13. Rao, A.G.; Van den Oudenalder, F.S.C.; Klein, S.A. Natural gas displacement by wind curtailment utilization in combined-cycle power plants. *Energy* **2019**, *168*, 477–491. [CrossRef]

14. Kotowicz, J.; Brzęczek, M. Analysis of increasing efficiency of modern combined cycle power plant: A case study. *Energy* **2018**, *153*, 90–99. [CrossRef]

15. Pattanayak, L.; Sahu, J.N.; Mohanty, P. Combined cycle power plant performance evaluation using exergy and energy analysis. *Env. Progr. Sust. Energy* **2017**, *36*, 1180–1186. [CrossRef]

16. Okubo, M.; Kuwahara, T. *New Technologies for Emission Control in Marine Diesel Engines*; Butterworth-Heinemann: Oxford, UK, 2020.

17. Sartomo, A.; Santoso, B.; Muraza, O. Recent progress on mixing technology for water-emulsion fuel: A review. *Energy Convers. Manag.* **2020**, *213*, 112817. [CrossRef]

18. Senčić, T.; Mrzljak, V.; Blecich, P.; Bonefačić, I. 2D CFD simulation of water injection strategies in a large marine engine. *J. Mar. Sci. Eng.* **2019**, *7*, 296. [CrossRef]

19. Lamas Galdo, M.I.; Castro-Santos, L.; Rodriguez Vidal, C.G. Numerical analysis of NOx reduction using ammonia injection and comparison with water injection. *J. Mar. Sci. Eng.* **2020**, *8*, 109. [CrossRef]

20. Fernández, I.A.; Gómez, M.R.; Gómez, J.R.; Insua, Á.B. Review of propulsion systems on LNG carriers. *Renew. Sustain. Energy Rev.* **2017**, *67*, 1395–1411. [CrossRef]

21. Ammar, N.R. Environmental and cost-effectiveness comparison of dual fuel propulsion options for emissions reduction onboard LNG carriers. *Shipbuilding* **2019**, *70*, 61–77. [CrossRef]

22. Altosole, M.; Benvenuto, G.; Zaccone, R.; Campora, U. Comparison of Saturated and Superheated Steam Plants for Waste-Heat Recovery of Dual-Fuel Marine Engines. *Energies* **2020**, *13*, 985. [CrossRef]

23. Altosole, M.; Benvenuto, G.; Campora, U.; Laviola, M.; Trucco, A. Waste heat recovery from marine gas turbines and diesel engines. *Energies* **2017**, *10*, 718. [CrossRef]

24. Grzesiak, S.; Adamkiewicz, A. Application of Steam Jet Injector for Latent Heat Recovery of Marine steam Turbine Propulsion Plant. *New Trend. Prod. Eng.* **2018**, *1*, 235–244. [CrossRef]

25. Marques, C.H.; Caprace, J.D.; Belchior, C.R.; Martini, A. An Approach for Predicting the Specific Fuel Consumption of Dual-Fuel Two-Stroke Marine Engines. *J. Mar. Sci. Eng.* **2019**, *7*, 20. [CrossRef]

26. Mrzljak, V.; Poljak, I.; Mrakovčić, T. Energy and exergy analysis of the turbo-generators and steam turbine for the main feed water pump drive on LNG carrier. *Energy Convers. Manag.* **2017**, *140*, 307–323. [CrossRef]

27. Behrendt, C.; Stoyanov, R. Operational characteristic of selected marine turbounits powered by steam from auxiliary oil-fired boilers. *New Trend. Prod. Eng.* **2018**, *1*, 495–501. [CrossRef]

28. Mrzljak, V. Low power steam turbine energy efficiency and losses during the developed power variation. *Tech. J.* **2018**, *12*, 174–180. [CrossRef]

29. Tanuma, T. *Advances in Steam Turbines for Modern Power Plants*; Woodhead Publishing: Cambridge, MA, USA, 2017.

30. Sun, L.; Hua, Q.; Shen, J.; Xue, Y.; Li, D.; Lee, K.Y. Multi-objective optimization for advanced superheater steam temperature control in a 300 MW power plant. *Appl. Energy* **2017**, *208*, 592–606. [CrossRef]

31. Szargut, J. *Exergy Method—Technical and Ecological Applications*; WIT Press: Southampton, UK, 2005.

32. Kanoglu, M.; Çengel, Y.A.; Dincer, I. *Efficiency Evaluation of Energy Systems*; Springer Briefs in Energy; Springer: Berlin/Heidelberg, Germany, 2012.

33. Ahmadi, G.R.; Toghraie, D. Energy and exergy analysis of Montazeri Steam Power Plant in Iran. *Renew Sustain. Energy Rev.* **2016**, *56*, 454–463. [CrossRef]

34. Si, N.; Zhao, Z.; Su, S.; Han, P.; Sun, Z.; Xu, J.; Cui, X.; Hu, S.; Wang, Y.; Jiang, L.; et al. Exergy analysis of a 1000 MW double reheat ultra-supercritical power plant. *Energy Convers. Manag.* **2017**, *147*, 155–165. [CrossRef]

35. Ibrahim, T.K.; Basrawi, F.; Awad, O.I.; Abdullah, A.N.; Najafi, G.; Mamat, R.; Hagos, F.Y. Thermal performance of gas turbine power plant based on exergy analysis. *Appl. Therm. Eng.* **2017**, *115*, 977–985. [CrossRef]

36. Aghbashlo, M.; Tabatabaei, M.; Hosseini, S.S.; Dashti, B.B.; Soufiyan, M.M. Performance assessment of a wind power plant using standard exergy and extended exergy accounting (EEA) approaches. *J. Clean. Prod.* **2018**, *171*, 127–136. [CrossRef]

37. AlZahrani, A.A.; Dincer, I. Energy and exergy analyses of a parabolic trough solar power plant using carbon dioxide power cycle. *Energy Convers. Manag.* **2018**, *158*, 476–488. [CrossRef]

38. Abuelnuor, A.A.A.; Saqr, K.M.; Mohieldein, S.A.A.; Dafallah, K.A.; Abdullah, M.M.; Nogoud, Y.A.M. Exergy analysis of Garri "2" 180 MW combined cycle power plant. *Renew. Sustain. Energy Rev.* **2017**, *79*, 960–969. [CrossRef]

39. Zhao, Z.; Su, S.; Si, N.; Hu, S.; Wang, Y.; Xu, J.; Jiang, L.; Chen, G.; Xiang, J. Exergy analysis of the turbine system in a 1000 MW double reheat ultra-supercritical power plant. *Energy* **2017**, *119*, 540–548. [CrossRef]

40. Medica-Viola, V.; Mrzljak, V.; Anđelić, N.; Jelić, M. Analysis of Low-Power Steam Turbine with One Extraction for Marine Applications. *Our Sea* **2020**, *67*, 87–95. [CrossRef]

41. Presciutti, A.; Asdrubali, F.; Baldinelli, G.; Rotili, A.; Malavasi, M.; Di Salvia, G. Energy and exergy analysis of glycerol combustion in an innovative flameless power plant. *J. Clean. Prod.* **2018**, *172*, 3817–3824. [CrossRef]

42. Szablowski, L.; Krawczyk, P.; Badyda, K.; Karellas, S.; Kakaras, E.; Bujalski, W. Energy and exergy analysis of adiabatic compressed air energy storage system. *Energy* **2017**, *138*, 12–18. [CrossRef]

43. Arshad, A.; Ali, H.M.; Habib, A.; Bashir, M.A.; Jabbal, M.; Yan, Y. Energy and exergy analysis of fuel cells: A review. *Therm. Sci. Eng. Progr.* **2019**, *9*, 308–321. [CrossRef]

44. Lorencin, I.; Anđelić, N.; Mrzljak, V.; Car, Z. Exergy analysis of marine steam turbine labyrinth (gland) seals. *Sci. J. Mar. Res.* **2019**, *33*, 76–83. [CrossRef]

45. Kavian, S.; Aghanajafi, C.; Mosleh, H.J.; Nazari, A.; Nazari, A. Exergy, economic and environmental evaluation of an optimized hybrid photovoltaic-geothermal heat pump system. *Appl. Energy* **2020**, *276*, 115469. [CrossRef]

46. Nami, H.; Anvari-Moghaddam, A. Geothermal driven micro-CCHP for domestic application–Exergy, economic and sustainability analysis. *Energy* **2020**, *207*, 118195. [CrossRef]

47. Liu, X.; Yang, X.; Yu, M.; Zhang, W.; Wang, Y.; Cui, P.; Zhu, Z.; Ma, Y.; Gao, J. Energy, exergy, economic and environmental (4E) analysis of an integrated process combining CO_2 capture and storage, an organic Rankine cycle and an absorption refrigeration cycle. *Energy Convers. Manag.* **2020**, *210*, 112738. [CrossRef]

48. Sun, K.; Wu, X.; Xue, J.; Ma, F. Development of a new multi-layer perceptron based soft sensor for SO_2 emissions in power plant. *J. Proc. Control* **2019**, *84*, 182–191. [CrossRef]

49. Hamed, W.; Salim, N. Use Data Mining Techniques to Identify Parameters That Influence Generated Power in Thermal Power Plant. *JECS* **2017**, *17*, 52–64. Available online: http://journal.sustech.edu/index.php/JECS/article/view/165 (accessed on 7 October 2020).

50. Lorencin, I.; Anđelić, N.; Mrzljak, V.; Car, Z. Genetic algorithm approach to design of multi-layer perceptron for combined cycle power plant electrical power output estimation. *Energies* **2019**, *12*, 4352. [CrossRef]

51. Khademi, M.; Moadel, M.; Khosravi, A. Power prediction and technoeconomic analysis of a solar PV power plant by MLP-ABC and COMFAR III, considering cloudy weather conditions. *Int. J. Chem. Eng.* **2016**, *2016*, 1031943. [CrossRef]

52. Demirdelen, T.; Aksu, I.O.; Esenboga, B.; Aygul, K.; Ekinci, F.; Bilgili, M. *A New Method for Generating Short-Term Power Forecasting Based on Artificial Neural Networks and Optimization Methods for Solar Photovoltaic Power Plants*; Springer Nature Singapore Pte Ltd.: Singapore, 2019. [CrossRef]

53. Wahid, F.; Ghazali, R.; Shah, A.S.; Fayaz, M. Prediction of energy consumption in the buildings using multi-layer perceptron and random forest. *IJAST* **2017**, *101*, 13–22. [CrossRef]

54. Tahan, M.; Tsoutsanis, E.; Muhammad, M.; Karim, Z.A. Performance-based health monitoring, diagnostics and prognostics for condition-based maintenance of gas turbines: A review. *Appl. Energy* **2017**, *198*, 122–144. [CrossRef]

55. Lorencin, I.; Anđelić, N.; Mrzljak, V.; Car, Z. Multilayer perceptron approach to condition-based maintenance of marine CODLAG propulsion system components. *Sci. J. Mar. Res.* **2019**, *33*, 181–190. [CrossRef]

56. Ferrero Bermejo, J.; Gómez Fernández, J.F.; Pino, R.; Crespo Márquez, A.; Guillén López, A.J. Review and Comparison of Intelligent Optimization Modelling Techniques for Energy Forecasting and Condition-Based Maintenance in PV Plants. *Energies* **2019**, *12*, 4163. [CrossRef]

57. Dixit, S.; Verma, N.K. Intelligent Condition Based Monitoring of Rotary Machines with Few Samples. *IEEE Sens. J.* **2020**, *2020*. [CrossRef]

58. Baressi Šegota, S.; Lorencin, I.; Musulin, J.; Štifanić, D.; Car, Z. Frigate Speed Estimation Using CODLAG Propulsion System Parameters and Multilayer Perceptron. *Our Sea* **2020**, *67*, 117–125. [CrossRef]

59. Dhini, A.; Kusumoputro, B.; Surjandari, I. Neural network based system for detecting and diagnosing faults in steam turbine of thermal power plant. In Proceedings of the 2017 IEEE 8th International Conference on Awareness Science and Technology (iCAST), Taichung, Taiwan, China, 8–10 November 2017; pp. 149–154. [CrossRef]

60. Tian, D.; Deng, J.; Vinod, G.; Santhosh, T.V.; Tawfik, H. A Neural Networks Design Methodology for Detecting Loss of Coolant Accidents in Nuclear Power Plants. In *Applications of Big Data Analytics*; Springer: Cham, Switzerland, 2018; pp. 43–61. [CrossRef]

61. Ayo-Imoru, R.M.; Cilliers, A.C. Continuous machine learning for abnormality identification to aid condition-based maintenance in nuclear power plant. *Ann. Nucl. Energy* **2018**, *118*, 61–70. [CrossRef]

62. Strušnik, D.; Avsec, J. Artificial neural networking and fuzzy logic exergy controlling model of combined heat and power system in thermal power plant. *Energy* **2015**, *80*, 318–330. [CrossRef]

63. Strušnik, D.; Golob, M.; Avsec, J. Artificial neural networking model for the prediction of high efficiency boiler steam generation and distribution. *Simul. Model. Pract. Theory* **2015**, *57*, 58–70. [CrossRef]

64. Agrež, M.; Avsec, J.; Strušnik, D. Entropy and exergy analysis of steam passing through an inlet steam turbine control valve assembly using artificial neural networks. *Int. J. Heat Mass Transf.* **2020**, *156*, 119897. [CrossRef]

65. *Marine Steam Turbine MS40-2—Instruction Book for Marine Turbine Unit*; Hyundai-Mitsubishi, Hyundai Heavy Industries, Co., Ltd.: Ulsan, Korea, 2004.

66. Mrzljak, V.; Poljak, I.; Medica-Viola, V. Dual fuel consumption and efficiency of marine steam generators for the propulsion of LNG carrier. *Appl. Therm. Eng.* **2017**, *119*, 331–346. [CrossRef]

67. Koroglu, T.; Sogut, O.S. Conventional and advanced exergy analyses of a marine steam power plant. *Energy* **2018**, *163*, 392–403. [CrossRef]

68. Çiçek, A.N. Exergy Analysis of a Crude Oil Carrier Steam Plant. Master's Thesis, Istanbul Technical University, Istanbul, Turkey, 2009. (In Turkish)

69. Mrzljak, V.; Poljak, I.; Medica-Viola, V. Thermodynamical analysis of high-pressure feed water heater in steam propulsion system during exploitation. *Shipbuilding* **2017**, *68*, 45–61. [CrossRef]

70. Taylor, D.A. *Introduction to Marine Engineering*, 2nd ed.; Elsevier Butterworth-Heinemann: Oxford, UK, 1996.

71. Škopac, L.; Medica-Viola, V.; Mrzljak, V. Selection Maps of Explicit Colebrook Approximations according to Calculation Time and Precision. *Heat Transf. Eng.* **2020**, *2020*, 1–15. [CrossRef]

72. Carlton, J. *Marine Propellers and Propulsion*, 4th ed.; Butterworth-Heinemann: Oxford, UK, 2019.

73. Kocijel, L.; Poljak, I.; Mrzljak, V.; Car, Z. Energy Loss Analysis at the Gland Seals of a Marine Turbo-Generator Steam Turbine. *Tech. J.* **2020**, *14*, 19–26. [CrossRef]

74. Moran, M.; Shapiro, H.; Boettner, D.D.; Bailey, M.B. *Fundamentals of Engineering Thermodynamics*, 7th ed.; John Wiley and Sons, Inc.: Hoboken, NJ, USA, 2011.

75. Fernández, I.A.; Gómez, M.R.; Gómez, J.R.; López-González, L.M. H_2 production by the steam reforming of excess boil off gas on LNG vessels. *Energy Convers Manag.* **2017**, *134*, 301–313. [CrossRef]

76. Medica-Viola, V.; Baressi Šegota, S.; Mrzljak, V.; Štifanić, D. Comparison of conventional and heat balance based energy analyses of steam turbine. *Sci. J. Mar. Res.* **2020**, *34*, 74–85. [CrossRef]

77. Dincer, I.; Rosen, M.A. *Exergy: Energy, Environment and Sustainable Development*, 2nd ed.; Elsevier: Oxford, UK, 2013.

78. Baldi, F.; Ahlgren, F.; Nguyen, T.V.; Thern, M.; Andersson, K. Energy and exergy analysis of a cruise ship. *Energies* **2018**, *11*, 2508. [CrossRef]

79. Kumar, V.; Pandya, B.; Matawala, V. Thermodynamic studies and parametric effects on exergetic performance of a steam power plant. *Int. J. Ambient. Energy* **2019**, *40*, 1–11. [CrossRef]

80. Mrzljak, V.; Blecich, P.; Anđelić, N.; Lorencin, I. Energy and exergy analyses of forced draft fan for marine steam propulsion system during load change. *J. Mar. Sci. Eng.* **2019**, *7*, 381. [CrossRef]

81. Ray, T.K.; Datta, A.; Gupta, A.; Ganguly, R. Exergy-based performance analysis for proper O&M decisions in a steam power plant. *Energy Convers. Manag.* **2010**, *51*, 1333–1344. [CrossRef]

82. Aljundi, I.H. Energy and exergy analysis of a steam power plant in Jordan. *Appl. Therm. Eng.* **2009**, *29*, 324–328. [CrossRef]

83. Mrzljak, V.; Senčić, T.; Žarković, B. Turbogenerator Steam Turbine Variation in Developed Power: Analysis of Exergy Efficiency and Exergy Destruction Change. *Model. Simul. Eng.* **2018**, *2018*, 2945325. [CrossRef]

84. Tan, H.; Shan, S.; Nie, Y.; Zhao, Q. A new boil-off gas re-liquefaction system for LNG carriers based on dual mixed refrigerant cycle. *Cryogenics* **2018**, *92*, 84–92. [CrossRef]

85. Noroozian, A.; Mohammadi, A.; Bidi, M.; Ahmadi, M.H. Energy, exergy and economic analyses of a novel system to recover waste heat and water in steam power plants. *Energy Convers. Manag.* **2017**, *144*, 351–360. [CrossRef]

86. Nanaki, E.A.; Xydis, G. *Exergetic Aspects of Renewable Energy Systems: Insights to Transportation and Energy Sector for Intelligent Communities*; CRC Press: Boca Raton, FL, USA, 2019.

87. Erdem, H.H.; Akkaya, A.V.; Cetin, B.; Dagdas, A.; Sevilgen, S.H.; Sahin, B.; Teke, I.; Gungor, C.; Atas, S. Comparative energetic and exergetic performance analyses for coal-fired thermal power plants in Turkey. *Int. J. Therm. Sci.* **2009**, *48*, 2179–2186. [CrossRef]

88. Adibhatla, S.; Kaushik, S.C. Energy and exergy analysis of a super critical thermal power plant at various load conditions under constant and pure sliding pressure operation. *Appl. Therm. Eng.* **2014**, *73*, 51–65. [CrossRef]

89. Goodfellow, I.; Bengio, Y.; Courville, A. *Deep Learning*; The MIT Press: Cambridge, MA, USA, 2016.

90. Moon, T.; Hong, S.; Choi, H.Y.; Jung, D.H.; Chang, S.H.; Son, J.E. Interpolation of greenhouse environment data using multilayer perceptron. *Comput. Electron. Agric.* **2019**, *166*, 105023. [CrossRef]

91. Lorencin, I.; Anđelić, N.; Španjol, J.; Car, Z. Using multi-layer perceptron with Laplacian edge detector for bladder cancer diagnosis. *Artif. Intell. Med.* **2020**, *102*, 101746. [CrossRef]

92. Car, Z.; Baressi Šegota, S.; Anđelić, N.; Lorencin, I.; Mrzljak, V. Modeling the Spread of COVID-19 Infection Using a Multilayer Perceptron. *Comput. Math. Methods Med.* **2020**, *2020*, 5714714. [CrossRef] [PubMed]

93. Khalid, A.; Sundararajan, A.; Acharya, I.; Sarwat, A.I. Prediction of li-ion battery state of charge using multilayer perceptron and long short-term memory models. In Proceedings of the 2019 IEEE Transportation Electrification Conference and Expo (ITEC), Detroit, MI, USA, 19–21 June 2019; pp. 1–6. [CrossRef]

94. Bisong, E. *The Multilayer Perceptron (MLP). Building Machine Learning and Deep Learning Models on Google Cloud Platform*; Apress: Berkeley, CA, USA, 2019; pp. 401–405. [CrossRef]

95. Eger, S.; Youssef, P.; Gurevych, I. Is it time to swish? Comparing deep learning activation functions across NLP tasks. *arXiv* **2019**, arXiv:1901.02671.

96. Jagtap, A.D.; Kawaguchi, K.; Karniadakis, G.E. Locally adaptive activation functions with slope recovery term for deep and physics-informed neural networks. *Proc. R. Soc. A* **2020**, *476*, 20200334. [CrossRef]

97. Dureja, A.; Pahwa, P. Analysis of non-linear activation functions for classification tasks using convolutional neural networks. *Recent Pat. Comput. Sci.* **2019**, *12*, 156–161. [CrossRef]

98. Hastie, T.; Tibshirani, R.; Friedman, J. *The Elements of Statistical Learning: Data Mining, Inference, and Prediction*, 2nd ed.; Springer Science & Business Media: New York, NY, USA, 2009.

99. Pedregosa, F.; Varoquaux, G.; Gramfort, A.; Michel, V.; Thirion, B.; Grisel, O.; Blondel, M.; Prettenhofer, P.; Weiss, R.; Dubourg, V.; et al. Scikit-learn: Machine learning in Python. *J. Mach. Learn. Res.* **2011**, *12*, 2825–2830.

100. Abraham, A.; Pedregosa, F.; Eickenberg, M.; Gervais, P.; Mueller, A.; Kossaifi, J.; Gramfort, A.; Thirion, B.; Varoquaux, G. Machine learning for neuroimaging with scikit-learn. *Front. Neuroinf* **2014**, *8*, 14. [CrossRef]

101. Géron, A. *Hands-On Machine Learning with Scikit-Learn, Keras, and TensorFlow: Concepts, Tools, and Techniques to Build Intelligent Systems*, 2nd ed.; O'Reilly Media: Sebastopol, CA, USA, 2019.

102. Liashchynskyi, P.; Liashchynskyi, P. Grid Search, Random Search, Genetic Algorithm: A Big Comparison for NAS. *arXiv* **2019**, arXiv:1912.06059.

103. Bari, A.H.; Gavrilova, M.L. Multi-layer perceptron architecture for kinect-based gait recognition. In Proceedings of the Computer Graphics International Conference, Calgary, AB, Canada, 17–20 June 2019; Springer: Cham, Siwtzerland; pp. 356–363. [CrossRef]

104. Sakar, C.O.; Polat, S.O.; Katircioglu, M.; Kastro, Y. Real-time prediction of online shoppers' purchasing intention using multilayer perceptron and LSTM recurrent neural networks. *Neural Comput. Appl.* **2019**, *31*, 6893–6908. [CrossRef]

105. Nagelkerke, N.J. A note on a general definition of the coefficient of determination. *Biometrika* **1991**, *78*, 691–692. [CrossRef]

106. Nakagawa, S.; Johnson, P.C.; Schielzeth, H. The coefficient of determination R^2 and intra-class correlation coefficient from generalized linear mixed-effects models revisited and expanded. *J. R. Soc. Interface* **2017**, *14*, 20170213. [CrossRef]

107. Qi, J.; Du, J.; Siniscalchi, S.M.; Ma, X.; Lee, C.H. On mean absolute error for deep neural network based vector-to-vector regression. *IEEE Signal Proc. Lett.* **2020**, *27*, 1485–1489. [CrossRef]

108. Štifanić, D.; Musulin, J.; Miočević, A.; Baressi Šegota, S.; Šubić, R.; Car, Z. Impact of COVID-19 on Forecasting Stock Prices: An Integration of Stationary Wavelet Transform and Bidirectional Long Short-Term Memory. *Complexity* **2020**, *2020*, 1846926. [CrossRef]

109. Berrar, D. Cross-validation. *Encycl. Bioinform. Comput. Biol.* **2019**, *1*, 542–545.

110. Bishop, C.M. *Pattern Recognition and Machine Learning*; Springer: New York, NY, USA, 2006.

111. Moayedi, H.; Osouli, A.; Nguyen, H.; Rashid, A.S.A. A novel Harris hawks' optimization and k-fold cross-validation predicting slope stability. *Eng. Comput.* **2019**, *2019*, 1–11. [CrossRef]

112. BURA Supercomputer, Computing Resources. Available online: https://cnrm.uniri.hr/bura/ (accessed on 30 October 2020).

113. Anaconda Software Distribution. Anaconda Documentation. Anaconda Inc. 2020. Available online: https://docs.anaconda.com/ (accessed on 30 October 2020).

114. Lemmon, E.W.; Huber, M.L.; McLinden, M.O. *Reference Fluid Thermodynamic and Transport Properties-REFPROP*; Version 9.0, User's Guide; NIST: Gaithersburg, MD, USA, 2010.

115. Mrzljak, V.; Poljak, I.; Žarković, B. Exergy analysis of steam pressure reduction valve in marine propulsion plant on conventional LNG carrier. *Our Sea* **2018**, *65*, 24–31. [CrossRef]

116. SUITABLE PT100 MEASURING PROBE (4-WIRE). Available online: https://www.greisinger.de/files/upload/en/produkte/kat/k16_011_EN_oP.pdf (accessed on 3 October 2020).

117. JTG Series of Pressure Transmitters. Available online: http://smte.kr/product/data/pdf/pdf_100812100836_552363.pdf (accessed on 4 October 2020).

118. JTD Series of Differential Pressure Transmitters. Available online: http://www.krtproduct.com/krt_Picture/sample/1_spare%20part/yamatake/Fi_ss01/SS2-DST100-0100.pdf (accessed on 3 October 2020).

Publisher's Note: MDPI stays neutral with regard to jurisdictional claims in published maps and institutional affiliations.

Article

Fault Tree Analysis and Failure Diagnosis of Marine Diesel Engine Turbocharger System

Vlatko Knežević [1,*], Josip Orović [1], Ladislav Stazić [2] and Jelena Čulin [1]

[1]   Maritime Department, University of Zadar, Mihovila Pavlinovića 1, 23000 Zadar, Croatia; jorovic@unizd.hr (J.O.); jculin@unizd.hr (J.Č.)

[2]   Faculty of Maritime Studies, University of Split, R. Boškovića 37, 21000 Split, Croatia; lstazic@pfst.hr

*   Correspondence: vknezevi1@unizd.hr

Received: 10 November 2020; Accepted: 7 December 2020; Published: 9 December 2020

Abstract: The reliability of marine propulsion systems depends on the reliability of several sub-systems of a diesel engine. The scavenge air system is one of the crucial sub-systems of the marine engine with a turbocharger as an essential component. In this paper, the failures of a turbocharger are analyzed through the fault tree analysis (FTA) method to estimate the reliability of the system and to predict the cause of failures. The quantitative method is used for assessing the probability of faults occurring in the turbocharger system. The main failures of a scavenge air sub-system, such as air filter blockage, compressor fouling, turbine fouling (exhaust side), cooler tube blockage and cooler air side blockage, are simulated on a Wärtsilä-Transas engine simulator for a marine two-stroke diesel engine. The results obtained through the simulation can provide improvement in the maintenance plan, reliability of the propulsion system and optimization of turbocharger operation during exploitation time.

Keywords: reliability; fault tree analysis; failure diagnosis; diesel engine turbocharger; maintenance

1. Introduction

The reliability and safety of marine propulsion have a major role during the exploitation period which cannot be neglected. The safe operation of the vessel depends on the reliability of the main engine propulsion system. The reliability of any component or system is defined as the probability that a component or system will perform a required function for a given period of time when used under stated operating conditions [1]. The safety factor of a system is usually related to reliability and it can be defined as the avoidance of conditions that can cause injury, loss of life or damage to equipment and the surrounding environment [2]. Due to the complexity of the marine main engines and their sub-systems, it is difficult to predict when and how many failures will occur during a voyage.

For early detecting and avoiding unnecessary failures, the method of fault diagnosis is used. The main objectives of fault diagnosis are detection, isolation and fault analyses. The main tasks of fault diagnosis are to determine the type of faults, size, time of failure and localization of faults [3]. It is necessary during the exploitation of the ship to continuously monitor and record the technical conditions and parameters of all main and auxiliary equipment. Once the operating parameters have been assessed, the reliability and availability of any component can be estimated and measures for reducing the risk of failure can be considered. Furthermore, with a detailed failure diagnosis, the maintenance plan of any component of the system can be optimized and enhanced. An improved maintenance plan can reduce life-cycle costs such as the cost of preventive and corrective maintenance, cost of materials and energy and cost of spare parts transport and installation. Nowadays, the main challenge for turbocharger manufacturers is to increase efficiency in terms of fuel economy and environmental performance.

Failure diagnosis in this paper is focused on the turbocharger of a marine diesel two-stroke MAN 6S60MC-C engine. The purpose of this paper is to diagnose the most frequent symptoms of turbocharger faults using a deductive method fault tree analysis (FTA) and to simulate these failures on a Wärtsilä-Transas engine simulator to optimize operating conditions and improve the reliability of the turbocharger by avoiding undesirable events.

Literature Overview

Because turbochargers are the most important part of the scavenge system, they must have high reliability to ensure the reliability of the main engine, which is also concluded in research papers [4] and [5], where the reliability of the main engine subsystems is estimated, including turbocharger failure. It is concluded that turbocharger failure can have a major impact on the main engine operation and proper matching of the turbocharger and main engine is highly important. The matching method with an electric turbo compound for a two-stroke marine engine is proposed in article [6] and the method for the effective mistuning identification of marine turbocharger bladed discs is discussed in article [7]. These two methods can improve marine turbocharger efficiency but mostly in the manufacturing period.

Adamkiewicz [8] analyzed the relations between cause and effect of operation faults in a few turbocharger models with a method based on expert knowledge and operational diagnostic experience. Monieta [9] used a method with acceleration vibration signals for assessing the technical condition of turbochargers in three four-stroke engines. The research has shown that the diagnostic parameters of technical condition are more reliable with this method than the resource of the operating hours of the engine. In [10], marine propulsion system reliability is estimated using fault tree analysis. The failure probability of the entire ship propulsion system is hard to estimate due to the complexity of the system and each component has a specified life-cycle and maintenance interval. This method is suitable for the main engine subsystems or individual components of the engine.

The FTA method is used in a research study [11] for the risk assessment of the container terminal operations. The results have shown that human factors were the most common cause of accidents due to negligence in operating with equipment or vehicles. This method is also used in research paper [12] as a tool for modeling the marine main engine reliability.

The research paper [13] recommends reliability-centered maintenance (RCM) methodology to optimize the failure database of marine diesel engines. This methodology is useful to obtain an accurate and reliable database for predicting failures. However, this research is done on a four-stroke marine diesel generator with a power value of 1200 kW.

The analysis of failures during the early operation period of a ship is presented in article [14]. The observed marine engine was two-stroke, low speed and turbocharged, belonging to a bulk cargo ship and, moreover, failure analysis was conducted only during the first year of operation.

The various failures of marine engine operations are simulated in study [15] on a Konsberg Maritime engine simulator. The simulated incorrect engine operations were: worn and clogged injector nozzle, exhaust valve leakage, early injection timing. The importance of early-stage fault detection and efficiency management (planned maintenance) is emphasized.

In most cases, failure analysis and reviews are lacing simulation of faults during the operating conditions of a vessel. With this simulator-based methodology, it is possible to achieve more efficient operation of the engine, predict possible faults of the turbocharger system and develop enhanced maintenance intervals for the system.

2. Two-Stroke Marine Diesel Engine Turbocharger

The turbocharger in marine diesel engines is an essential element of the scavenge air system, which directly influences the power output, engine efficiency and emission of exhaust gases. The main components of the turbocharger are: turbine wheel (rotor with turbine blades), turbine nozzle ring, steel shaft (turbine wheel on one end and compressor impeller assembled on the other), air compressor, silencer, diffuser, air filter and cooler.

In this paper, the focus is on the air filter, compressor impeller, turbine wheel and air cooler, because the engine room crew regularly inspects these components. Additionally, one of the important components of the turbocharger shaft that affects its reliability and durability are bearings. The key role of the bearing system is to control the radial and axial motion of the shaft and to reduce friction losses that have an impact on fuel efficiency. Furthermore, with new stringent emission regulations and demands, the lubricating oil viscosities become lower, so manufacturers must produce bearings that can maintain the stability of the rotor and avoid increased wear [16].

Leading manufacturers of turbochargers in the shipping industry are MAN Diesel & Turbo, ABB Turbocharging and Mitsubishi [17]. One of the new MAN TCR turbocharger models is shown in Figure 1. The new TCR turbocharger series has a wide range of applications, with engine power outputs from 390 to 7000 kW. The upcoming series is TCT which is specifically optimized for IMO Tier III engines, with a lighter design, superior charging efficiency and high air pressure.

Figure 1. The turbocharger of two-stroke marine engine (MAN, TCR 18) [18].

Some turbochargers are designed with variable turbine area technology which enables the volume of charge air to be precisely matched to the quantity of fuel injected at all points regarding an engine's load and speed range [18]. Aside from mechanical innovations, there is research on using digital services that can provide reliable and simplified monitoring of turbocharger parameters to reduce maintenance costs.

The turbocharger of a simulated engine in this paper is a single stage type and it uses a charging system with constant pressure. Nowadays, manufacturers are trying to increase the energy efficiency of engines with two-stage turbochargers that use low- and high-pressure stages to deliver the charge air to cylinders at high pressure. Two-stage turbochargers can improve engine efficiency and reduce specific fuel oil consumption (SFOC), and these developments are important to satisfy new IMO NOx Tier regulations [19].

3. Fault Tree Analysis Method

The applied method for fault detection in this paper is FTA. The FTA method is a graphical model of the various combinations of faults that will result in the occurrence of the predefined undesired event [20]. With the FTA, the reliability of the marine propulsion system or any component of the sub-system can be estimated by calculating failure probability. The main purpose of FTA studies is to develop comprehensive technology for early fault detection, system life prediction and enhanced maintenance intervals.

When creating a fault tree model, it is necessary to define the causal connections between events (failures) of the analyzed system, identify all possible faults that can cause the top event to happen and consider appropriate corrective measures. The reliability of the system (turbocharger) depends on the occurrence probability of undesired failures of its sub-units. In this case, it is the exhaust and air side of the turbine, air cooler and turbine shaft. Figure 2 shows a fault tree structure for the analyzed turbocharger.

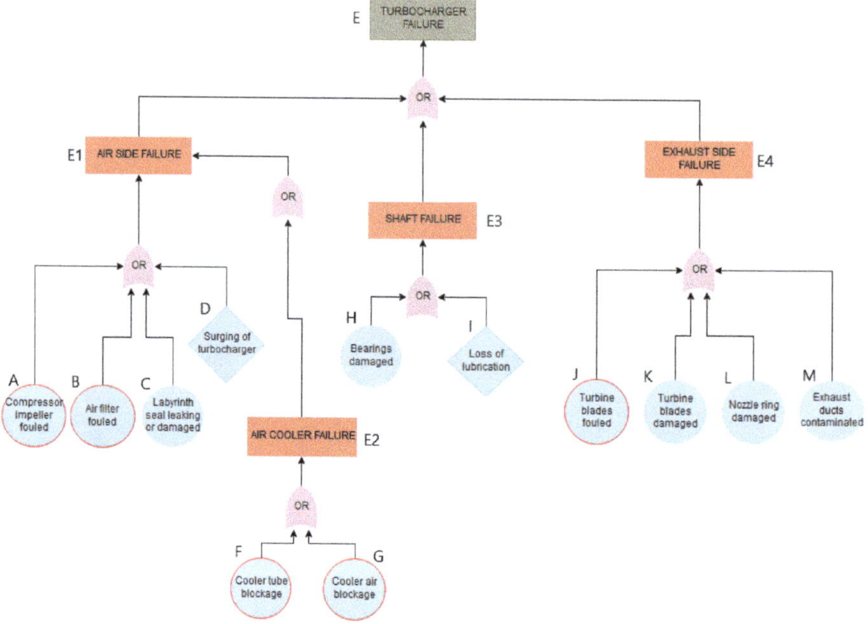

Figure 2. The fault tree of the turbocharger system.

The first step in the FTA is to define the top event, which is the most unwanted event in the system. Furthermore, it is important to determine all the events and conditions that leads to the top event. The fault events that lead to the top event are at the bottom and intermediate events are connected with logic gates. Logic gates represent the branches of the fault tree with their multiple inputs and just one output.

The primary events with their related symbols included in this fault tree model are:

- Basic events are faults events due to excessive operational stress resulting in the system element being out of operation [20]. These events do not need any further development and they are presented graphically with a circle (bottom events of fault tree). Basic events with a red outline are the ones that are simulated in this research.

- Intermediate events are faults that occur as a result of the combination through logic gates. They are symbolized by a rectangle and they pass through logic gates to the top event.

- Undeveloped events are specific faults that are not further developed, either because the event is of insufficient consequence or because information for the event is unavailable. They are graphically presented with a diamond.

- An OR gate is a logic gate that indicates that an output event occurs if one or more input events occur.

When the fault tree is constructed it can be used for assessing the probability of the basic events that are key parameters to determine the probability of the occurrence of the top event. For a better understanding of the interactive connections between basic and intermediate events, quantitative analysis is used and expressed with Boolean algebra. Using the probability theory, the fault tree model of the turbocharger can be expressed as:

$$P(E) = P(E1) + P(E3) + P(E4) \tag{1}$$

$$P(E1) = P(A) + P(B) + P(C) + P(D) + P(E2)$$

$$P(E2) = P(F) + P(G) - P(F \cap G)$$

$$P(E3) = P(H|I) = P(H)$$

$$P(E4) = P(J) + P(K) + P(L) + P(M) - P(J)P(K)P(L)P(M)$$

The occurrence of the top event P(E), is obtained as the sum of the fault probabilities of E1, E3 and E4. Basic events connected with an OR logic gate are considered as mutually exclusive (faults cannot occur at same time) or independent (occurrence of one event does not affect the occurrence of other events). The probability of event E1 can be calculated as the sum of mutually exclusive events (A, B, C and D) and the fault probability of event E2 that consists of two independent events. In the case of event E3, the probability of a fault is equal to the probability of basic event H, because event I is completely dependent on event H (loss of lubrication is unlikely to cause shaft failure without affecting bearing temperature). The occurrence of event E4 is defined as the probability fault sum of basic events that are all independent.

The probability of the occurrence of a fault event output from the "OR" gate can be calculated with the formula [21]:

$$P_{(y0)} = 1 - \prod_{i=1}^{k}\{1 - P(yi)\} \tag{2}$$

where:

$P(y_0)$: the probability of the occurrence of the OR gate output event

k: the number of input events in the OR gate

$P(yi)$: the probability of the input event in the OR gate. The input event is $yi = 1, 2, 3, \dots, k$

4. Fault Simulation and Diagnosis

The faults (air filter blockage, compressor fouled, cooler tube and air side fouled, turbine fouled) are simulated using a Wärtsilä-Transas 5000 engine room simulator. The simulator provides a detailed copy of the vessel system and engine room models with interactive parameters and features for simulating exploitation conditions such as machinery faults or environment effects (wind, waves, hull fouling).

In this study, a propulsion plant of Tanker LCC (Aframax) with the main engine—MAN B&W 6S60 MC-C—is used (Figure 3). The main engine is two-stroke, low speed, reversible, crosshead type with six cylinders and constant pressure turbocharging. One turbocharger is fitted with equipment for washing the compressor and turbine side. Additionally, the engine is equipped with an air cooler for a fresh water cooling system. The type of fuel used for simulation is marine diesel oil (MDO) with a defined maximum sulfur content according to new IMO regulations. All mentioned faults are simulated while the engine is operating at 85% (nominal continuous rating) of maximum output with ambient air temperature set to 22 °C and humidity of 60%.

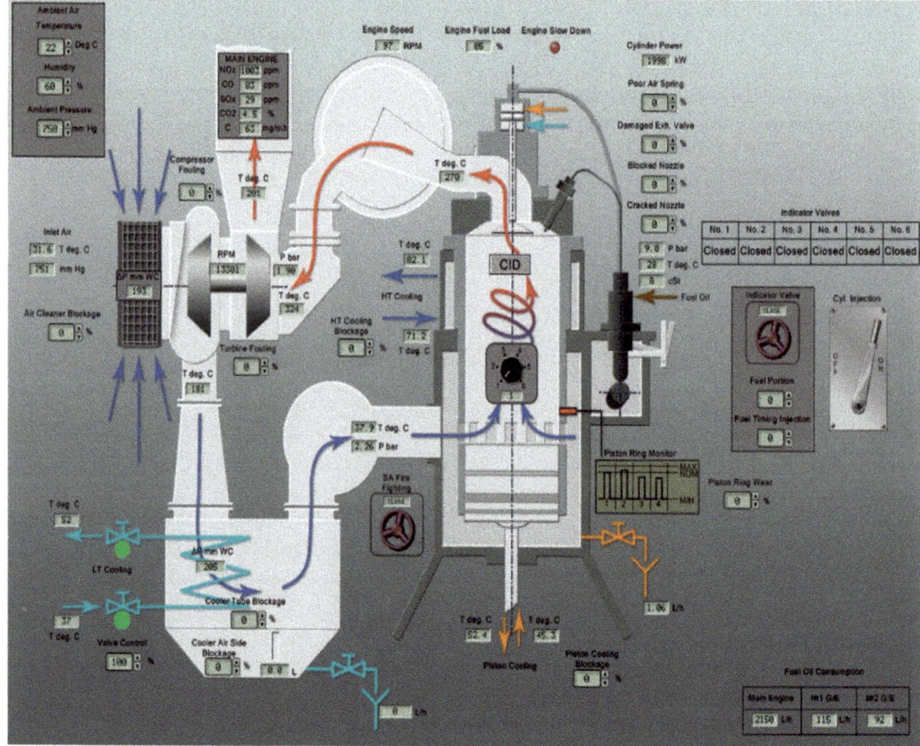

Figure 3. The interface of the main engine (cylinder) on Wärtsilä-Transas simulator [22].

4.1. Air Filter Fault

The blockage of the air filter is one of the most common faults associated with engine turbocharger systems. Fouling of the air filter and air flow ducts will significantly affect the quality of energy conversion and also it could cause an increase in fuel consumption. During operation, the air filter will eventually get contaminated and display the following inefficiencies [9]:

- Increase in the flow resistance
- Loss of filtering properties
- Loss of tightness

The amount of filter blockage can be set in the range of 0–100%, however, due to the automatic start of "slow down" operating mode (after 40% of filter blockage), faults of 10%, 20%, 30% and 40% of the fouled filter were simulated. The main engine parameters are presented in Table 1 for each fault.

Table 1. Air filter fouled.

Main Engine Parameters	Fault—Air Filter Fouled				
	0%	10%	20%	30%	40%
Main engine inlet temperature (°C)	31.60	31.70	31.80	32.40	32.70
Turbocharger air filter pressure drop (mm WC)	192.92	248.57	285.50	285.66	286.25
Scavenge air inlet temperature (°C)	181.12	173.49	165.57	141.51	124.46
Scavenge air pressure drop (mm WC)	205.27	200.66	194.62	153.53	110.47
Scavenge air temperature manifold (°C)	37.90	36.50	36.20	35.90	35.90
Main engine scavenge air pressure (bar)	2.26	1.99	1.77	1.41	1.31
Average cylinder exhaust temperature (°C)	258.49	274.44	292.48	353.55	484.33
Exhaust gas outlet temperature (°C)	200.82	219.88	241.47	315.62	456.35
Turbocharger turbine inlet temperature (°C)	323.49	336.48	351.48	405.67	530.47
Turbocharger turbine inlet pressure (bar)	1.90	1.70	1.50	1.20	0.91
Turbocharger rpm (r/min)	13,296	13,240	13,442	12,537	11,478
Main engine rpm (r/min)	96.60	96.60	96.60	96.60	96.50
Main engine fuel load (%)	84.48	84.48	84.48	84.51	85.55

The increase in filter fouling percentage results in less air supplied on the air side of the turbocharger, which is shown by these indicators in Table 1:

- slightly increased main engine inlet temperature
- increased air filter pressure drop
- reduced scavenge air inlet temperature
- reduced turbocharger scavenge air pressure drop
- reduced scavenge air temperature and pressure
- increased exhaust gas temperature on all cylinders
- increased turbocharger turbine inlet temperature
- reduced turbine inlet pressure
- reduced turbocharger rpm
- slightly increased main engine fuel load

A fault of the air filter could drastically affect the operation of the main engine if the amount (percentage) of fouling increases. Since deposits on the air filter increase, the pressure drop on the air filter is also increased, which results in ineffective compressor operation. The compressor supplies less fresh air into the cylinders due to the fouled filter, thus the main engine scavenge air pressure is reduced (2.26 bar to 1.31 bar) and the scavenge air temperature (181.12 °C to 124.46 °C) before the air cooler is also reduced. Furthermore, this fault has an impact on the increase in exhaust temperatures in the cylinders and exhaust temperature at the turbine inlet. Insufficient air supply also reduces pressure at the turbine inlet (1.90 bar to 0.90 bar), therefore, turbocharger revolutions (rpm) are reduced. The impact of an air filter fault on the main engine combustion process is explained in Section 5.1.

4.2. Air Cooler Faults

Turbochargers increase the temperature of the air in intake manifold, so it is important to reduce these excessive temperatures to achieve an efficient combustion process and lower exhaust emissions. For this purpose, engines are equipped with a scavenge air cooler which is usually constructed of bronze alloy tubes for cooling water circulation and aluminum fins for necessary air flow [23]. Some of the air coolers are cooled by sea water, but in this case, a low-temperature fresh water circuit is used.

The loss of the cooling efficiency of the air cooler is related to insufficient air flow (blockage of cooler air side) and ineffective cooling (cooler tube blockage). These two main faults are simulated with the amount of fouling from 0–50%. The main engine monitored parameters during cooler tube blockage are presented in Table 2.

Table 2. Cooler tube blockage.

Main Engine Parameters	Fault—Cooler Tube Blockage					
	0%	10%	20%	30%	40%	50%
Main engine inlet temperature (°C)	31.60	31.50	31.50	31.50	31.40	31.40
Turbocharger air filter pressure drop (mm WC)	192.92	197.50	203.51	211.51	220.53	232.52
Scavenge air inlet temperature (°C)	181.12	183.54	188.52	194.50	201.51	210.50
Scavenge air pressure drop (mm WC)	205.27	206.50	207.53	210.53	212.53	216.51
Scavenge air temperature manifold (°C)	37.90	47.90	60.00	74.90	92.90	115.50
Main engine scavenge air pressure (bar)	2.26	2.31	2.38	2.47	2.58	2.71
Average cylinder exhaust temperature (°C)	258.49	264.51	272.51	284.50	298.51	315.53
Exhaust gas outlet temperature (°C)	200.82	204.51	209.52	217.51	226.51	237.50
Turbocharger turbine inlet temperature (°C)	323.49	329.53	338.53	351.50	365.53	384.50
Turbocharger turbine inlet pressure (bar)	1.90	2.00	2.00	2.10	2.20	2.30
Turbocharger rpm (r/min)	13,296	13,407	13,564	13,785	14,070	14,432
Main engine rpm (r/min)	96.60	96.60	96.60	96.60	96.60	96.60
Main engine fuel load (%)	84.48	84.48	84.48	84.48	84.50	84.50

A blockage of cooler tubes will affect the cooling water flow in tubes which will result in a loss of cooling efficiency, as well as these unwanted indications:

- increased air filter pressure drop
- increased scavenge air inlet temperature
- increased turbocharger scavenge air pressure drop
- increased scavenge air temperature manifold
- increased average exhaust gas temperature on all cylinders and exhaust outlet temperature
- increased turbocharger turbine inlet temperature
- increased turbocharger rpm

With reduced cooling efficiency, temperatures of exhaust gases and scavenge air increase, especially the temperature of charge air (temperature manifold) before entering the cylinders. The reasons and effects of fouled cooler tubes are discussed in Section 5.5.

In Table 3, the main engine parameters for a cooler air side fault are presented. Fouling of the air side of the cooler will reduce the amount and quality of combustion air entering the cylinders, leading to these indications:

- slightly increased main engine inlet temperature
- reduced air filter pressure drop
- reduced scavenge air inlet temperature
- increased turbocharger scavenge air pressure drop
- reduced main engine scavenge air inlet pressure
- increased average exhaust gas temperature on all cylinders and exhaust outlet temperatures
- increased turbine inlet temperature
- reduced turbine inlet pressure
- reduced turbocharger rpm
- slightly increased main engine fuel load

Table 3. Cooler air side blockage.

Main Engine Parameters	Fault—Cooler Air Side Blockage				
	0%	10%	20%	30%	40%
Main engine inlet temperature (°C)	31.60	31.70	31.90	32.30	32.90
Turbocharger air filter pressure drop (mm WC)	192.92	166.47	133.48	99.35	71.48
Scavenge air inlet temperature (°C)	181.12	174.49	165.49	154.40	144.47
Scavenge air pressure drop (mm WC)	205.27	215.59	221.45	223.57	236.26
Scavenge air temperature manifold (°C)	37.90	36.60	35.90	35.90	35.90
Main engine scavenge air pressure (bar)	2.26	2.00	1.68	1.34	1.04
Average cylinder exhaust temperature (°C)	258.49	274.47	300.52	347.63	411.51
Exhaust gas outlet temperature (°C)	200.82	219.45	250.58	302.51	370.52
Turbocharger turbine inlet temperature (°C)	323.49	336.45	359.27	400.65	459.50
Turbocharger turbine inlet pressure (bar)	1.90	1.70	1.40	1.10	0.80
Turbocharger rpm (r/min)	13,296	12,688	11,841	10,840	9793
Main engine rpm (r/min)	96.60	96.60	96.60	96.60	96.60
Main engine fuel load (%)	84.48	84.48	84.48	84.51	85.53

The amount of fouling is set to a maximum of 40%, due to the automatic start of "slow down" mode when exhaust temperatures exceed their set limit point. Moreover, less air supplied to the cylinders reduces the pressure of the scavenge air before entering the cylinders and turbine inlet pressure.

4.3. Compressor Wheel Fault

The purpose of the turbocharger compressor is to draw air from the engine room in an axial direction and then expel it in a radial direction with high velocity. Three essential components of the compressor that ensure high performance are: compressor wheel, casing and diffuser.

Maintenance of the compressor side is highly important to avoid fouling of the compressor blades which can lead to excessive air flow resistance. The simulation of the fouled compressor wheel is shown in Table 4. The amount of fouling is set to 25%, 50%, 75% and 90%, unlike faults of the air filter and cooler where operating mode "slow down" automatically starts after a certain percentage of fouling.

Table 4. Compressor wheel fouled.

Main Engine Parameters	Fault—Compressor Wheel Fouled				
	0%	25%	50%	75%	90%
Main engine inlet temperature (°C)	31.60	31.60	31.60	31.60	31.70
Turbocharger air filter pressure drop (mm WC)	192.92	189.38	185.52	183.54	182.50
Scavenge air inlet temperature (°C)	181.12	199.54	225.50	262.50	292.54
Scavenge air pressure drop (mm WC)	205.27	223.16	247.51	284.55	317.51
Scavenge air temperature manifold (°C)	37.90	37.20	38.50	40.00	41.30
Main engine scavenge air pressure (bar)	2.26	2.22	2.19	2.17	2.16
Average cylinder exhaust temperature (°C)	258.49	258.60	261.50	264.50	267.52
Exhaust gas outlet temperature (°C)	200.82	201.52	204.53	207.50	210.50
Turbocharger turbine inlet temperature (°C)	323.49	323.58	325.51	328.50	331.51
Turbocharger turbine inlet pressure (bar)	1.90	1.90	1.90	1.90	1.90
Turbocharger rpm (r/min)	13,296	13,240	13,152	13,103	13,084
Main engine rpm (r/min)	96.60	96.60	96.60	96.60	96.60
Main engine fuel load (%)	84.48	84.48	84.48	84.48	84.50

Table 4 presents indications which will help to establish these shortcomings of the compressor wheel/impeller:

- reduced air filter pressure drop
- increased scavenge air inlet temperature
- increased turbocharger scavenge air pressure drop
- increased scavenge air temperature manifold
- slightly reduced main engine scavenge air pressure
- increased average cylinder temperatures and outlet exhaust temperature
- increased turbocharger turbine inlet temperature
- reduced turbocharger rpm

While the amount of deposits on the compressor wheel is increasing it affects compressor efficiency and main engine parameters. With the reduced efficiency of the compressor, the scavenge air pressure is insufficient and the most affected parameter is the temperature of the scavenge air inlet (charge air temperature after the compressor), which increases significantly (181.12 °C to 292.54 °C). The increase in the fouling percentage on the compressor blades also affects main engine performance in terms of increased exhaust temperatures in the cylinders and at the turbine inlet. Furthermore, with an excessive charge air temperature, the air cooler cannot efficiently reduce this temperature, so the quality of air entering the cylinders is inadequate for proper combustion processes. The simulated scenario with a high percentage (90%) of fouling could even lead to severe damage to the compressor impeller. Compressor maintenance and optimization are discussed in Section 5.2.

4.4. Turbine Blades Fault

The turbocharger turbine side, which consists of a turbine casing and turbine wheel, is a crucial part for converting exhaust gas energy into mechanical energy (shaft power) to drive the compressor. Because the high-velocity and high-temperature exhaust gas is directed onto the turbine blades, without preventive maintenance, the exhaust side of the turbocharger can easily get contaminated with carbon deposits and soot from the combustion process. Fouled turbine blades cause an increase in the exhaust gas flow resistance, which leads to a reduction in turbine efficiency and an increase in the specific fuel consumption. The main engine parameters during fouled turbine wheel simulation are presented in Table 5.

Table 5. Turbine wheel fouled.

Main Engine Parameters	Fault—Turbine Wheel Fouled				
	0%	25%	50%	75%	90%
Main engine inlet temperature (°C)	31.60	31.60	31.90	32.20	32.50
Turbocharger air filter pressure drop (mm WC)	192.92	171.68	144.50	114.47	93.57
Scavenge air inlet temperature (°C)	181.12	164.49	146.61	129.50	118.55
Scavenge air pressure drop (mm WC)	205.27	201.67	195.61	184.50	172.47
Scavenge air temperature manifold (°C)	37.90	36.30	35.90	35.90	35.90
Main engine scavenge air pressure (bar)	2.26	1.98	1.66	1.34	1.13
Average cylinder exhaust temperature (°C)	258.49	269.49	290.48	324.50	358.50
Exhaust gas outlet temperature (°C)	200.82	223.49	255.53	300.51	340.50
Turbocharger turbine inlet temperature (°C)	323.49	332.46	349.52	379.51	410.39
Turbocharger turbine inlet pressure (bar)	1.90	1.70	1.40	1.10	0.90
Turbocharger rpm (r/min)	13,296	12,370	11,989	11,087	10,327
Main engine rpm (r/min)	96.60	96.60	96.60	96.60	96.60
Main engine fuel load (%)	84.48	84.48	84.48	84.51	84.51

Simulated faults in the range of 25–90% reduce turbine output capacity and result in these important indications:

- reduced air filter pressure drop
- reduced turbocharger scavenge air pressure drop
- reduced scavenge air inlet temperature
- reduced main engine scavenge air pressure
- increased turbocharger turbine inlet temperature
- increased average exhaust gas temperature on all cylinders
- increased exhaust gas outlet temperature
- reduced turbocharger rpm

The fault (fouling) of turbine blades mostly depends on the quality of the fuel used and the combustion process in the cylinders. Incomplete fuel burning causes layers of deposits on the turbine blades which result in excessive exhaust gas temperatures. An increased amount of fouling percentage on the turbine blades reduces the pressure of exhaust gases at the turbine inlet stage (1.90 bar to 0.90 bar), thus the turbine does not have the necessary output power to provide charge air with constant pressure.

5. Results and Discussion

5.1. Turbocharger Air Side Results

The required amount and quality of intake air for proper combustion processes depends on the efficient and optimized operating conditions of the compressor wheel and air filter. The fouling of the air filter and compressor wheel are the two most common faults on the air intake side during engine operation and they could easily be detected by monitoring engine parameters. These changes in main engine parameters are presented in the previous section, however, the results of faults can also be shown on the cylinder indicator diagram. Indicator diagrams are used to assess the performance of each cylinder to detect any differences in the combustion process during the voyage.

The results of air filter fault simulation are presented with an indicator diagram (cylinder pressure/crank angle) in Figure 4. Cylinder indicator diagrams (Figures 4–7) are obtained using the simulator's built-in option for recording the pressure in each engine cylinder and an analyzing option for a comparison of recorded indicator diagrams according to the fault percentage. In the indicator diagram, the horizontal axis represents the cylinder crankshaft angle and the vertical axis is pressure in the cylinder.

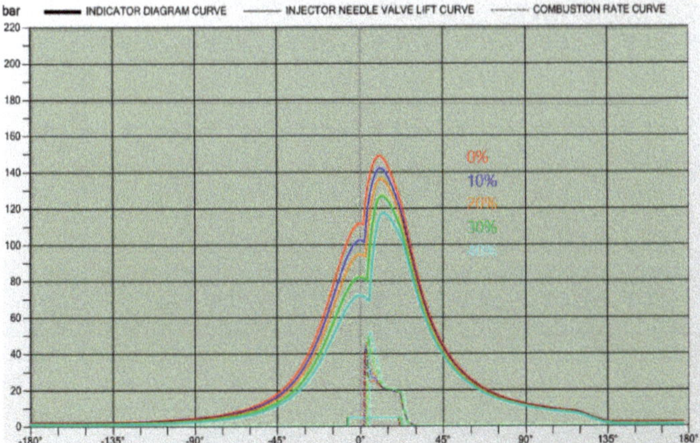

Figure 4. Indicator diagram during air filter faults.

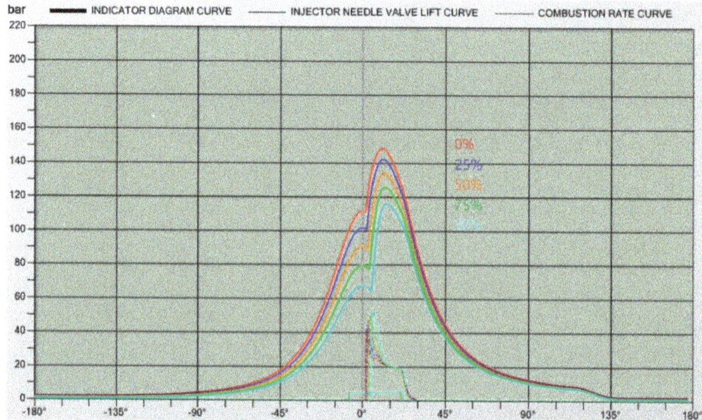

Figure 5. Indicator diagram for turbine blade faults.

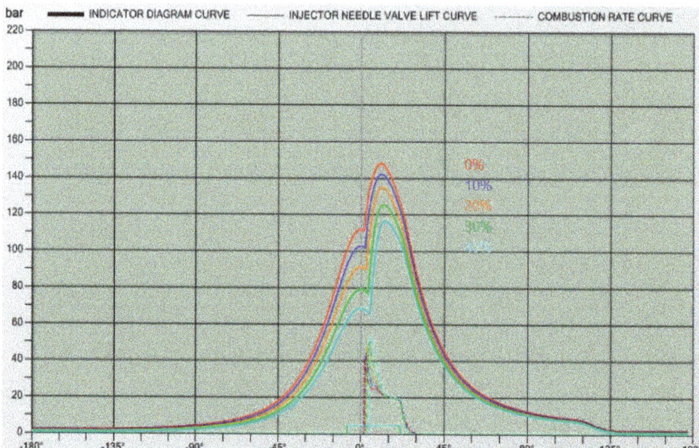

Figure 6. Indicator diagram of the fouled cooler—air side.

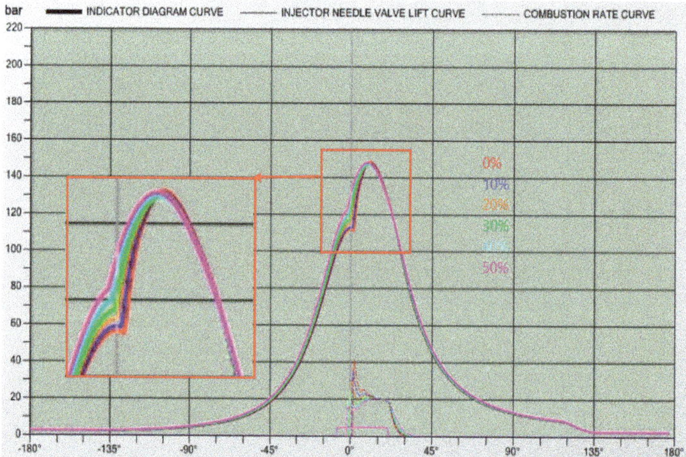

Figure 7. Indicator diagram of blocked cooler tubes.

As the contamination of the air filter increases, it affects the combustion pressures and injection timing crank angles. Each fault in the diagram is presented with different colors according to the fouling percentage, while the red line (0%) is a simulation in normal operating conditions. Analyzing the indicator diagram, the main differences during the combustion process in the cylinder are:

- reduced maximum combustion pressure from 150 bar to 119 bar
- reduced compression pressure from 112 bar to 73 bar
- increased angle of combustion start (1.5° to 4.7°), resulting in a late combustion process
- reduced (late) timing angle of fuel injection (−7.4° to −6.5°)

The fouling of the air filter and compressor impeller is usually from fuel and oil vapors, cargo residues and dust from the engine room. These are the reasons for increasing the flow resistance on the air side of the turbocharger, which eventually reduces the amount of air supplied for the combustion process and overall efficiency of the turbocharger.

With insufficient air supplied to the cylinders, the combustion process is improper, resulting in black smoke from the exhaust and increased fuel consumption. Because the amount of air for each cylinder is reduced, increased fuel injection causes an increase in exhaust gas temperatures and, eventually, activation of the alarm for high exhaust gas temperature and main engine slow down operating mode.

5.2. Optimization and Maintenance of Air Filter and Compressor Wheel

The adequate maintenance of each turbocharger component is highly important, not only for preventing failures but also to extend the durability and reliability of the system. Failures due to improper maintenance are usually caused by human errors and they are hard to predict or avoid. The assessment of human reliability during turbocharger maintenance procedures is presented in [24] and an evaluation of human error probabilities in [25], with emphasis on contributing factors for making errors such as a high level of noise and vibration, weather conditions, level of ship motion and stress. The probability of human error during the cleaning of the air filter is very low because it is an easy task to perform and an old filter with a silencer can be replaced or washed onboard.

Turbocharger manufacturers recommend maintenance and inspection intervals for each component according to engine operating hours. For an air filter on a two-stroke engine, the maintenance interval (clean air filter/depending on condition) is set to every 250 h [18]. Controlling the contamination of the air filter should be a crucial part of the inspection interval. The results of an air filter fault have shown that it drastically affects the main engine and turbocharger parameters, so the recommendation is to enhance the maintenance interval by checking the fouling of filter or unusual vibrations more often and to compare scavenge air pressure differences on the manometer.

The contamination of the compressor wheel and air intake casing is usually caused by lubricating oil vapors (entering through the labyrinth seal) or particles of fuel contained in the air. Moreover, due to the high speed of the turbocharger, foreign objects, parts of the equipment or even small particles in air flow ducts could lead to serious damage to the compressor wheel. A compressor wheel with damaged impeller blades affects the aerodynamics of the air flow and it results in insufficient scavenge air pressure and a reduction of compressor efficiency.

Another problem that can also affect compressor performance is surging (specific fault/event on fault tree). Surging occurs when the air pressure charged by the compressor is higher than the pressure inside the compressor and it creates a reverse air flow towards the inlet of the compressor. This deviation of the pressure is hard to predict and it could be due to sudden changes in the main engine load or imbalanced or damaged blades. In [26], a detailed measurement of engine performance during compressor surge is analyzed, with the conclusion that in marine engines with intake manifold of large volumes, it is hard to prevent the surging effect.

To optimize compressor working performance, it is necessary to evaluate the compressor efficiency. Compressor efficiency is defined as the ratio of the work a compressor performs under insulated conditions to that of an compressor under actual conditions, and it is expressed with Equation (3) [27]:

$$\eta_{is,c} = \frac{\Delta h_{is,c}}{\Delta h_c} = \frac{C_{p,c} \cdot (T_{2,is} - T_1)}{C_{p,c} \cdot (T_2 - T_1)} \tag{3}$$

The fouling of the compressor wheel does not affect the combustion pressures or timing and duration of fuel injection as an air filter fault. However, it significantly increases the compressor outlet temperature (in Table 3: scavenge air inlet temperature). This temperature is indicated as T_2 in Equation (3), and when it increases, the compressor efficiency is reduced.

To prevent this from happening, period maintenance is required at scheduled intervals. The most common methods for cleaning the compressor or turbine blades are wet and dry cleaning. The dry cleaning method is carried out with compressed air blown at the compressor wheel. However, this method is not recommended for heavier deposits. Manufacturers recommend wet cleaning (fresh water) for compressor blades while the engine is at full load. Usually, compressor maintenance is neglected or postponed until dry-docking rather than performing maintenance during operating hours. Sometimes neglected washing routines can lead to an increase in dirt deposits on both the compressor and turbine blades and this could cause an imbalance of the rotor or even bearing damage. The recommendation is to continuously monitor scavenge air temperature and perform the cleaning at intervals adjusted to the amount of contamination.

5.3. Turbocharger Exhaust Side Results

The exhaust side of the turbocharger consists of the gas inlet and outlet casing, nozzle ring and turbine rotor with blades for converting exhaust gas kinetic energy into mechanical energy. Turbine blades are directly exposed to a high temperature of exhaust gases, therefore, their condition depends on the quality of the combustion process and fuel used. Due to improper combustion, unburnt carbon and soot particles in exhaust gas can cause fouling and damage to the nozzle ring and blades. Moreover, severe damage to turbine blades can be caused by pieces of broken piston rings or valves.

The faults (fouling) of turbine blades also cause differences in the combustion process and they are presented in Figure 5.

The fouling of turbine blades leads to a reduction of turbine output capacity and overall efficiency. The main changes in the combustion process from the cylinder diagram are:

- reduced mean effective pressure (20.1 to 19.3 bar)
- reduced maximum combustion pressure (150 to 117 bar)
- reduced compression pressure from (112 to 68 bar)
- increased angle of combustion start (1.5° to 4.6°), resulting in late combustion process

In the case of a high fouling level, exhaust gas temperatures (at turbine inlet) are increased and the turbocharger consequently supplies less charged air in the cylinders, so the combustion process starts later and lasts longer.

5.4. Maintenance and Optimization of the Turbocharger Exhaust Side

To ensure the durability and efficiency of the turbocharger, maintenance of the turbine rotor, nozzle ring and blades at regular intervals is highly important. For cleaning the deposits on turbine blades, two methods that can be used without the need to stop the engine are the wet and dry cleaning methods. Usually, for turbine cleaning, the wet method is applied due to better cleaning effects and longer maintenance intervals [18].

For two-stroke engines, a maintenance interval of 150 operating hours is recommended [6], however, it should be adapted according to the quality of the fuel used. The fresh water for cleaning should be without any chemical additives and sprayed into the exhaust gas casing before the turbine at a pressure of 2 or 3 bar. In a scenario with high exhaust temperatures during wet cleaning, it is necessary to reduce engine load to avoid the thermal stress of turbine materials. The advantage of dry cleaning is that it can be carried out during operation at full load, however, heavier deposits are harder to remove and maintenance intervals are shortened.

The efficiency of the turbine depends on the energy in exhaust gases which is converted into turbine power output for the intake of air mass flow from the compressor side.

Turbine power is expressed with Equation (4) [27]:

$$P_t = \dot{m}_t \cdot C_{p,t} \cdot (T_3 - T_4) \tag{4}$$

In Equation (4), temperature T_3 represents the turbine inlet temperature and T_4 is exhaust outlet temperature. The results of the turbine fault have shown that both temperatures simultaneously increase proportionally to the amount of fouling. While these temperatures increase, the turbine power is reduced due to lower temperature difference (ΔT) and turbine gas flow rate ($\dot{m}_t$). With insufficient turbine outlet power, turbocharger revolutions (rpm) are also reduced and consequently less fresh air is supplied to the cylinders. Furthermore, less air has a negative effect on the scavenge air pressure (reduced from 2.26 to 1.13 bar) and turbine inlet pressure (reduced from 1.90 to 0.9 bar).

The presented results of turbine fouling indicate the importance of preventive maintenance. The high amount of deposits in the exhaust side of the turbocharger will eventually lead to damage to the turbine. Moreover, when the turbine wheel is damaged, it is imbalanced and it could cause serious issues for the shaft and bearings. Corrective measures for repairing turbocharger damage are not an easy task in large two-stroke marine engines during voyage. The reasons are usually related to missing spare parts or insufficient crew maintenance knowledge and experience. To avoid failure of the exhaust side, it is recommended to adjust the maintenance interval (according to contamination level) and to regularly observe exhaust gas temperatures (before and after the turbine).

5.5. Air Cooler Results

Turbocharger efficiency also depends on the operating condition of the scavenge air cooler. The results of two simulated faults (fouling of air side and tubes) have shown that it could significantly affect the combustion process and these faults should not be neglected. The effects of fouling include loss of heat transfer (tube blockage) and a pressure drop decrease (air side fouled).

Fouling of the air side occurs due to dust and atmospheric particulates contained in the air and the fins of the cooler have a role as a filter where particulates can deposit. As the layers of deposits increase, less air is supplied to the cylinders, leading to reduced inlet air pressure (from 2.26 bar to 1.04 bar) and turbine inlet pressure (from 1.90 bar to 0.80 bar). With reduced turbine inlet pressure, the turbocharger does not have enough output capacity for the compressor side, so the air filter pressure drop is also reduced (from 192 to 71 mm WC). Therefore, less supplied air will increase the exhaust gas temperatures from the cylinders and fuel consumption. Differences in the combustion process for this fault are presented in Figure 6.

The results shown in the indicator diagram (Figure 6) are similar to the results of the air filter fault due to insufficient charged air flow. The symptoms of this fault in the combustion process are:

- reduced mean effective pressure (20.3 to 19.5 bar)
- reduced maximum combustion pressure (150 to 118 bar)
- reduced compression pressure from (112 to 70 bar)
- increased angle of combustion start (1.5° to 4.6°)

Regular maintenance for the air cooler involves the injection of cleaning additives with water. This mist of water and solvents is necessary for cleaning the air fin deposits which can reduce air flow. The washing down process is highly important to ensure all the contamination is flushed out and to improve scavenging efficiency and heat transfer.

The second simulated fault of the air cooler (tube blockage) also has a negative effect on the combustion process. The blockage of water flow through the tubes is related to the inadequate treatment of cooling water, which has corrosive and sediment-laden properties. When cooler tubes are blocked, the temperatures of scavenge air and exhaust gases are increased. The temperature of scavenge air during normal operating conditions is 37.90 °C, while during 40% tube blockage, it is increased to 92.90 °C (Table 2). Moreover, the combustion process in the cylinders is inefficient, therefore exhaust temperatures and turbine inlet pressure are increased from 1.90 to 2.20 bar. With higher pressure at the turbine inlet and excessive temperatures, turbocharger shaft revolutions are increased, which could lead to turbocharger overspeeding. The results of the combustion process in the indicator diagram are presented in Figure 7.

A high temperature of scavenge air in the combustion process causes these changes in the indicator diagram:

- increased compression pressure from (112 to 124 bar)
- reduced angle of combustion start (1.5° to −2.0°), resulting in the early start of the combustion process

To avoid mechanical damage to the fins and tubes of the air cooler, it is recommended to detect the level of fouling in the early stage. The easiest way to detect this is to measure and control the scavenge air temperature and pressure difference. Sometimes maintenance of the air cooler is neglected during voyage and it is postponed until major overhaul. The efficiency and reliability of the air cooler depends on regular maintenance intervals and the control of fouling layers in tubes and fins, especially in large two-stroke marine engines with an enormous mass of air flow charged into cylinders.

6. Conclusions

The reliability of the marine turbocharger system is crucial to ensure efficient performance during the exploitation period. High reliability of the turbocharger system depends on preventing failures by using a fault diagnosis method, engine performance evaluation and appropriate maintenance intervals. Usually, fault diagnosis during engine operation depends on the engine crew's experience, which can lead to false conclusions and improper corrective actions.

Before analyzing turbocharger faults, it is necessary to evaluate relations between the cause and symptoms of all the possible faults that can occur during operation. In this paper, the most common faults in each turbocharger component are simulated and analyzed. The main conclusions of this research are:

- Fouling of the air filter can significantly affect the main engine performance and efficiency of the turbocharger, moreover, it could also result in an increase in fuel consumption. Regular maintenance intervals of the air filter should not be neglected and it is recommended to control the amount of fouling more often. Replacing an air filter with a new one or washing an old one is considered as an easy maintenance task and maintenance costs are negligible when compared with potential losses. For compressor wheel faults, it is recommended to perform a maintenance interval according to manufacturer instructions and to continuously monitor scavenge air temperature at the compressor outlet.
- With a high fouling level of the turbine wheel, exhaust gas temperatures are increased and this could lead to damage to the turbine blades. Furthermore, turbine capacity, power and efficiency of the combustion process are reduced. It is necessary to monitor the exhaust temperatures at the turbine inlet and outlet (exhaust ducts). The maintenance interval should be adjusted according to quality of fuel used and the level of contamination.
- The results of the simulation of air cooler faults have indicated that it is a highly important component of the turbocharger system and it should not be neglected in terms of inspection and preventive maintenance. The efficiency of the combustion process and reliability of turbocharging depend on the operating condition of the air cooler.

This method for diagnosing and simulating failures during the operating period is useful to provide analysis of failure causes and to improve the experience in early failure detection. However, turbochargers in low-speed marine engines are complex systems and many unpredictable factors can affect their efficiency. Some problems that are not analyzed in this paper but could also occur during operation are: insufficient ventilation around the engine, lack of lubrication or incorrect lubricating oil, turbocharger shaft misalignment, high ambient air humidity, surging effect, mismatching of the operating engine with the turbocharger system.

More stringent emission regulations and fuel economy are forcing manufacturers to adapt turbocharger performance to new exhaust technologies. With new technologies, there is still much uncertainty in terms of achieving optimal turbocharger efficiency and reliability, such as: how do new alternative fuels impact turbocharger performance? Does two-stage turbocharging have a long-term future? Will turbocharger emission reduction technologies add to maintenance costs? Does slow-steaming reduce turbocharger efficiency?

Although there will be many influences on turbocharger design in the near future, the main priorities will be the reliability of the system, energy efficiency and maintenance costs. For achieving these demands, the methodology presented in this paper is highly useful and practical. Operational efficiency of vessels could be enhanced by this methodology, with real-time scenarios and even simulation of joint bridge-engine operation. The advantage of a marine engine simulator is that failures can be simulated without any consequences for the engine or equipment. Otherwise, this faulty operation of marine engine and turbocharger can be dangerous and impossible under real conditions during voyage. Simulation results and diagnosis of failures can be used as a new educational method for students and useful information for ship owners and engine crew. The presented results can be further used for scientific research in the field of optimization of the main engine and turbochargers. Additionally, the results can be used for the evaluation of the fuel oil consumption and reduction of exhaust gas emissions.

Author Contributions: Conceptualization, V.K. and J.O.; methodology, V.K.; research simulation, V.K. and J.O.; validation, V.K., J.O. and L.S.; formal analysis, V.K., J.O. and J.Č.; writing—original draft preparation, V.K.; writing—review and editing, V.K., J.O., J.Č. and L.S. All authors have read and agreed to the published version of the manuscript.

Funding: This research received no external funding.

Conflicts of Interest: The authors declare no conflict of interest.

Abbreviations

The following abbreviations are used in the manuscript:

$C_{p,c}$	specific compressor enthalpy
$C_{p,t}$	specific heat of turbine gas
Δh_c	specific compressor enthalpy
$\Delta h_{is,c}$	isentropic compressor enthalpy
$\dot{m}_t$	turbine gas flow rate
P(E)	probability of event
P_t	turbine power
T_1	compressor inlet temperature
T_2	compressor outlet temperature
$T_{2,is}$	isentropic compressor outlet temperature
T_3	turbine inlet gas temperature
T_4	turbine outlet gas temperature

References

1. Smith, D.J. Reliability, Maintainability and Risk—7th Edition. Available online: https://www.elsevier.com/books/reliability-maintainability-and-risk/smith/978-0-7506-6694-7 (accessed on 17 September 2020).
2. Ebeling, C.E. *An Introduction to Reliability and Maintainability Engineering*, 3rd ed.; Waveland Press: Long Groove, IL, USA, 2019; ISBN 978-1-4786-3933-6.
3. Golub, I.; Antonić, R.; Dobrota, Đ.; Golub, I.; Antonić, R. Dobrota, Optimization of heavy fuel oil separator system by applying diagnostic inference methods. *Pomorstvo* **2011**, *25*, 173–188.
4. Anantharaman, M.; Khan, F.; Garaniya, V.; Lewarn, B. Reliability Assessment of Main Engine Subsystems Considering Turbocharger Failure as a Case Study. *TransNav* **2018**, *12*, 271–276. [CrossRef]
5. Anantharaman, M.; Islam, T.M.; Khan, F.; Garaniya, V.; Lewarn, B. Data Analysis to Evaluate Reliability of a Main Engine. *TransNav* **2019**, *13*, 403–407. [CrossRef]
6. Yang, M.; Hu, C.; Bai, Y.; Deng, K.; Gu, Y.; Qian, Y.; Liu, B. Matching method of electric turbo compound for two-stroke low-speed marine diesel engine. *Appl. Therm. Eng.* **2019**, *158*, 113752. [CrossRef]
7. Píštěk, V.; Kučera, P.; Fomin, O.; Lovska, A. Effective Mistuning Identification Method of Integrated Bladed Discs of Marine Engine Turbochargers. *J. Mar. Sci. Eng.* **2020**, *8*, 379. [CrossRef]
8. Adamkiewicz, A. An analysis of cause and effect relations in diagnostic relations of marine Diesel engine turbochargers. *Zeszyty Naukowe/Akademia Morska w Szczecinie* **2012**, *31*, 5–13.
9. Monieta, J. Fundamental Investigations of Marine Engines Turbochargers Diagnostic with Use Acceleration Vibration Signals. In Proceedings of the 2nd International Conference on Chemistry, Chemical process and Engineering, Yogyakarta, Indonesia, 14–15 August 2018; p. 020044.
10. Ta, T.V.; Vu, N.H.; Triet, M.A.; Thien, D.M.; Cang, V.T. Assesment of Marine Propulsion System Reliability Based on Fault Tree Analysis. *Int. J. Trans. Eng. Technol.* **2017**, *2*, 55. [CrossRef]
11. Budiyanto, M.A.; Fernanda, H. Risk Assessment of Work Accident in Container Terminals Using the Fault Tree Analysis Method. *J. Mar. Sci. Eng.* **2020**, *8*, 466. [CrossRef]
12. Laskowski, R. Fault Tree Analysis as a tool for modelling the marine main engine reliability structure. *Sci. J. Mar. Univ. Szczec.* **2015**, *41*, 71–77.
13. Vera-García, F.; Pagán Rubio, J.A.; Hernández Grau, J.; Albaladejo Hernández, D. Improvements of a Failure Database for Marine Diesel Engines Using the RCM and Simulations. *Energies* **2020**, *13*, 104. [CrossRef]
14. Chybowski, L.; Gawdzińska, K.; Laskowski, R. Assessing the Unreliability of Systems during the Early Operation Period of a Ship—A Case Study. *J. Mar. Sci. Eng.* **2019**, *7*, 213. [CrossRef]
15. Karatuğ, Ç.; Arslanoğlu, Y. Importance of early fault diagnosis for marine diesel engines: A case study on efficiency management and environment. *Ships Offshore Struct.* **2020**, 1–9. [CrossRef]
16. Turbocharger Fundamentals. Available online: https://dieselnet.com/tech/air_turbocharger.php#turbine (accessed on 9 October 2020).
17. Monieta, J.; Sendecki, A. Database and Knowledge about Essential Manufacturers of Marine Self-Ignition Engines. *J. Mar. Sci. Eng.* **2020**, *8*, 239. [CrossRef]
18. MAN Diesel & Turbo. *TCR Turbocharger, Project Guide Book, Augsburg*; MAN Diesel & Turbo SE: Augsburg, Germany, 2014.
19. IMO—Nitrogen Oxides (NOx)—Regulation 13. Available online: https://www.imo.org/en/OurWork/Environment/Pages/Nitrogen-oxides-(NOx)-%E2%80%93-Regulation-13.aspx (accessed on 29 October 2020).
20. Vesely, W.; Goldberg, F.; Roberts, N.; Haasl, D. *Fault Tree Handbook*; NUREG-0492; Nuclear Regulatory Commission: Rockville, MD, USA, 1981.
21. Brocken_Improving_The_Reliability_Of_Ship_Machinery.pdf|Ships|Oil Tanker. Available online: https://www.scribd.com/document/372482692/Brocken-Improving-The-Reliability-Of-Ship-Machinery-pdf (accessed on 9 October 2020).
22. MAN B & W 6S60MC-C Diesel Engine—Tanker LCC (Aframax). *Trainee Manual Book*; Transas Marine Ltd.: Saint-Petersburg, Russia, 2017.
23. Charge Air Cooler Maintenance. Available online: https://manualzilla.com/doc/6016061/charge-air-cooler-maintenance (accessed on 2 November 2020).
24. Islam, R.; Abbassi, R.; Garaniya, V.; Khan, F. Development of a human reliability assessment technique for the maintenance procedures of marine and offshore operations. *J. Loss Prev. Process. Ind.* **2017**, *50*, 416–428. [CrossRef]

25. Demirel, H. An Evaluation of Human Error Probabilities for Critical Failures in Auxiliary Systems of Marine Diesel Engines. *J. Mar. Sci. Appl.* **2020**. [CrossRef]
26. Vrettakos, N.A. Analysis and characterization of a marine turbocharger's unstable performance. *Proc. IMechE* **2018**, *232*, 293–306. [CrossRef]
27. Chung, J.; Chung, J.; Kim, N.; Lee, S.; Kim, G. An Investigation on the Efficiency Correction Method of the Turbocharger at Low Speed. *Energies* **2018**, *11*, 269. [CrossRef]

Publisher's Note: MDPI stays neutral with regard to jurisdictional claims in published maps and institutional affiliations.

Journal of
*Marine Science
and Engineering*

MDPI

Article

Verification of Vibration Isolation Effectiveness of the Underwater Vehicle Power Plant

Yang Yang [1,2,*], Guang Pan [1], Shaoping Yin [2], Ying Yuan [3,*] and Qiaogao Huang [1]

1   School of Marine Science and Technology, Northwestern Polytechnical University, Xi'an 710072, China; panguang@nwpu.edu.cn (G.P.); huangqiaogao@nwpu.edu.cn (Q.H.)
2   The 705 Research Institute, China Shipbuilding Industry Corporation, Xi'an 710077, China; summertalentyy@163.com
3   School of Physics and Optoelectronic Engineering, Xidian University, Xi'an 710071, China
*   Correspondence: allen-yang1988@mail.nwpu.edu.cn (Y.Y.); yuanying@xidian.edu.cn (Y.Y.)

Abstract: In order to enhance the vibration isolation effectiveness of an underwater vehicle power plant, and alleviate the mechanical vibration of the outer housing, initially discrete vibration isolators were improved, and three new types of ring vibration isolators designed, i.e., ring metal rubber isolators, magnesium alloy isolators and modified ultra-high polyethylene isolators (MUHP). A vibrator excitation test was carried out, and the isolation effectiveness of the three types of vibration isolators was evaluated, adopting insertion loss and vibration energy level drop. The results showed that compared with the initial isolators and the other two new types of isolators, MUHP showed the most significant vibration isolation effectiveness. Furthermore, its effectiveness was verified by a power vibration test of the power plant. To improve the vibration isolation effectiveness, in addition to vibration isolators, it is essential to carry out investigations on high-impedance housings.

Keywords: underwater vehicle; isolation; flexible foundation; vibration mitigation

Citation: Yang, Y.; Pan, G.; Yin, S.; Yuan, Y.; Huang, Q. Verification of Vibration Isolation Effectiveness of the Underwater Vehicle Power Plant. *J. Mar. Sci. Eng.* **2021**, *9*, 382. https://doi.org/10.3390/jmse9040382

Academic Editor: Igor Poljak

Received: 11 March 2021
Accepted: 31 March 2021
Published: 3 April 2021

Publisher's Note: MDPI stays neutral with regard to jurisdictional claims in published maps and institutional affiliations.

1. Introduction

The low-vibration performance design of underwater vehicles is critical to their concealment and navigation performance. For a thermodynamic underwater vehicle, the mechanical noise of the power system is the main noise source when it sails at a relatively deep depth [1]. The source of mechanical noise is mainly excited by the operation of the power plant, which contains multiple vibration sources. These vibration may be transmitted to the outer housing of the underwater vehicle through different paths, resulting in structural vibration and noise radiation. As one of the most effective measures to mitigate vibration transmission, vibration isolators are generally utilized in the structural design of underwater vehicle power plants. Traditional vibration isolators frequently use rubber as the vibration isolation material. To facilitate installation, rubber is commonly vulcanized with the metal frame to form an independent vibration isolator. Engineering practice shows that the transmission of vibration in the power system can be effectively mitigated by selecting appropriate rubber materials. However, rubber is prone to aging and shows poor impact resistance during long-term storage. In addition, the installation space of a vibration isolator is extremely limited due to the cramped space inside an underwater vehicle. The metal skeleton in the rubber vibration isolator occupies a part of the installation space of the elastic material, thereby reducing the volume of vibration isolation materials. Therefore, it is urgent to develop some new types of vibration isolators that could employ more vibration isolation materials. In addition, a reasonable and accurate evaluation of the vibration isolation effectiveness of isolators in underwater vehicles is also extremely important.

The dynamic vibration isolation system of an underwater vehicle is a typical elastic foundation vibration isolation. The power plant occupies most of the mass of a power

137

system. Compared with the power plant, the outer housing of the vehicle is lighter and thinner. There is an inevitable vibration coupling between the power plant, the vibration isolator and the outer housing. The coupling, on the one hand, affects the vibration isolation effectiveness; on the other hand, it produces external sound radiation. The classic model of a vibration isolation system is the mass-spring-foundation model. In this model, the basic vibration isolation structure is commonly regarded as a rigid structure with infinite impedance. However, actual infrastructures often do not meet the above assumptions. For a basic structure that does not meet the rigidity assumption, an elastic foundation vibration isolation model has been developed. Scholars considered the actual structural form and analyzed the vibration characteristics of a vibration isolation system with beams [2,3], plates [4–6] and even cylindrical shells [7–10] by simplifying the basic structure. Some scholars considered the problem of non-linearity [11,12]. Flotow [13] summarized and proposed the elastic foundation–rigid equipment approximate modeling method by analyzing and collating the modeling methods of mechanical equipment–vibration isolation system basic structure. The finite element method is widely adopted to build a vibration isolation system model for more complex infrastructure forms [14]. Experimental research occupies an important part in the research and design of vibration isolators [15–19]. There are many evaluation methods for the vibration effectiveness of elastic foundation isolation system, which include power flow transfer spectrum [20], the maximum and minimum singular values of the effective ratio matrix of force and velocity [21], the transfer rate matrix of force and velocity [22] and the force transmission rate [23–25]. In underwater vehicles, the purpose of an isolation system is to mitigate the vibration transmission of the equipment to the foundation and reduce the vibration energy level of the foundation. For an isolator system in engineering terms, the biggest concern is the vibration energy level transmitted to the foundation after the isolator, that is, the vibration response of the foundation in the isolation system. Therefore, the insertion loss and vibration level drop are often utilized as indexes in the evaluation of the effectiveness of an underwater vehicle isolation system.

Discrete metal rubber vibration isolators were employed as initial vibration isolators for underwater vehicle power plants. However, the vibration isolation effectiveness was insignificant. Three new types of vibration isolators, namely ring metal rubber vibration isolators, magnesium alloy vibration isolators and modified ultra-high polyethylene vibration isolators (MUHP) were designed with the goal of improving vibration isolation effectiveness. Compared with ordinary rubber, MUHP has a lower density, higher heat distortion temperature, and good toughness and impact resistance. In addition, it has excellent damping performance with a loss factor of about 10^{-1}, which can be well adapted to the working environment of underwater vehicles. Magnesium alloy has the advantages of high strength, low density and a low modulus of elasticity, while metal rubber has high damping characteristics and is widely utilized in aerospace vehicles. Different from general engineering structures, a vibration isolator is commonly designed as a ring structure due to the structural characteristics of an underwater vehicle. Therefore, the vibration effectiveness of isolators in underwater vehicles need to be well explored. To provide more comprehensive experimental evidence that testifies to the effectiveness of isolators for the mitigation of power-plant-generated vibration, a test method was designed and carried out in a laboratory. In the test, three types of vibration isolators were installed in an experimental device, in which the power plant was replaced by a simulator of the same weight. By picking up the average vibration response of the power plant and outer housing of the vehicle, the insertion loss and vibration level drop of different vibration isolators were obtained to evaluate vibration isolation effectiveness.

2. Problem Description and Test Methodology

The power plant vibration isolator system consists of a load, vibration isolators and the outer housing, which is schematically shown in Figure 1. The housing is a thin-walled cylindrical shell, the material is aluminum and the load is an aluminum solid cylinder structure of which the length is the same as the length of the housing. The initial vibration isolators are divided into two groups: the front composition and the rear composition. Each group is composed of 6 identical metal rubber vibration isolation elements, which are arranged approximately evenly along the circumference, and are shown in Figure 1.

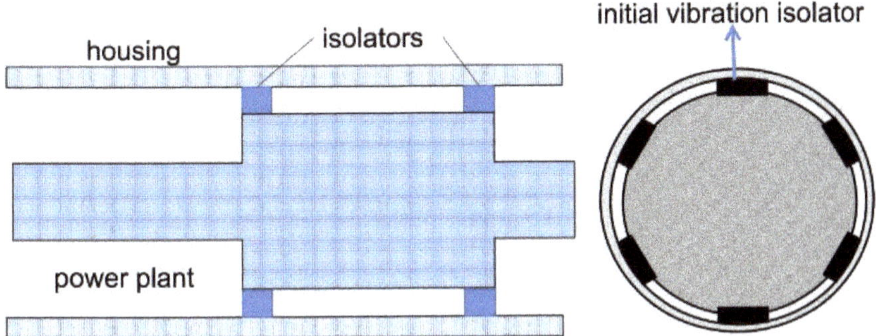

Figure 1. Schematic diagram of the dynamic vibration isolation device.

The insertion loss and vibration energy level drop of the vibration isolation system were tested and a schematic diagram of the test system is shown in Figure 2. The test device for the vibration isolation effectiveness of the vibration isolator was composed of a load, vibration isolators and an outer housing. The entire device was suspended by an elastic rope to simulate a free–free boundary. The load was a solid cylinder and the housing was an aluminum thin-walled cylindrical shell. Moreover, the length of the load was the same as the length of the housing. The isolators were divided into two annular isolators, located in the front and rear. The structural parameters of the systems are shown in Table 1.

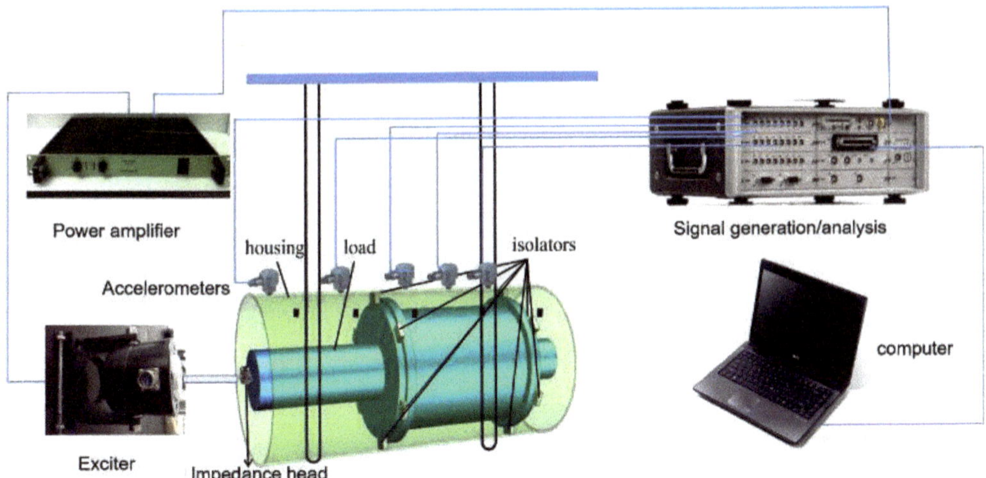

Figure 2. Schematic diagram of the vibration isolation system test.

Table 1. The structural parameters of the test system.

	The Housing			The Load		
Material	Length (mm)	Thickness (mm)	Inner Diameter (mm)	Material	Length (mm)	Inner Diameter (mm)
aluminum	500	5	313	aluminum	500	293

In the test system, the exciter model was BK4808, the power amplifier was BK2712, the type of acceleration sensor was PCB353B04, the data acquisition system was LMS SCADAS Mobile SCM05 and LMS.Test.Lab software was utilized for excitation control and acceleration test. In the test, the exciter acted as a steady-state excitation source to generate white noise excitation. The impedance head was connected to the exciter through the excitation rod to pick up the acceleration response signal and force signal of the excitation point. The exciter was attached to the beam by a flexible cord and the excitation rod extended on the axial direction of the power plant simulator.

In the test model, a total of 20 radial acceleration sensors were located on the outer surface of the housing, which are shown in Figure 3. The 20 sensors were divided into 5 groups, which were respectively arranged in the axial position with equal space. In addition, 4 acceleration sensors in the radial direction were installed at the front and rear ends of the power simulator, respectively.

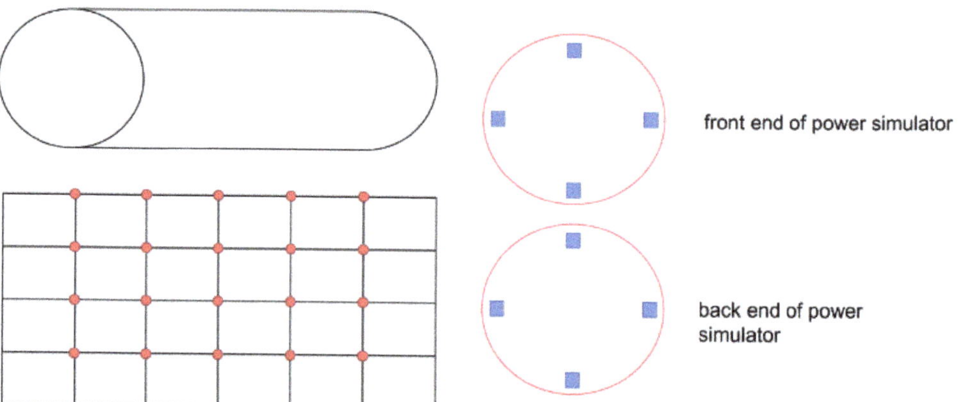

Figure 3. Acceleration measuring point distribution diagram of the housing and power plant simulator.

The average acceleration responses of the housing and power plant simulator were determined by:

$$a = \left(\frac{1}{N} \sum_{i=1}^{N} a_i^2 \right)^{\frac{1}{2}} \tag{1}$$

where a_i and N was the acceleration value of the i-th measuring point and the number of acceleration sensors on the housing or power plant simulator.

The vibration energy levels of the housing and power plant simulator were calculated by:

$$L = 20 \log_{10} a \tag{2}$$

The insertion loss of the vibration isolation system was obtained by the following formula:

$$L_I = L_{h1} - L_{h2} \tag{3}$$

where L_{h1}, L_{h2} represented the average vibration levels of the housing before isolation and after isolation, respectively.

The vibration level drop of the vibration isolation system was obtained by the following formula:

$$L_D = L_p - L_h \tag{4}$$

where L_p, L_h represented the average vibration levels of the power plant simulator and housing.

The larger L_I and L_D, the lower the response of the housing after isolation, and the more significant the vibration isolation effectiveness of the vibration isolation system was.

The test procedure was:

(a) Without installing vibration isolators, a rigid aluminum ring was placed between the load and the housing; thus, the load was directly in contact with the housing through the aluminum ring, the system was excited by the exciter and then, the frequency response of each acceleration measuring point was picked up;

(b) Step (a) was repeated and the final acceleration response of each measuring point was obtained by linearly averaging the results of the two repeated measurements, represents the acceleration before isolation;

(c) Replacing the aluminum ring of Step (a) with initial vibration isolators, the load and the housing were connected through the vibration isolators, and the shock was excited by the exciter and the data acquisition system picked up the frequency response of each acceleration measuring point;

(d) Repeating Step (b), the final acceleration response of each measuring point was obtained by linearly averaging the results of the two repeated measurements and represents the acceleration after isolation;

(e) Repeating Step (c) and Step (d) the vibration data of the three new design vibration isolators was obtained.

3. Results of Initial Isolators

Figure 4 compares the vibration response of the housing with the background noise (the vibration response obtained by the sensors when the vibrator was not working). It can be seen from the figure that, at the peak of the housing's response, the signal-to-noise ratio reached 70 dB, and outside of the response peak frequency points, the signal-to-noise ratio at other frequencies reached more than 40 dB. This signal-to-noise ratio was sufficient for evaluating the vibration isolation effectiveness of isolators.

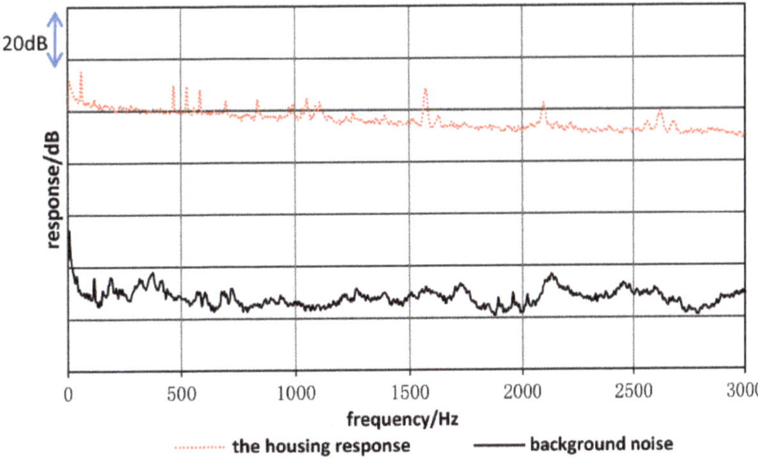

Figure 4. Signal-to-noise ratio of the housing vibration response.

The results of insertion loss and vibration energy level drop of initial isolators are shown in Figures 5 and 6, respectively. It can be seen that, in the wide frequency range, the vibration isolation system did not have a good vibration isolation effectiveness, and the insertion loss was always less than 5 dB. As the frequency increased, the vibration isolation effectiveness did not improve, which could not satisfy the vibration isolation requirements.

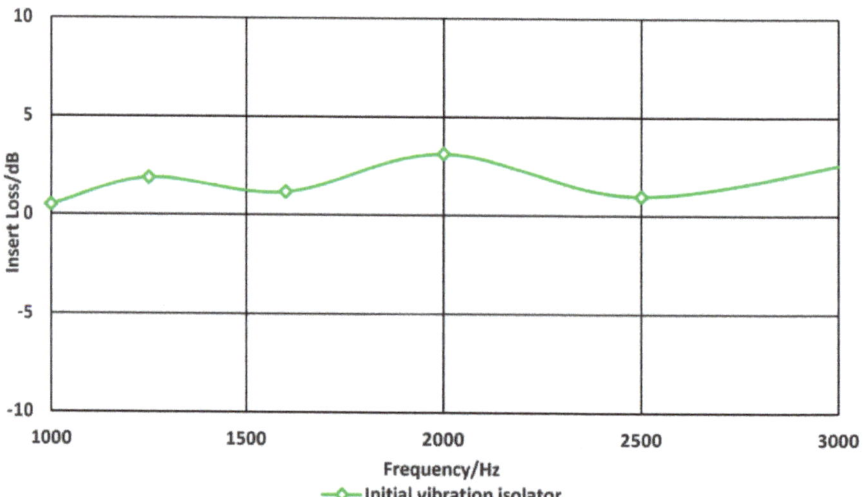

Figure 5. Insertion loss test results of initial isolator.

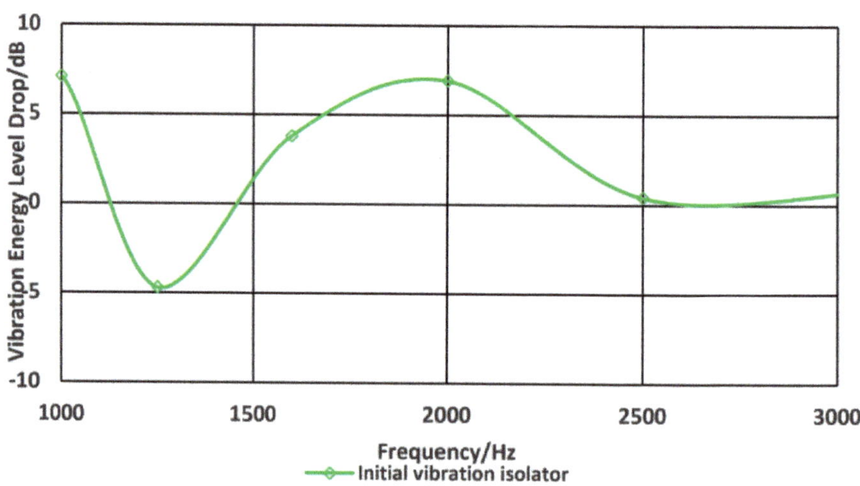

Figure 6. Vibration level drop test results of initial isolator.

Analysis and research show that [1]: the primary reason for the poor vibration isolation effectiveness was that the impedance of the isolator was greater than the housing. In order to improve the vibration isolation effectiveness of the system, this article employed the above analysis results as a guide to design a variety of vibration isolators using different materials, and select vibration isolators that meet the engineering requirements through experimental tests.

4. Improved Isolator Design

The appearance of the newly designed vibration isolators and their installation in the power plant are shown in Figure 7. The three newly designed types of vibration isolators are shown in Figure 8, and their material parameters are shown in Table 2. They were a ring metal rubber vibration isolator, magnesium alloy vibration isolator and MUHP. Each type of the vibration isolator contained two annular vibration isolators, namely a front vibration isolator and a rear vibration isolator. It should be noted that the overall dimensions of the three new types of vibration isolators were kept the same, and the front vibration isolators and the rear vibration isolators were connected to the dynamic simulation device by screws.

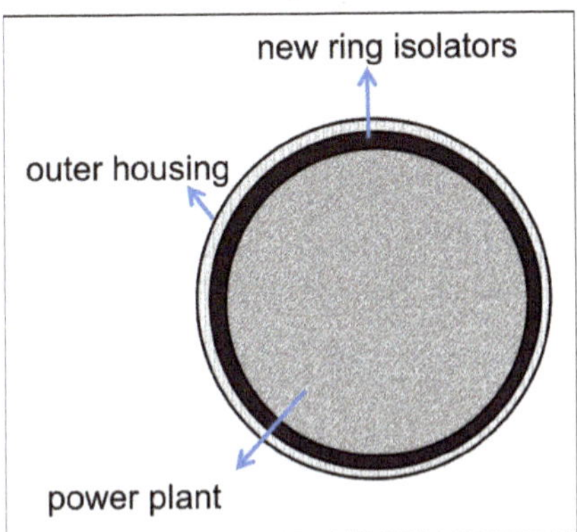

Figure 7. Schematic diagram of the new vibration isolators.

(**a**) The ring metal rubber isolator

(**b**) MUHP

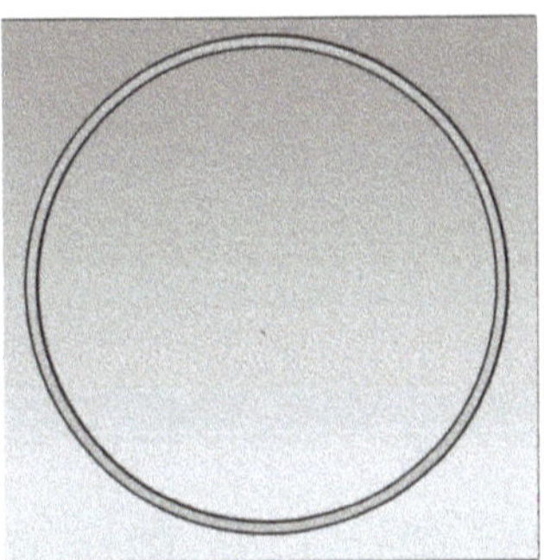

(**c**) The magnesium alloy isolator

Figure 8. Photos of the three new isolators.

Table 2. Material parameters of the isolators.

	Metal Rubber	MUHP	Magnesium Alloy
elasticity modulus (pa)	1.2×10^7	4.2×10^8	4.5×10^{10}
material density (kg/m^3)	1752	940	1820

The ring metal rubber vibration isolator included an inner metal ring, an outer metal ring and several arc-shaped metal rubber vibration damping strips. The inner metal ring and the outer metal ring were enclosed to form a cavity, and a plurality of arc-shaped metal rubber vibration damping strips was arranged in the cavity. The arc-shaped metal rubber vibration damping strips were clamped between the inner and outer metal rings in a pre-tightened manner. Both the magnesium alloy isolators and MUHP were integrated vibration isolators, that is, the entire body of the vibration isolator was an independent element.

5. Discussion

The same test method as the initial isolators was employed and the insertion losses of the three types of vibration isolators are shown in Figure 9. It can be seen from the figure that the ring metal rubber vibration isolators showed a vibration effectiveness of 3–5 dB at an interval of 1–1.5 kHz. However, the insertion loss decreased as the frequency continued to increase. The insertion loss of the magnesium alloy vibration isolators was relatively low, especially in the vicinity of 1 kH; within the range of 1.5–2 kHz, the insertion loss was negative. This also meant that the presence of the magnesium alloy isolators made the vibration response of the housing greater compared to the rigid connection between the power plant and the housing in these frequency bands. MUHP showed a more significant and stable vibration isolation effectiveness in the interest frequency band whose insertion loss was stable at 3–5 dB compared with the other three types of vibration isolators, and there was a trend of continuous improvement as the frequency increased.

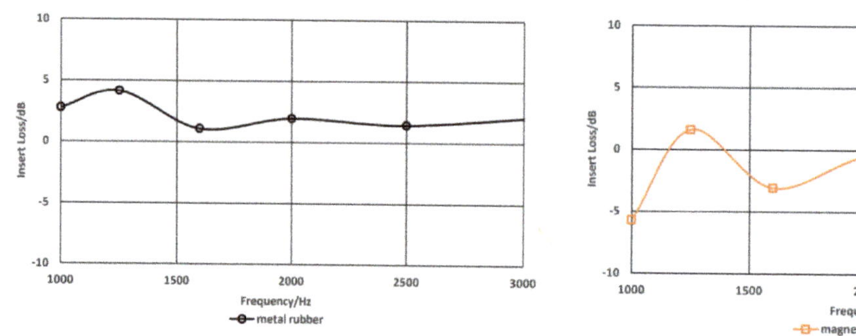
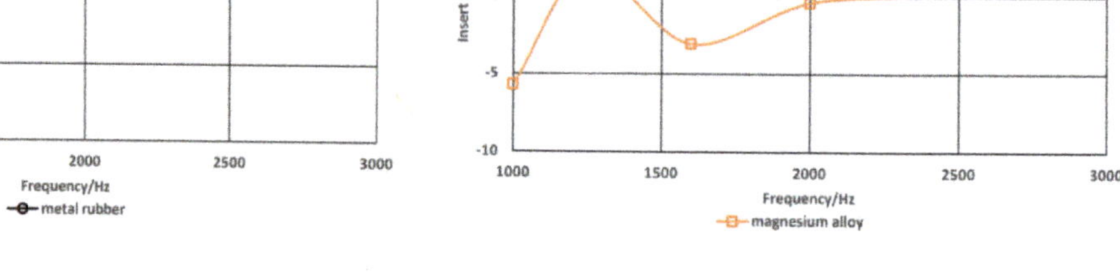

(**a**) Metal rubber

(**b**) Magnesium alloy

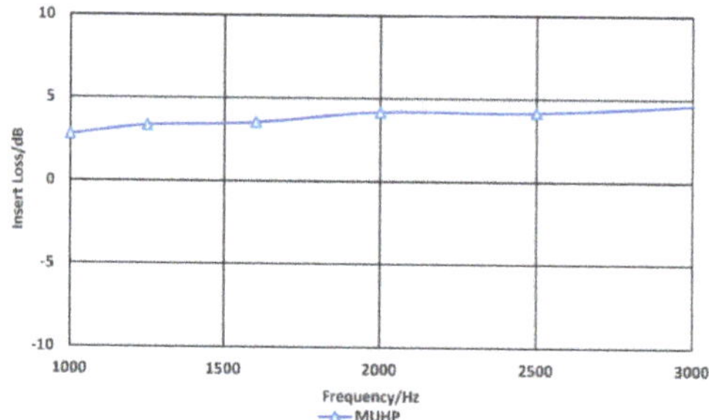

(**c**) MUHP

Figure 9. Insertion losses of the three new isolators.

From the above comparison, it can be seen that with insertion loss as the evaluation criterion, the vibration isolation effectiveness of MUHP was more significant than that of the initial vibration isolators among the three newly designed vibration isolators. The vibration isolation effectiveness of metal rubber vibration isolators was similar to that of the initial vibration isolators, while the vibration isolation effectiveness of magnesium alloy vibration isolators was insignificant and the phenomenon of vibration amplification appeared.

The vibration level drops for the three types of vibration isolators are shown in Figure 10. It can be seen from the figure that the vibration level drop of MUHP was greater than 4 dB in the full frequency band, and the maximum value reached 12 dB, which appeared at about 1.6 kHz. The vibration level drop of the ring metal rubber isolators was 0–7 dB, and the maximum value appeared at 1 kHz and 2 kHz. In addition, the maximum value of the vibration level drop of the magnesium alloy vibration isolators also appeared at 2 kHz, but were all less than 10 dB. Observing that the vibration level drop curves of the four types of vibration isolators include the initial isolators, it was found that the vibration level drop showed a downward trend with the increase in frequency after the maximum value of the vibration level drop was reached, which was related to the elasticity of the housing. The outer housing of the underwater vehicle was a typical elastic structure and thus the power plant vibration isolation device was the elastic foundation vibration isolation. As the excitation frequency increased, the modal density of the housing increases and more modes were excited, resulting in the deterioration of the vibration isolation effectiveness of the vibration isolators.

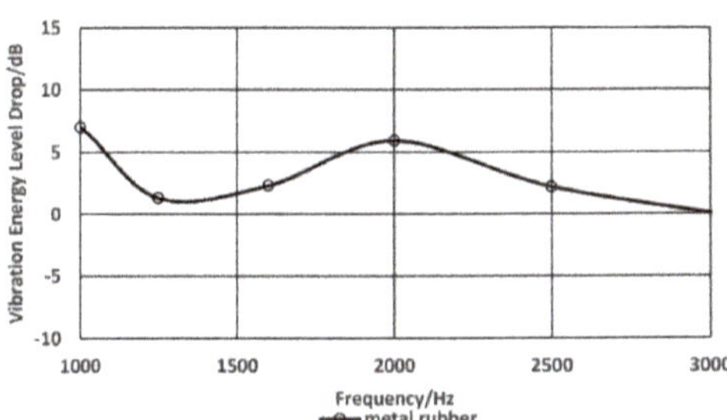

(**a**) Metal rubber

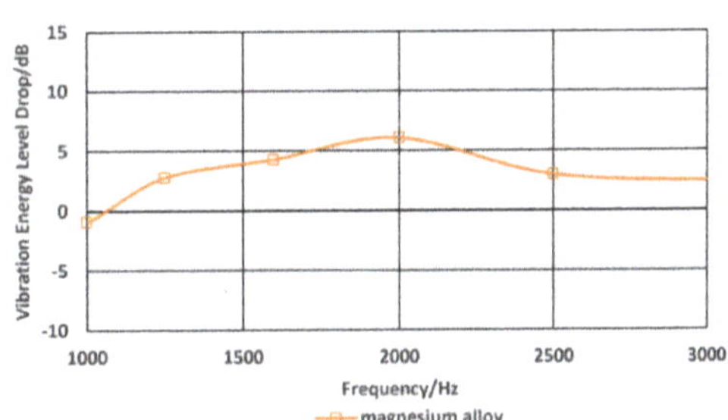

(**b**) Magnesium alloy

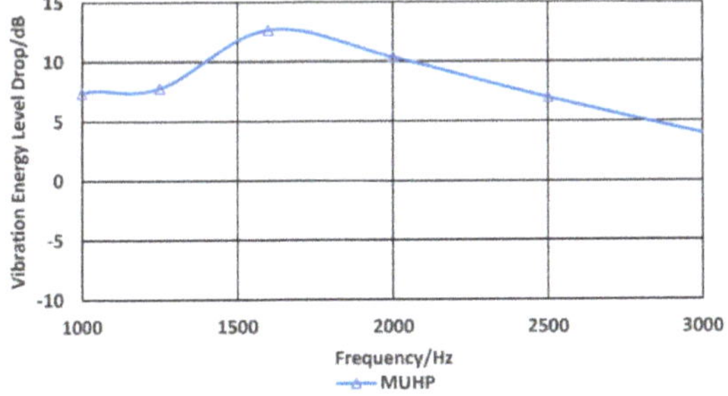

(**c**) MUHP

Figure 10. Vibration energy level drop of the three new isolators.

The insertion losses and vibration energy level drop of the four types of vibration isolators were compared in the exciter excitation test. The results showed that the vibration isolation effectiveness of MUHP was more significant than that of the initial vibration isolators and the other two new isolators in the frequency band. To further verify the vibration isolation effectiveness of MUHP in the actual working condition of the power plant, the power vibration test of the vehicle, including MUHP, was carried out and compared with the initial vibration isolators. The result is shown in Figure 11. It can be clearly seen from the figure that the vibration energy level of the housing surface was mitigated by about 3 dB after installing the MUHP compared to that of the initial isolators.

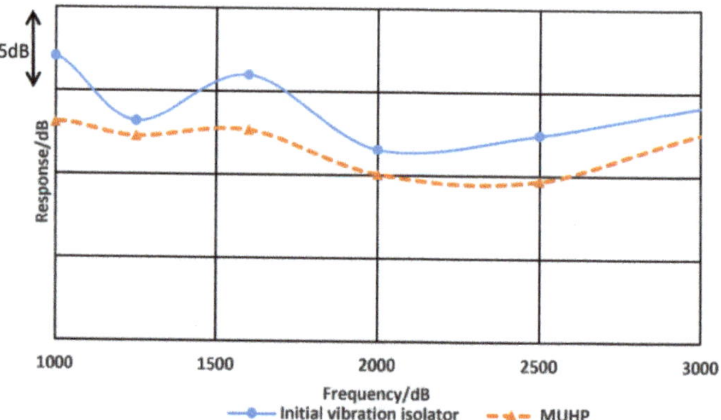

Figure 11. The housing response when two types of vibration isolators were installed.

The vibration isolation effectiveness of the underwater vehicle power plant has been effectively improved through the development of a variety of new vibration isolators. However, the reduction of the vibration energy level of the underwater vehicle housing was very limited. The reason is that in addition to the vibration isolator, the housing is also an important factor affecting the vibration isolation effectiveness of the power plant.

Supposing M, k, c represents the load mass, the stiffness and damping of the isolator, respectively, k_1 is the stiffness of the elastic support system, thus the mechanical impedance of the load, the isolator and the foundation are $Z_m = -M\omega^2$, $Z_k = k + ic\omega$ and $Z_s = 1/R_m$, the impedance of the support system is $Z_{k1} = k_1$, Let $Z_1 = \frac{Z_k}{Z_{k1}+Z_s}, Z_2 = \frac{Z_k}{Z_m}$, then the insertion loss can be written as [1]:

$$L_I = 20\lg(1 + \frac{1}{Z_1 + Z_2})$$ (5)

It can be clearly seen from the calculation (5) that the insertion loss of the vibration isolation system is mainly related to the impedance ratio of the isolator to the housing and the elastic support and the impedance ratio of the isolator to the load, and is determined by the larger of the two impedance ratios.

The expression of the vibration level drop can be written as:

$$L_D = 20\lg\left(1 + \frac{Z_{k1} + Z_s}{Z_k}\right)$$ (6)

The magnitude of the vibration energy level drop depends on the quotient of the sum of the impedance of the housing and the supporting boundary with the impedance of the isolator and the load has no effect on it. If the support stiffness is 0, the greater the ratio of the impedance of the housing to the vibration isolator, the greater the vibration level drop and the better the vibration isolation effectiveness. When the mechanical impedance of the load is much larger than that of the sum of the housing and the supporting boundary, the insertion loss is approximately equal to the vibration level drop.

It can be seen from the above comparison and analysis that the other reason for the insignificant vibration isolation effectiveness of the power plant was that the impedance of the isolator was approximately equivalent to the housing. Reducing the mechanical impedance of the vibration isolator is only one of the means to improve the vibration isolation effectiveness. To further alleviate the vibration of the underwater vehicle housing and improve the vibration isolation effectiveness of the vibration isolation device, it is necessary to simultaneously carry out the investigation on the high impedance housing.

6. Conclusions

In this article, the vibration isolation effectiveness of three newly designed vibration isolators, i.e., ring metal rubber isolators, magnesium alloy isolators and MUHP, was evaluated by insertion loss and vibration energy level drop. Our test results showed that the vibration isolation effectiveness of MUHP had a superior vibration isolation performance compared with the original vibration isolators and the other two new vibration isolators. In order to further verify the performance of MUHP in the working condition of the power plant, the power vibration test of the vehicle, when the initial isolators and MUHP were installed, was carried out. The results showed that the vibration response of the outer housing of the vehicle was mitigated by about 3 dB compared with the initial vibration isolators installed. To further alleviate the mechanical vibration of the outer housing, it is essential to carry out investigations on high-impedance housings.

Author Contributions: Contributed to synthesis, testing, data analysis and writing the manuscript: Y.Y. (Yang Yang); supervised: G.P. S.Y. and Q.H.; contributed to revising the language of the manuscript and suggested the work: Y.Y. (Ying Yuan). All authors have read and agreed to the published version of the manuscript.

Funding: This research was funded by the National Natural Science Foundation of China (62005204) and the Fundamental Research Funds for the Central Universities.

References

1. Yang, Y.; Guang, P.; Shaoping, Y. Experimental investigation on vibration isolation effectiveness of two sets of underwater vehicle dynamic systems. *J. Ship Res.* **2020**, *64*, 226–233. [CrossRef]
2. Goyder, H.; White, R. Vibrational power flow from machines into built-up structures, Part II: Wave propagation and power flow in beam-stiffened plates. *J. Sound Vib.* **1980**, *68*, 77–96. [CrossRef]
3. Pinnington, R.; White, R. Power flow through machine isolators to resonant and non-resonant beams. *J. Sound Vib.* **1981**, *75*, 179–197. [CrossRef]
4. Pan, J.; Hansen, C.H. Total power flow from a vibrating rigid body to a thin panel through multiple elastic mounts. *J. Acoust. Soc. Am.* **1992**, *92*, 895–907. [CrossRef]
5. Gardonio, P.; Pinnington, R. Active isolation of structural vibration on a multiple-degree-of-freedom system, part I: The dynamics of the system. *J. Sound Vib.* **1997**, *207*, 61–93. [CrossRef]
6. Li, W.; Lavrich, P. Prediction of power flows through machine vibration isolators. *J. Sound Vib.* **1999**, *224*, 757–774. [CrossRef]
7. Li, W.; Daniels, M.; Zhou, W. Vibrational power transmission from a machine to its supporting cylindrical shell. *J. Sound Vib.* **2002**, *257*, 283–299. [CrossRef]
8. Ma, X.; Jin, G.; Liu, Z. Active structural acoustic control of an elastic cylindrical shell coupled to a two-stage vibration isolation system. *Int. J. Mech. Sci.* **2014**, *79*, 182–194. [CrossRef]
9. Oliazadeh, P.; Farshidianfar, A. Analysis of different techniques to improve sound transmission loss in cylindrical shells. *J. Sound Vib.* **2017**, *389*, 276–291. [CrossRef]
10. Guo, W.; Li, T.; Zhu, X. Vibration and acoustic radiation of a finite cylindrical shell submerged at finite depth from the free surface. *J. Sound Vib.* **2017**, *393*, 338–352. [CrossRef]
11. Lu, Z.; Brennan, M.J. On the transmissibilities of nonlinear vibration isolation system. *J. Sound Vib.* **2016**, *375*, 28–37. [CrossRef]
12. Liu, X.-L.; Shangguan, W.-B.; Jing, X. Vibration isolation analysis of clutches based on trouble shooting of vehicle accelerating noise. *J. Sound Vib.* **2016**, *382*, 84–99. [CrossRef]
13. Flotow, V. An expository overview of active control of machinery mounts. In Proceedings of the 27th IEEE Conference on Decision and Control, Austin, TX, USA, 7–9 December 1988; IEEE: Piscataway, NJ, USA, 1988; pp. 2029–2032.

14. Hong, J.; He, X.; Zhang, D. Vibration isolation design for periodically stiffened shells by the wave finite element method. *J. Sound Vib.* **2018**, *419*, 90–102. [CrossRef]
15. Syam, W.P.; Jianwei, W.; Zhao, B. Design and analysis of strut-based lattice structures for vibration isolation. *Precis. Eng.* **2018**, *52*, 494–506. [CrossRef]
16. Pan, P.; Shen, S.; Shen, Z. Experimental investigation on the effectiveness of laminated rubber bearings to isolate metro generated vibration. *Measurement* **2018**, *122*, 554–562. [CrossRef]
17. Zhou, W.; Li, D. Experimental research on a vibration isolation platform for momentum wheel assembly. *J. Sound Vib.* **2013**, *332*, 1157–1171. [CrossRef]
18. Li, H.; Li, H.Y.; Chen, Z.B.; Tzou, H.S. Experiments on active precision isolation with a smart conical adapter. *J. Sound Vib.* **2016**, *374*, 17–28. [CrossRef]
19. Lu, L.-Y.; Chen, P.-R.; Pong, K.-W. Theory and experiment of an inertia-type vertical isolation system for seismic protection of equipment. *J. Sound Vib.* **2016**, *366*, 44–61. [CrossRef]
20. Goyder, H.G.D.; White, R.G. Vibrational power flow on machines into built-up structures, part I: Introduction and approximate analyses of beam and plate-like foundations. *J. Sound Vib.* **1980**, *68*, 59–75. [CrossRef]
21. Singh, R.; Kim., S. Examination of multi-dimensional vibration isolation measures and their correlation to sound radiation over a broad frequency range. *J. Sound Vib.* **2003**, *262*, 419–455. [CrossRef]
22. Swanson, D.; Miller., L.; Norris., M. Multidimensional mount effectiveness for vibration isolation. *J. Aircr.* **1994**, *31*, 188–196. [CrossRef]
23. Ibrahim, R.A. Recent advances in nonlinear passive vibration isolators. *J. Sound Vib.* **2008**, *314*, 371–452. [CrossRef]
24. Carrella, A.; Brennan, M.J.; Waters, T.P. Static analysis of a passive vibration isolator with quasi-zero stiffness characteristic. *J. Sound Vib.* **2007**, *01*, 678–689. [CrossRef]
25. Zhang, J.; Guo, Z.; Zhang, Y. Dynamic characteristics of vibration isolation platforms considering the joints of the struts. *Acta Astronaut.* **2016**, *126*, 120–137. [CrossRef]

Journal of
*Marine Science
and Engineering*

Article

Use of Genetic Programming for the Estimation of CODLAG Propulsion System Parameters

Nikola Anđelić [1], Sandi Baressi Šegota [1], Ivan Lorencin [1], Igor Poljak [2], Vedran Mrzljak [1,*] and Zlatan Car [1]

[1] Faculty of Engineering, University of Rijeka, Vukovarska 58, 51000 Rijeka, Croatia; nandelic@riteh.hr (N.A.); sbaressisegota@riteh.hr (S.B.Š.); ilorencin@riteh.hr (I.L.); car@riteh.hr (Z.C.)
[2] Maritime Department, University of Zadar, Mihovila Pavlinovića 1, 23000 Zadar, Croatia; ipoljak1@unizd.hr
* Correspondence: vmrzljak@riteh.hr; Tel.: +385-98-174-5205

Abstract: In this paper, the publicly available dataset for the Combined Diesel-Electric and Gas (CODLAG) propulsion system was used to obtain symbolic expressions for estimation of fuel flow, ship speed, starboard propeller torque, port propeller torque, and total propeller torque using genetic programming (GP) algorithm. The dataset consists of 11,934 samples that were divided into training and testing portions in an 80:20 ratio. The training portion of the dataset which consisted of 9548 samples was used to train the GP algorithm to obtain symbolic expressions for estimation of fuel flow, ship speed, starboard propeller, port propeller, and total propeller torque, respectively. After the symbolic expressions were obtained the testing portion of the dataset which consisted of 2386 samples was used to measure estimation performance in terms of coefficient of correlation (R^2) and Mean Absolute Error (MAE) metric, respectively. Based on the estimation performance in each case three best symbolic expressions were selected with and without decay state coefficients. From the conducted investigation, the highest R^2 and lowest MAE values were achieved with symbolic expressions for the estimation of fuel flow, ship speed, starboard propeller torque, port propeller torque, and total propeller torque without decay state coefficients while symbolic expressions with decay state coefficients have slightly lower estimation performance.

Keywords: CODLAG; data-driven modelling; genetic programming; decay state coefficients

Citation: Anđelić, N.; Baressi Šegota, S.; Lorencin, I.; Poljak, I.; Mrzljak, V.; Car, Z. Use of Genetic Programming for the Estimation of CODLAG Propulsion System Parameters. *J. Mar. Sci. Eng.* **2021**, *9*, 612. https://doi.org/10.3390/jmse9060612

Academic Editor: Tie Li

Received: 25 May 2021
Accepted: 30 May 2021
Published: 2 June 2021

Publisher's Note: MDPI stays neutral with regard to jurisdictional claims in published maps and institutional affiliations.

1. Introduction

The marine propulsion systems are used to generate thrust to propel a ship across the water, with various types of marine prime movers being used [1,2]. The gas turbines are often used in combination with other types of propulsion systems due to their poor thermal efficiency at low power output. The other key factor for using such propulsion systems is to allow a reduction of emissions in sensitive environmental areas or while in port [3]. In some cases, ships have steam turbines which are also used to improve the efficiency of gas turbines in a combined cycle, where waste heat from gas turbine exhaust is used to boil water and create steam.

The combined diesel-electric and gas (CODLAG) is a modified diesel and gas propulsion system for ships. In it, the electric motors which are powered by diesel generators are connected to the propeller shafts. To achieve higher speed, the gas turbine is used to power shafts over a cross-connecting gearbox. For cruise speed, the drive train of the turbine is disengaged with clutches. Since electric motors work efficiently over a wide range of revolutions they can be directly connected to the propeller shaft so simpler gearboxes are used for combining the mechanical output of the turbine and diesel-electric system.

Literature Review

The most commonly used maintenance approach was to repair systems as necessary [4]. This approach in the long run proved to be very expensive especially when gathering data from the field is cheaper and breakdown-related costs may overcome the

149

asset value [5]. Condition-based maintenance (CBM) is triggering maintenance activities as they are indicated by the condition of the system [4]. This approach tracks the condition of system parts which is used to predict their potential degradation and to plan when maintenance activities will be performed. To perform accurate fault prognosis the CBM requires real-time tracking and diagnosis of the target system.

The comprehensive approach in the simulation of CODLAG propulsion system behavior during transients and off-design conditions is presented by Altosole et al. (2010) [6]. With this model, the authors were able to capture the unbalance of the shaft line during a turning maneuver. The influence of the deterioration of the main components (gas turbine, propellers, and ship hull) on the behavior of the CODLAG propulsion system was performed in [7]. The different detailed simulation models of the CODLAG propulsion system were developed by Martelli (2017) [8] to investigate the system performance under different operational conditions. The publicly available dataset has been developed using numerical simulation of CODLAG propulsion plant [9], where the performance advantages of exploiting machine learning (ML) methods in modeling the degradation of the propulsion plant over time are tested. In [10], the multi-layer perceptron (MLP) was applied on data available dataset in the prediction of the gas turbine and turbo compressor decay state coefficients. In the case of gas turbine decay state coefficient prediction, the lowest mean relative error of 0.622% was achieved while in the case of turbo compressor decay state coefficient, the lowest mean relative error of 1.094% was achieved. In [11], the MLP was again used for the estimation of the frigate speed. The results showed that MLP could estimate the shipping speed with an error of just 3.4485×10^{-5} knots. In [12], the publicly available CODLAG dataset was used to train genetic programming algorithm to obtain symbolic expressions for estimation gas turbine shaft torque and fuel flow. The three best symbolic expressions obtained for gas turbine shaft torque estimation generated R^2 scores of 0.999201, 0.999296, and 0.999374, respectively. The three best symbolic expressions obtained for fuel flow estimation generated R^2 scores of 0.995495, 0.996465, and 0.996487, respectively.

Beyond the aforementioned papers, many researchers opted for an application of AI-based modeling techniques in the application in propulsion system research area. Cheliotis et al. (2020) [13] demonstrate the application of Exponentially Weighted Moving Average (EWMA) for fault detection in maritime systems. The proposed research achieves an R^2 score of 0.96 in both observed cases. Uyanik et al. (2020) [14] proposed an ML approach to the prediction of a container vessel fuel consumption. Through the application of multiple algorithms, such as Multiple Linear Regression, Ridge and LASSO Regression, Support Vector Regression, Tree-Based Algorithms, and Boosting Algorithms are applied and evaluated using R^2. The best results are achieved through multiple linear regression and ridge regression with an R^2 value of 0.999. Berghout et al. (2021) [15] applied an Extreme Learning Machine in combination with other techniques in the application for prediction of condition-based maintenance of naval propulsion systems. The newly proposed approach demonstrates not only higher accuracy, but also better generalization under different training paradigms. Tsaganos et al. (2020) [16] demonstrated the application of AdaBoost classifier for the improvement of engine fault detection. Based on the achieved performance, with an accuracy of 96.5%, the authors concluded that the ensemble methods such as used are an appropriate choice for the given problem. Bachmayer et al. (2020) [17] discussed ML applications in underwater propulsion systems, concluding that such approaches are fast enough for use in the real-time system for detection of soft and hard errors.

GP is an Artificial Intelligence (AI) method for evolving expressions such as computer programs or equations. The roots of GP can be traced back to Alan Turing [18] but the computational limitations of that time prevented further development. After almost 30 years the small programs were successfully evolved, as reported in [19]. The genetic algorithm (GA) for evolving programs was officially introduced by Koza in 1988 [20]. The algorithm can be used to develop symbolic ecpressions which allow for direct modelling of various tasks [21–23].

Based on an extensive literature review the following questions arise:

- does the correlation exist, and how strong is the correlation between the parameters of CODLAG propulsion system dataset [9], and
- is it possible to obtain the symbolic expressions using GP algorithm for fuel flow estimation, ship speed estimation, starboard and port propeller torque, and total torque-with and without decay state coefficients.

The correlation analysis will give a better insight into the CODLAG propulsion system dataset [9] which will be a good starting point for GP algorithm implementation. After the symbolic expressions were obtained and tested the results of correlation analysis will provide sufficient information in further investigation of symbolic expressions.

The novelty of the research lies in multiple elements. The authors have applied the correlation analysis to determine the parameter importance of individual dataset parameters, in order to improve the results of the AI-based methods. The main novelty of the paper is the generation of equations which can be applied to the prediction of the aformentioned parameters (fuel flow, ship speed, as well as starboard, port and total propeller torque) by the future researchers. As a final research novelty, the influence of decay coefficients has been tested.

First, the researchers will present the used dataset, with methods applied to the analysis of it. Then, a short description of the GP algorithm is provided, along with the used hyperparameters and evaluation metrics. The results are presented and discussed; following that, providing information on the correlation coefficients of the parameters in the dataset, metrics achieved with the trained models along with the used hyperparameters and regressed equations. Drawn conclusions, addressing the posed research questions, are given in the end.

2. Materials and Methods

In this section, the publicly available dataset [9] is described in detail as well as the correlation analysis, genetic programming algorithm, and metric used to evaluate obtained symbolic expressions.

2.1. Dataset Description

The dataset that was used in this paper is a publicly available dataset available at the UCI machine learning repository [9]. The dataset was obtained using a numerical simulator of a naval vessel (Frigate) characterized by a Gas Turbine (GT) propulsion plant. The simulator that was used to obtain the dataset consists of different blocks such as propeller, hull, GT, gearbox, and controller. These components were developed and fine-tuned on several similar real propulsion plants. This dataset also incorporates the performance decay over time of the GT components such as turbo compressors and turbines. The two propellers are driven from power generated with GT and two electric motors which are transmitted using a system that consists of three gearboxes and four clutches. The scheme of the CODLAG propulsion system is shown in Figure 1.

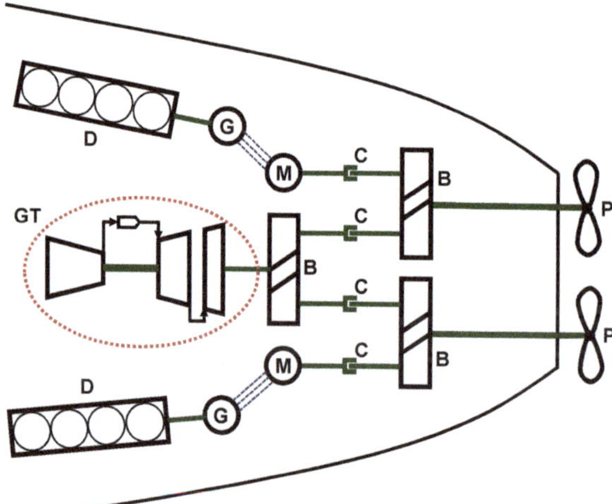

Figure 1. The scheme of CODLAG propulsion system (B-gear box, C-clutch, D-diesel engine, G-electrical generator, GT- gas turbine, M-electrical motor, P-frigate propeller).

The GT shown in Figure 1 consists of a turbo compressor, combustion chamber, high pressure (HP), and low pressure (LP) gas turbines. It should be noted that the power produced in HP gas turbine is used only for turbo compressor drives (gas generator) while the power produced by LP gas turbine is used for ship propulsion in combination with power produced by electric motors. The detailed scheme of GT used in the CODLAG propulsion system is shown in Figure 2.

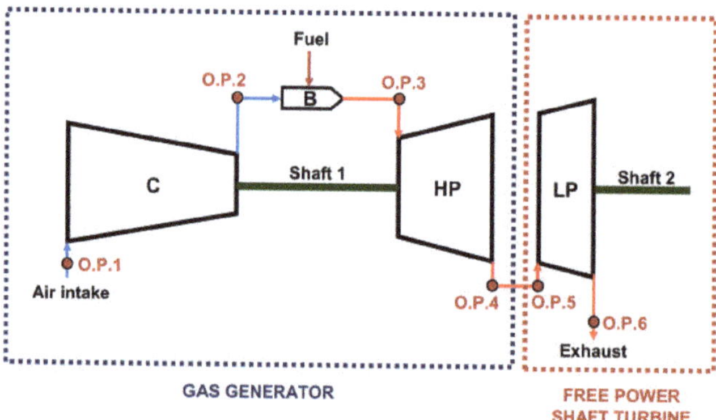

Figure 2. The scheme of GT component used in CODLAG propulsion system (C-turbo compressor; B-combustion chamber; HP-high pressure turbine; LP-low pressure turbine, O.P.-Operating Point).

As seen in Figure 2, the HP gas turbine together with turbo compressor (C) and combustion chamber (B) represents the gas generator. The only connection between HP gas turbine and LP gas turbine is achieved by flue gases that go from HP gas turbine to LP gas turbine. The LP gas turbine is a free power shaft turbine. System maintenance is an important factor of complex propulsion systems. To describe the gas turbine and turbo compressor the decay state coefficient is used as the numerical indicator of their condition. In this dataset, the decay state coefficients of gas turbine and turbo compressor are simulated in the MatLab software package as the consequence of fouling. The source of fouling is the exhaust gases and oil vapors that produce impurities on gas turbine blades and impurities of intake air of turbo compressor. The fouling in the gas turbine is simulated as the gas flow rate decrease while in the turbo compressor the fouling is simulated as a decrease of airflow rate M_c and isentropic efficiency η_c. In Table 1 the dataset parameters with corresponding values range and units are provided, while Figure 3 shows the T-s diagram of the Gas turbine for the CODLAG system.

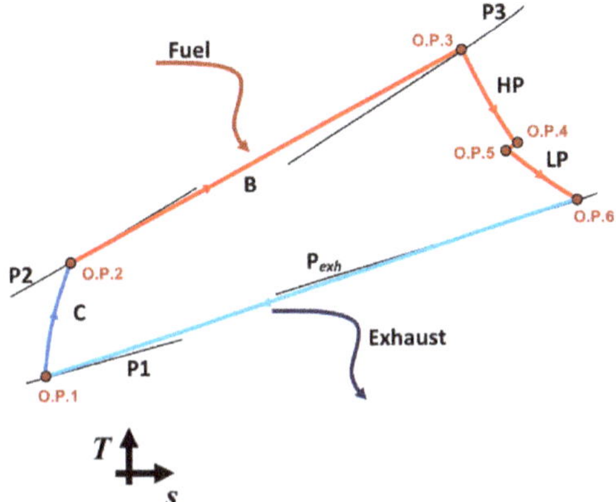

Figure 3. Thermodynamic process of the gas turbine from the analyzed CODLAG propulsion system in T-s diagram (O.P.-Operating Point).

Table 1. The list of physical values in CODLAG dataset with corresponding range of values and units.

Physical Variable	Range	Unit
Lever position (l_p)	1.138–9.3	-
Ship speed (v)	3–27	kn
Gas turbine shaft torque (GTT)	253.547–72,784.872	kNm
GT rate of revolutions (GTn)	1307.675–3560.741	rpm
Gas generator rate of revolutions (GGn)	6589.002–9797.103	rpm
Starboard propeller torque (Ts)	5.304–645.249	kN
Port propeller torque (Tp)	5.304–645.249	kN
High pressure turbine exit temperature (T48)	442.364–1115.797	°C
Turbo compressor inlet air temperature (T1)	288	°C
Turbo compressor outlet air temperature (T2)	540.442–789.094	°C
HP turbine exit pressure (P48)	1.093–4.56	bar
Turbo compressor inlet air pressure (P1)	0.998	bar
Turbo compressor outlet air pressure (P2)	5.828–23.14	bar
GT exhaust gas pressure (P_{exh})	1.019–1.052	bar
Turbine injection control (TIC)	0–92.556	%
Fuel flow (m_f)	0.068–1.832	kg/s
Turbo compressor decay state coefficient	0.95–1	-
Turbine decay state coefficient	0.975–1	-

2.2. Correlation Analysis

In this paper, two types of correlation analysis will be applied to the CODLAG propulsion system dataset to determine the correlation between input and output variables i.e., Pearsons and Spearman correlation analysis.

The Pearson's product-moment correlation coefficient r measures the linear relationship between two continuous variables [24]. For example, let x and y represent the quantitative measures of two random variables on the same sample of n. The Pearson's correlation coefficient r can be written in the following form [25]:

$$r = \frac{\sum_{i=1}^{n}(x_i - \bar{x})(y_i - \bar{y})}{\sqrt{\sum_{i=1}^{n}(x_i - \bar{x})}\sqrt{\sum_{i=1}^{n}(y_i - \bar{y})}} \tag{1}$$

where

$$\bar{x} = \frac{1}{n}\sum_{i=1}^{n} x_i \quad \text{and} \quad \bar{y} = \frac{1}{n}\sum_{j=1}^{n} y_i \tag{2}$$

are the mean values of variable x and y, respectively. Assuming that the sample variances of x and y are positive i.e., $s_x^2 > 0$ and $s_y^2 > 0$ the linear correlation coefficient r can be written as the ratio of the sample covariance of the two variables to the product of their respective standard deviations s_x and s_y as [26,27]:

$$r = \frac{\text{Cov}(x,y)}{s_x s_y}, \tag{3}$$

where Cov represents covariance. The range of correlation measurement r is between -1 and $+1$. There are three different cases of correlation measurement between x and y and these are:

- $r > 0$-the linear correlation between x and y are positive i.e., higher absolute levels of one variable are associated with lower levels of the other,
- $r = 0$-indicates the absence of any association between x and y, and
- $r < 0$-the linear correlation between x and y is negative i.e., higher absolute levels of one variable are associated with lower levels of the other.

The magnitude of the correlation coefficient indicates the strength of association, while the sign of the linear correlation coefficient indicates the direction of the association. For example, if the value of the correlation coefficient is equal to $+1$ the variables have a perfect linear positive correlation which means that if one variable increases, the second increases proportionally in the same direction. On the other hand, if the correlation coefficient value is equal to -1, the variables have a negative correlation and move in the opposite direction of each other. If the value of one variable increases the value of the other variable decreases proportionally. When two variables x and y are normally distributed, the population Pearson's product-moment correlation coefficient can be determined as [28]:

$$\rho = \frac{\text{Cov}(x,y)}{\sigma_x \sigma_y}, \tag{4}$$

where σ_x and σ_y are the population standard deviations of x and y, respectively. It should be noted that if both variables are normally distributed the coefficient ρ is not significant since it is affected by extreme values.

Spearman's correlation coefficient evaluates the monotonic relationship between two continuous variables [29]. In a monotonic relationship, the variables tend to change together, but not at constant rate. For two variables x and y the Spearman's rank correlation coefficient computes the correlation between the rank of two variables which can be written in the following form [30]:

$$r_s = \frac{\sum_{i=1}^{n}(x_i' - \bar{x}')(y_i' - \bar{y}')}{\sqrt{\sum_{i=1}^{n}(x_i' - \bar{x}')}\sqrt{\sum_{i=1}^{n}(y_i' - \bar{y}')^2}} \tag{5}$$

where x' and y' are ranks of x and y, respectively. The Spearman's correlation is basically the rank-based version of the Pearson's correlation coefficient. The range of Spearman's coefficient is from -1 up to $+1$. Similar to Pearson correlation coefficient, the Spearman's correlation coefficient is 0 for variables that are correlated in a non-monotonic way. An alternative formula used to calculate the Spearman rank correlation can be written in the following form [31]:

$$r_s = 1 - \frac{6\sum_{i=1}^{2} d_i}{n(n^2-1)},$$ (6)

where d_i is the difference between the ranks of corresponding values x_i and y_i. To avoid the step of determining the ranks of the variables, Equation (5) was used for the calculation of Spearman's correlation coefficients in this paper.

2.3. Genetic Programming

The genetic programming algorithm is a technique of evolving programs from an initial population of random, unfit programs from generation to generation and fits them for a particular task with the application of genetic operations (crossover and mutation) [32]. In GP computer programs are represented as three structures. The example of computer program $(X_1 + 2.7X_2) + (X_3 - 3.7X_4)$ is shown in Figure 4.

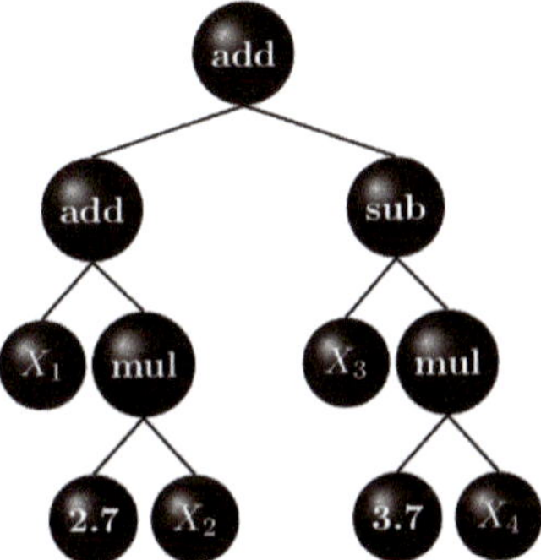

Figure 4. The example of computer program represented as three structure.

The variables and constants shown in Figure 4 are leaves of the tree and in GP they are called terminals while the arithmetic operations are internal nodes called functions. The set of functions and terminals together form the primitive set of a GP system.

As stated earlier the initial population consists of random, naive programs which are developed using a primitive set. Various methods can be used to initialize the population however in this paper the ramped-half-and-half method is used. This method is a combination of the full and grow method. In the full method, the nodes are taken at random from the function set until the maximum tree depth is reached. After the maximum tree depth is reached only terminals must be chosen. In grow method the nodes are selected from the whole primitive set until the depth limit is reached. Once the depth limit is reached only terminals may be chosen. Since both methods do not provide a very wide array of sizes and shape the ramped half-and-half method is used. In this method, half of the initial population is generated using the full method, and the other half using the grow method. This procedure is done using a range of depth limits to ensure that the variety of tree sizes and shapes in population. After the initial population is generated each population member must be evaluated to determine its fitness value. In this paper, the Mean Absolute Error is a fitness measure that will be used to evaluate each population member. The MAE formula can be written in the following form [33]:

$$MAE = \frac{\sum_{i=1}^{n} |y_i - x_i|}{n},$$ (7)

where y_i is prediction and x_i is the true value thus the difference between those two values represents an average of the absolute errors while n represents the number of samples. It should be noted that this measure will also be used later for further evaluation of symbolic expressions on the testing portion of the dataset. After the initial population has been created the selection must be performed to select population members that will represent parents of the next generation. There are various types of the selection procedure which can be used; however, in this paper, the tournament selection procedure was used. The tournament selection starts from a random selection of population members from all population members [34]. These population members are compared with each other and the best of them (tournament winner) is chosen to be the parent. For crossover operation two parents are needed so, two selection tournaments are made. However, for mutation operation, only one population member (tournament winner) is required so only one tournament selection is required. In GP the most commonly used form of crossover is the subtree crossover. This operation requires two parents and the crossover point or a node is randomly selected in each parent tree. The subtrees are swapped between those two parents to generate the members of the next generation. In GP there are three types of mutation operations and these are subtree mutation, hoist mutation, and point mutation. In each mutation case, only one tournament winner is needed. The subtree mutation starts by randomly selecting the subtree on the tournament winner and this subtree is replaced by a randomly generated subtree to form an offspring of the next generation. The hoist mutation operation starts by randomly selecting the subtree on the tournament winner. Then a random subtree of that subtree is selected and is then hoisted into the original subtree location to form the member of the next generation. The point mutation operation starts by selecting random nodes on the tournament winner which will be replaced. The terminals are then replaced by other terminals and functions are replaced by other functions.

To terminate the execution of the GP algorithm the stopping criteria are needed. Two different stopping criteria are usually used in GP and these are the maximum number of generations and the stopping criteria value. The maximum number of generations is the termination criteria that terminates the execution of GP after the maximum number of generations is reached. The stopping criteria value represents the lowest fitness function value which can be achieved by population members in a generation. If the lowest value is achieved the GP algorithm execution is terminated.

The other important parameter in the GP algorithm is the parsimony coefficient [35] which is responsible for penalizing large growth of symbolic expressions without improvement in their fitness value by making them less favorable for tournament selection.

2.4. Evaluation Metrics

After all symbolic expressions were obtained with the GP algorithm on the training portion of the dataset these symbolic expressions are then evaluated on the testing portion of the dataset. In this paper, two metrics are used for the evaluation of estimation performance of symbolic expressions and these are the R^2 and MAE metric. Since the MAE was already described in the previous section here only the R^2 metric will be described.

The R^2 metric or the coefficient of determination is the proportion of the variance in the dependent variable that is predictable from the independent variable. The formula for calculating the R^2 metric can be written in the following form

$$R^2 = 1 - \frac{S_{RESIDUAL}}{S_{TOTAL}} = 1 - \frac{\sum_{i=0}^{m}(y_i - \hat{y}_i)^2}{\sum_{i=0}^{m}(y_i - \frac{1}{m}\sum_{i=0}^{m} y_i)^2} \tag{8}$$

Two sets of solutions i.e., the real data y and the data obtained by the model $\hat{y}$ are compared by this metric in terms of variance. The result of R^2 metric can be in the range from 0 to 1. If the R^2 value is equal to 1.0 means that there is no variance between the real data and the data obtained by the model. The R^2 value of 0 means none of the variances in the real data are explained in the model data.

3. Results and Discussion

In this section, the preparatory steps for implementation of GP are described as well as the results obtained using correlation analysis and symbolic expressions obtained for estimation of the fuel flow, ship speed, starboard, and propeller torque, and total torque, respectively. After extensive research, the obtained results are discussed in detail.

3.1. Results

Before presenting the best symbolic expressions for estimation of specific output values the two types of correlation analysis were performed and these are Pearsons and Spearman's correlation analyses. The results of Pearsons and Spearman's correlation analyses are shown in Figures 5 and 6.

Pearson Correlation Coefficient Values

	I_p	v	GTT	GTn	GGn	Ts	Tp	T48	T1	T2	P48	P1	P2	Pexh	TIC	m_f	GCDSC	GTDSC
I_p	1.0	0.9999	0.961	0.9621	0.986	0.9592	0.9592	0.9612	0.0	0.9827	0.9631	0.0	0.9691	0.9534	0.9136	0.9314	0.0	0.0
v	0.9999	1.0	0.9582	0.9604	0.9866	0.9564	0.9564	0.9588	0.0	0.9812	0.9606	0.0	0.967	0.9508	0.91	0.9278	0.0	0.0
GTT	0.961	0.9582	1.0	0.9897	0.933	0.9992	0.9992	0.9911	0.0	0.9902	0.9989	0.0	0.9976	0.996	0.9779	0.9951	0.003	0.0004
GTn	0.9621	0.9604	0.9897	1.0	0.943	0.9886	0.9886	0.9796	0.0	0.9893	0.9951	0.0	0.996	0.994	0.9623	0.9802	0.0014	-0.0
GGn	0.986	0.9866	0.933	0.943	1.0	0.9296	0.9296	0.9398	0.0	0.9667	0.9375	0.0	0.9459	0.9242	0.8791	0.897	-0.0188	0.01
Ts	0.9592	0.9564	0.9992	0.9886	0.9296	1.0	1.0	0.986	0.0	0.9874	0.998	0.0	0.9962	0.9962	0.9775	0.9944	0.0008	0.0001
Tp	0.9592	0.9564	0.9992	0.9886	0.9296	1.0	1.0	0.986	0.0	0.9874	0.998	0.0	0.9962	0.9962	0.9775	0.9944	0.0008	0.0001
T48	0.9612	0.9588	0.9911	0.9796	0.9398	0.986	0.986	1.0	0.0	0.9923	0.9894	0.0	0.9905	0.9801	0.9697	0.9863	-0.0396	-0.0385
T1	0.0	0.0	0.0	0.0	0.0	0.0	0.0	0.0	1.0	0.0	0.0	0.0	0.0	0.0	0.0	0.0	0.0	0.0
T2	0.9827	0.9812	0.9902	0.9893	0.9667	0.9874	0.9874	0.9923	0.0	1.0	0.9917	0.0	0.9944	0.9835	0.9587	0.9765	-0.0472	-0.0169
P48	0.9631	0.9606	0.9989	0.9951	0.9375	0.998	0.998	0.9894	0.0	0.9917	1.0	0.0	0.9994	0.9979	0.9757	0.9927	0.0082	-0.0027
P1	0.0	0.0	0.0	0.0	0.0	0.0	0.0	0.0	0.0	0.0	0.0	1.0	0.0	0.0	0.0	0.0	0.0	0.0
P2	0.9691	0.967	0.9976	0.996	0.9459	0.9962	0.9962	0.9905	0.0	0.9944	0.9994	0.0	1.0	0.9963	0.9721	0.9893	0.0083	-0.0183
Pexh	0.9534	0.9508	0.996	0.994	0.9242	0.9962	0.9962	0.9801	0.0	0.9835	0.9979	0.0	0.9963	1.0	0.9742	0.991	0.0353	0.0118
TIC	0.9136	0.91	0.9779	0.9623	0.8791	0.9775	0.9775	0.9697	0.0	0.9587	0.9757	0.0	0.9721	0.9742	1.0	0.9855	-0.032	-0.0189
m_f	0.9314	0.9278	0.9951	0.9802	0.897	0.9944	0.9944	0.9863	0.0	0.9765	0.9927	0.0	0.9893	0.991	0.9855	1.0	-0.0137	-0.0173
GCDSC	0.0	0.0	0.003	0.0014	-0.0188	0.0008	0.0008	-0.0396	0.0	-0.0472	0.0082	0.0	0.0083	0.0353	-0.032	-0.0137	1.0	0.0
GTDSC	0.0	0.0	0.0004	-0.0	0.01	0.0001	0.0001	-0.0385	0.0	-0.0169	-0.0027	0.0	-0.0183	0.0118	-0.0189	-0.0173	0.0	1.0

Figure 5. The result of Pearsons correlation analysis.

As seen in Figure 5 the highest positive correlation values are obtained for 14 out of 18 variables in the dataset. This means that if the value of these input variables increases the value with the output variable will also increase. However, the GCDSC (turbo compressor decay state coefficient) and GTDSC (turbine decay state coefficient) have positive, negative, and no correlation values with other variables in the dataset. Both decay state coefficients do not correlate with ship speed (v), have small positive correlation values (0.0008, 0.0001) with starboard and port propeller torque, and negative correlation values (-0.0137, -0.0173) with fuel flow. If the correlation value is negative this means that if the value of input variables increases the value of the output variable will decrease or vice versa. It should be noted that T1 (GT turbo compressor inlet air temperature) and P1 (GT turbo compressor inlet air pressure) have no correlation with any variable in the dataset except with itself. These two variables represent the ambient temperature and pressure which were set to constant values during the simulation of the CODLAG propulsion system. The variation of these two variables would not have any effect on the output variable. The results of Spearman's correlation analysis are shown in Figure 6.

Spearman Correlation Coefficient Values

	l_p	v	GTT	GTn	GGn	Ts	Tp	T48	T1	T2	P48	P1	P2	Pexh	TIC	m_f	GCDSC	GTDSC
l_p	1.0	1.0	0.9889	0.9719	0.9937	0.9938	0.9938	0.9696	0.0	0.9883	0.9892	0.0	0.9889	0.9913	0.9316	0.9746	0.0	0.0
v	1.0	1.0	0.9889	0.9719	0.9937	0.9938	0.9938	0.9696	0.0	0.9883	0.9892	0.0	0.9889	0.9913	0.9316	0.9746	0.0	0.0
GTT	0.9889	0.9889	1.0	0.9732	0.9885	0.9905	0.9905	0.9832	0.0	0.9938	0.9843	0.0	0.9865	0.9805	0.9307	0.9849	-0.0414	0.0264
GTn	0.9719	0.9719	0.9732	1.0	0.9766	0.9785	0.9785	0.9568	0.0	0.9725	0.9609	0.0	0.9623	0.9712	0.8886	0.9569	-0.0499	0.0407
GGn	0.9937	0.9937	0.9885	0.9766	1.0	0.9925	0.9925	0.9677	0.0	0.9904	0.9732	0.0	0.976	0.9795	0.9255	0.9706	-0.098	0.047
Ts	0.9938	0.9938	0.9905	0.9785	0.9925	1.0	1.0	0.9646	0.0	0.9857	0.9798	0.0	0.9811	0.9861	0.9136	0.9683	-0.0295	0.0174
Tp	0.9938	0.9938	0.9905	0.9785	0.9925	1.0	1.0	0.9646	0.0	0.9857	0.9798	0.0	0.9811	0.9861	0.9136	0.9683	-0.0295	0.0174
T48	0.9696	0.9696	0.9832	0.9568	0.9677	0.9646	0.9646	1.0	0.0	0.9906	0.9755	0.0	0.9843	0.959	0.9498	0.9989	-0.0717	-0.0629
T1	0.0	0.0	0.0	0.0	0.0	0.0	0.0	0.0	1.0	0.0	0.0	0.0	0.0	0.0	0.0	0.0	0.0	0.0
T2	0.9883	0.9883	0.9938	0.9725	0.9904	0.9857	0.9857	0.9906	0.0	1.0	0.9821	0.0	0.9897	0.9737	0.939	0.9916	-0.0933	-0.0309
P48	0.9892	0.9892	0.9843	0.9609	0.9732	0.9798	0.9798	0.9755	0.0	0.9821	1.0	0.0	0.9964	0.9886	0.9237	0.9797	0.0871	-0.0348
P1	0.0	0.0	0.0	0.0	0.0	0.0	0.0	0.0	0.0	0.0	0.0	1.0	0.0	0.0	0.0	0.0	0.0	0.0
P2	0.9889	0.9889	0.9865	0.9623	0.976	0.9811	0.9811	0.9843	0.0	0.9897	0.9964	0.0	1.0	0.9826	0.9328	0.9881	0.0318	-0.088
Pexh	0.9913	0.9913	0.9805	0.9712	0.9795	0.9861	0.9861	0.959	0.0	0.9737	0.9886	0.0	0.9826	1.0	0.9132	0.964	0.0778	0.0304
TIC	0.9316	0.9316	0.9307	0.8886	0.9255	0.9136	0.9136	0.9498	0.0	0.939	0.9237	0.0	0.9328	0.9132	1.0	0.9538	-0.0856	-0.0681
m_f	0.9746	0.9746	0.9849	0.9569	0.9706	0.9683	0.9683	0.9989	0.0	0.9916	0.9797	0.0	0.9881	0.964	0.9538	1.0	-0.0559	-0.0681
GCDSC	0.0	0.0	-0.0414	-0.0499	-0.098	-0.0295	-0.0295	-0.0717	0.0	-0.0933	0.0871	0.0	0.0318	0.0778	-0.0856	-0.0559	1.0	0.0
GTDSC	0.0	0.0	0.0264	0.0407	0.047	0.0174	0.0174	-0.0629	0.0	-0.0309	-0.0348	0.0	-0.088	0.0304	-0.0681	-0.0681	0.0	1.0

Figure 6. The result of Spearman's correlation analysis.

The results of performed Spearman's correlation analyses have similar results as in the case of Pearson's correlation analysis. The correlation analysis showed that 14 out of 18 variables have positive correlation values. The T1 and P1 are constant values throughout the entire dataset so they do not correlate with any other variable except with themselves i.e., the correlation values are zero. The results of correlation analyses also showed that two decay state coefficients (GCDSC and GTDSC) have positive, negative, or no correlation value. As in the case of Pearson's correlation analysis, the two decay state coefficients do not correlate with ship speed (0.0, 0.0), positive and negative correlation values with starboard (-0.0295, 0.0174) and port propeller torque (-0.0295, 0.0174), and negative correlation values with fuel flow (-0.0559, -0.0681).

As stated in the abstract and introduction of this paper there is a total of 10 different GP analyses performed and these are fuel flow, ship speed, starboard propeller torque, port propeller torque, and total propeller torque analysis with and without decay state coefficients. It should be noted that for starboard propeller torque analysis the port propeller torque variable will be excluded from the dataset. The same procedure was applied for port propeller torque analysis. For total propeller torque analysis, the starboard and port propeller torque values were added together and excluded from the dataset as input variables. Table 2 shows input and output variables for each of the analyses.

Table 2. The input and output variables used in the GP algorithm to obtain symbolic expressions for estimation of fuel flow, ship speed, starboard, port, and total propeller torque with and without decay coefficient.

Physical Variable	Representation of Variables in GP				
	Fuel Flow Analysis	Ship Speed Analysis	Starboard Propeller Torque Analysis	Port Propeller Torque Analysis	Total Propeller Torque Analysis
Lever position (l_p)	X_0	X_0	X_0	X_0	X_0
Ship speed (v)	X_1	y	X_1	X_1	X_1
Gas turbine shaft torque (GTT)	X_2	X_1	X_2	X_2	X_2
GT rate of revolutions (GTn)	X_3	X_2	X_3	X_3	X_3
Gas generator rate of revolutions (GGn)	X_4	X_3	X_4	X_4	X_4
Starboard propeller torque (Ts)	X_5	X_4	y	-	-
Port propeller torque (Tp)	X_6	X_5	-	y	-
High pressure turbine exit temperature (T48)	X_7	X_6	X_5	X_5	X_5
turbo compressor inlet air temperature (T1)	X_8	X_7	X_6	X_6	X_6
turbo compressor outlet air pressure (P2)	X_9	X_8	X_7	X_7	X_7
HP turbine exit pressure (P48)	X_{10}	X_9	X_8	X_8	X_8
Turbo compressor inlet air pressure (P1)	X_{11}	X_{10}	X_9	X_9	X_9
Turbo compressor outlet air pressure (P2)	X_{12}	X_{11}	X_{10}	X_{10}	X_{10}
GT exhaust gas pressure (P_{exh})	X_{13}	X_{12}	X_{11}	X_{11}	X_{11}
Turbine injection control (TIC)	X_{14}	X_{13}	X_{12}	X_{12}	X_{12}
Fuel flow (m_f)	y	X_{14}	X_{13}	X_{13}	X_{13}
Turbo compressor decay state coefficient	X_{15}	X_{15}	X_{14}	X_{14}	X_{14}
Trubine decay state coefficient	X_{16}	X_{16}	X_{15}	X_{15}	X_{15}
Total Propeller Torque (Ts+Tp)	-	-	-	-	y

As seen in Table 2 for fuel flow and ship speed analysis there is a total of 16 input variables and one output variable. In the case of fuel flow and ship speed without decay coefficients, there is a total of 14 input variables. In the case of starboard and port propeller torque analysis, there is a total of 13 input variables in the case without decay state coefficient while in the case with decay state coefficients there is a total of 15 input variables. The same number of input variables with and without decay state coefficients is applied for the total propeller torque but the output variable is the sum of starboard and port propeller torque values. The GP range of GP parameters that were used in all these analyses is shown in Table 3.

Table 3. The range of GP parameters used in all analyses.

GP Parameter	Lower Bound	Upper Bound
Population size	500	1000
Number of generations	100	500
Tournament selection size	50	100
Tree depth	(3–7)	(6–12)
Crossover coefficient	0.9	1
Subtree mutation coefficient	0.01	0.1
Hoist mutation coefficient	0.01	0.1
Point mutation coefficient	0.01	0.1
Stopping criteria value	1×10^{-6}	0.001
Maximum number of samples	0.9	1.0
Constant range	-0.1	0.1
Parsimony coefficient	1×10^{-4}	0.01

As seen in Table 3 the dominating genetic operator is crossover coefficient when compared to three mutation coefficient. The stopping criteria range is very small; however, in all GP algorithm execution, this value was never achieved so the GP algorithm execution was terminated when the maximum number generation was reached. The parsimony coefficient value is responsible for penalizing the large growth of population members without improvement in fitness value i.e., bloat phenomenon. The values of the parsimony coefficient in all analyses were small to allow the growth of population members from generation to generation.

3.1.1. The Symbolic Expressions for Fuel Flow Estimation with and without Decay State Coefficients

To obtain symbolic expressions for fuel flow estimation with decay state coefficients total of 16 input variables were used from the training dataset part and fuel flow was used as the output variable which is shown in Table 2. In the case of fuel flow estimation without decay state coefficients only 14 input variables were used. After multiple GP algorithm executions, the three best symbolic expressions with and without decay state coefficients were selected based on their performance in terms of R^2 and MAE values, respectively. The three best symbolic expressions with and without decay state coefficients for fuel flow estimation are presented in Tables 4 and 5.

Table 4. Three best symbolic expressions for fuel flow estimation with decay state coefficients with corresponding R^2 and MAE score.

GP Parameters Population, Generations, Selection Size, Tree Depth, Crossover Coef., Subtree Mutation Coef., Hoist Mutation Coef., Point Mutation Coef., Stopping Criteria, Samples, Constant Range, Parsimony Coef.	Symbolic Expression	R^2	MAE
[930, 243, 81, (3, 11), 0.91, 0.021, 0.015, 0.041, 0.0002, 0.95, (−0.043, 0.021), 0.0003]	$y_{mfDF1} = (\log(\min(\sqrt{\sin(\log(\frac{X_{12}}{X_{15}X_{16}}))},$ $\tan(\sin(\tan(\sin(\log(\frac{X_{12}}{X_{13}X_{15}}))))))))^{\frac{1}{2}}$	0.99398	0.02664
[742, 103, 92, (4, 11), 0.9, 0.026, 0.035, 0.02, 0.0002, 0.91, (−0.071, 0.02), 0.0038]	$y_{mfDF2} = \log(X_{10})\cos(\log(\cos(X_{16})))$ $\cos(\log(\tan(X_{11})))\max(X_{15},\log(X_{10}))$	0.993	0.03695
[927, 346, 80, (6, 9), 0.9, 0.032, 0.039, 0.019, 0.0002, 0.92, (−0.063, 0.056), 0.0008]	$y_{mfDF3} = \log\left(X_{10}X_{15}\cos\left(\frac{X_{13}+\sin(X_{16}+X_3)}{\sin(X_0)+3.35241}\right)\right)$	0.95526	0.08184

Table 5. Three best symbolic expressions for fuel flow estimation without decay state coefficients with corresponding R^2 and MAE score.

GP Parameters Population, Generations, Selection Size, Tree Depth, Crossover Coef., Subtree Mutation Coef., Hoist Mutation Coef., Point Mutation Coef., Stopping Criteria, Samples, Constant Range, Parsimony Coef.	Symbolic Expression	R^2	MAE
[962, 289, 52, (6, 8), 0.91, 0.017, 0.035, 0.03, 0.000524, 0.99, (−0.073, 0.0014), 0.0029]	$y_{mf1} = \dfrac{X_{10}}{\sqrt{\dfrac{\ln(X_2)\sqrt{\frac{\ln(X_2)}{\ln(X_{10})}}}{X_{10}}}}$	0.9964	0.02276
[1000, 141, 83, (5, 9), 0.9, 0.022, 0.012, 0.032, 0.000986, 0.98, (−0.049, 0.0943), 0.0013]	$y_{mf2} = \sqrt{\tan(X_1)}\sin(\sqrt{\tan(\max(X_1,\ln(X_4)))}$ $\sin(\sin(\sin((\sin(\sin(\sin(\sqrt{\sin(X_1)})))$ $\sqrt{\tan(\max(X_1,\ln(X_4)))})^{\frac{1}{2}}))))$	0.99591	0.02341
[582, 365, 85, (4, 7), 0.9, 0.022, 0.027, 0.018, 0.00046, 0.91, (−0.0103, 0.0905), 0.0003]	$y_{mf3} = \dfrac{\ln(X_{10})}{\tan\left(\sin\left(\frac{\frac{\ln(X_{10})}{X_{13}}}{X_{13}}+X_{11}\right)\right)}$	0.99578	0.023027

When Tables 4 and 5 are compared it can be noticed that decay state coefficients are decreasing the performance of fuel flow estimation in terms of R^2 and MAE values. The population size for each case was near 1000 except for the third case without decay state coefficients where population size is near the lower boundary of 500. The crossover coefficient was the dominating genetic operation for each case. All six symbolic expressions are small in size so the bloat phenomenon did not occur although the values of the parsimony coefficients in all six cases are extremely small. The fuel flow estimation performance of all six symbolic expressions is shown in Figure 7.

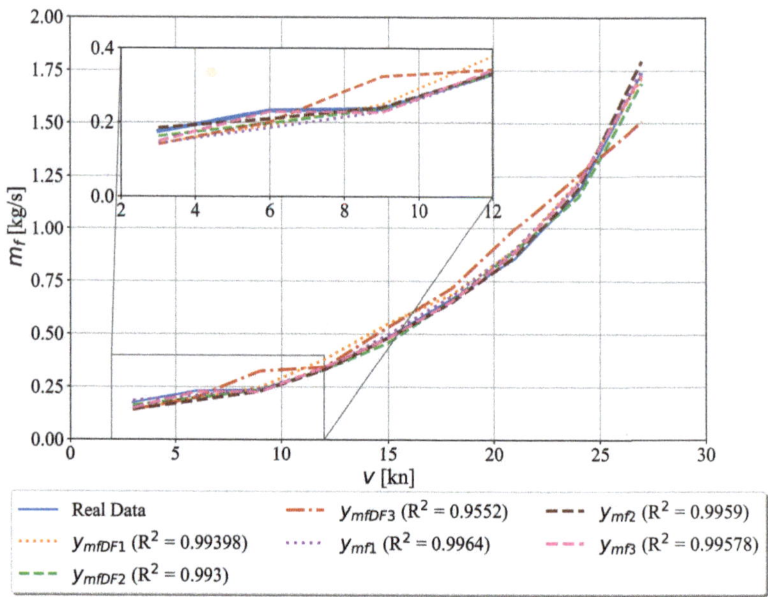

Figure 7. The comparison of estimated fuel flow with real data versus the ship speed.

As seen in Figure 7 all symbolic expressions are estimating the fuel flow with high accuracy except for y_{mfDF3} which has the highest deviation from the real data. When the estimation performance of symbolic expressions with decay state coefficients is compared to those without decay state coefficients it can be noticed that those symbolic expressions with decay state coefficients have slightly lower estimation accuracy. However, those symbolic expressions with decay state coefficients are more important symbolic expressions for CBM since they could indicate the potential degradation of system performance.

3.1.2. The Symbolic Expressions for Ship Speed Estimation with and without Decay State Coefficients

In the case of ship speed estimation using GP with decay state coefficients, the total of 16 input variables was considered while in the case without decay state coefficients the GCDSC and GTDSC input variables were omitted. The output variable in both cases was the shipping speed as indicated in Table 2. After multiple GP algorithm executions using the training dataset part, all symbolic expressions were tested on the testing dataset part to determine R^2 and MAE value. Based on the highest R^2 and MAE value the three best symbolic expressions with and without decay state coefficients were chosen and shown in Tables 6 and 7.

Table 6. Three best symbolic expressions for ship speed estimation with decay state coefficients with corresponding R^2 and *MAE* score.

GP Parameters - Population, Generations, Selection Size, Tree Depth, Crossover Coef., Subtree Mutation Coef., Hoist Mutation Coef., Point Mutation Coef., Stopping Criteria, Samples, Constant Range, Parsimony Coef.	Symbolic Expression	R^2	*MAE*
$[548, 311, 87, (3, 8),$ $0.91, 0.017, 0.017, 0.018,$ $0.000926, 0.92,$ $(-0.015, 0.044), 0.0013]$	$y_{ssDF1} = (X_{15} + X_{16})\left(\frac{X_0}{X_{10} + X_{12}} + X_0\right)$	0.99843	0.2858
$[784, 458, 77, (4, 7),$ $0.9, 0.015, 0.015, 0.06,$ $9.3 \times 10^{-5}, 0.9,$ $(-0.0083, 0.082), 0.0063]$	$y_{ssDF2} = X_0 X_{15} X_{16} + X_0 X_{15} + X_0 X_{16}$	0.99788	0.32584
$[585, 286, 69, (3, 12),$ $0.9, 0.024, 0.025, 0.023,$ $0.000191, 0.92,$ $(-0.00084, 0.018), 0.0053]$	$y_{ssDF3} = \|\|\log(X_{14})\| + \tan(X_{15} + X_{16})\| + \sqrt{X_4}$	0.99593	0.41067

Table 7. Three best symbolic expressions for ship speed estimation without decay state coefficients with corresponding R^2 and *MAE* score.

GP Parameters - Population, Generations, Selection Size, Tree Depth, Crossover Coef., Subtree Mutation Coef., Hoist Mutation Coef., Point Mutation Coef., Stopping Criteria, Samples, Constant Range, Parsimony Coef.	Symbolic Expression	R^2	*MAE*
$[732, 352, 86, (6, 10),$ $0.92, 0.012, 0.013, 0.023,$ $0.000231, 0.9,$ $(-0.073, 0.031), 0.003]$	$y_{sp1} = \frac{X_0 - 0.066}{X_{12}} + 2X_0 - 0.279$	0.9998925	0.06729
$[945, 479, 70, (6, 7),$ $0.91, 0.016, 0.016, 0.014,$ $9.4 \times 10^{-5}, 0.98,$ $(-0.085, 0.0049), 0.0097]$	$y_{sp2} = \sqrt{X_0(X_0 - X_{14}) \log\left(X_3 + X_4\sqrt{X_6}\right)}$	0.999825	0.08665
$[690, 152, 82, (6, 12),$ $0.9, 0.047, 0.01, 0.018,$ $3.6 \times 10^{-5}, 0.94,$ $(-0.023, 0.058), 0.0078]$	$y_{sp3} = X_{14} \cos(X_{12} - X_{14} \cos(X_0 - X_{10})) +$ $\log(X_0) + \sqrt{X_4}$	0.999541	0.11797

As seen in Tables 6 and 7 those symbolic expressions with decay state coefficients included in the analyses have slightly lower estimation accuracy in terms of R^2 and MAE values when compared to those symbolic expressions without decay state coefficients. As in the case of fuel flow estimations, both decay state coefficients are in all three symbolic expressions shown in Table 6. In this analysis, the crossover coefficient was the dominating genetic operation when compared to the remaining three mutation coefficient values, and the parsimony coefficient was extremely low. The tree depth range of the initial population was lower in the case of symbolic expressions with decay state coefficients. The stopping criteria value in all these analyses was never achieved due to the extremely low value, so the GP algorithm executions were terminated after the maximum number of generations was reached. The estimation performance of all six symbolic expressions is shown in Figure 8.

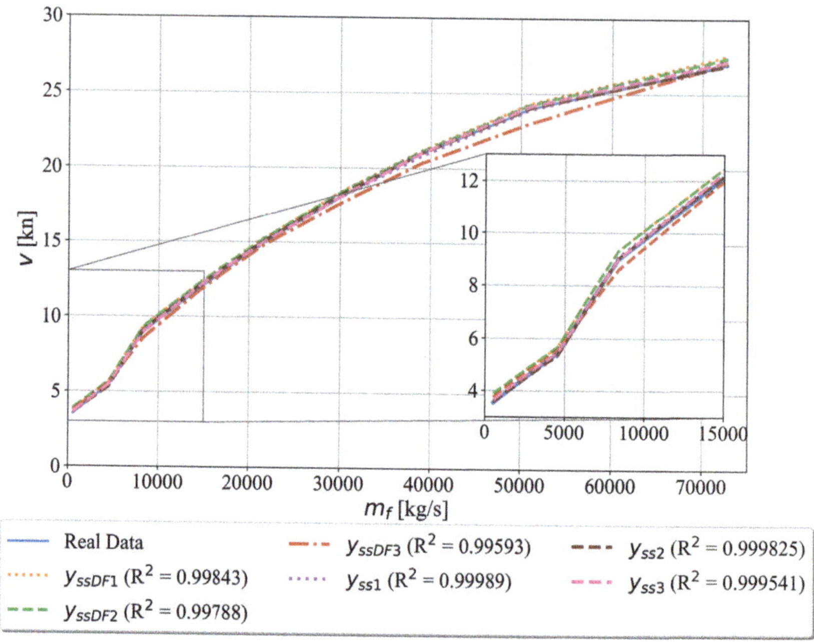

Figure 8. The Comparison of Estimated Ship Speed with Real Data Versus the Fuel Flow.

In Figure 8 the variation of ship speed versus the fuel flow is shown. The estimation accuracy of ship speed using symbolic expressions with decay state coefficients is slightly lower than those without decay state coefficients which are also indicated by achieved R^2 and MAE values.

3.1.3. The Symbolic Expressions for Starboard Propeller Torque Estimation with and without Decay State Coefficients

In the case of starboard propeller torque analysis using the GP algorithm, the port propeller torque was excluded from the analysis since it has almost identical values as the starboard propeller torque. Therefore, if the port propeller torque was included as an input variable in the GP algorithm this would result in early termination of GP algorithm execution. With the exclusion of port propeller torque from the analysis, the total number of input variables in the case of decay state coefficient is 15 while in the case without decay state coefficient the total number of input variables is 13. The list of input and output variables is shown in Table 2. After multiple GP algorithm executions using the training dataset part the obtained symbolic expressions were evaluated on the testing dataset part to determine the R^2 and MAE values, respectively. Based on the highest R^2 and lowest MAE values the three best symbolic expressions with and without decay state coefficients were selected and shown in Tables 8 and 9 with corresponding GP parameters.

Table 8. Three best symbolic expressions for starboard torque estimation with decay state coefficients with corresponding R^2 and MAE score.

GP Parameters - Population, Generations, Selection Size, Tree Depth, Crossover Coef., Subtree Mutation Coef., Hoist Mutation Coef., Point Mutation Coef., Stopping Criteria, Samples, Constant Range, Parsimony Coef.	Symbolic Expression	R^2	MAE
[996, 399, 69, (3, 10), 0.92, 0.013, 0.035, 0.018, 9.06×10^{-7}, 0.94, $(-0.078, 0.077)$, 0.0061]	$y_{stDF1} = X_0 + X_1 X_{10} + 5 X_{13} + X_8 + X_{SPTDF11} + X_{SPTDF12}$	0.99985	1.98477
[821, 418, 92, (4, 10), 0.909, 0.044, 0.018, 0.011, 1.46×10^{-7}, 0.98, $(-0.002, 0.07)$, 0.0022]	$y_{stDF2} = X_1 X_{10} \min(X_{11}, X_{SPTDF21})$	0.99959	3.16776
[598, 398, 63, (4, 11), 0.9, 0.033, 0.018, 0.032, 1.68×10^{-7}, 0.96, $(-0.0055, 0.014)$, 0.0016]	$y_{stDF3} = X_{12} X_{SPTDF31}$	0.99737	7.9579

Table 9. Three best symbolic expressions for starboard torque estimation without decay state coefficients with corresponding R^2 and MAE score.

GP Parameters - Population, Generations, Selection Size, Tree Depth, Crossover Coef., Subtree Mutation Coef., Hoist Mutation Coef., Point Mutation Coef., Stopping Criteria, Samples, Constant Range, Parsimony Coef.	Symbolic Expression	R^2	MAE
[554, 233, 81, (5, 11), 0.9, 0.052, 0.025, 0.017, 5.1×10^{-7}, 0.95, $(-0.087, 0.028)$, 0.0031]	$y_{st1} = \sqrt{X_{SPT11} X_{SPT12}}$	0.99994	1.0697
[792, 144, 63, (5, 8), 0.92, 0.039, 0.013, 0.025, 6.25×10^{-7}, 0.92, $(-0.07, 0.01)$, 0.0069]	$y_{st2} = X_0(X_1 + X_{SPT21})$	0.99989	1.3387
[824, 297, 57, (6, 7), 0.91, 0.014, 0.032, 0.031, 4.08×10^{-7}, 0.92, $(-0.08, 0.039)$, 0.0039]	$y_{st3} = \frac{X_1 X_{10} X_{SPT31}}{\tan(\tan(X_9))} + X_1 X_{10} + \log(X_{13}) + X_{SPT32}$	0.99981	1.8535

As seen in Tables 8 and 9 some new variables were introduced to shorten the size of symbolic expressions in the aforementioned tables. The full form of $X_{SPTDF11}$, $X_{SPTDF12}$, $X_{SPTDF21}$, $X_{SPTDF31}$, X_{SPT11}, X_{SPT12}, X_{SPT21}, X_{SPT31}, and X_{SPT32} is shown in Appendices A.1 and A.2, respectively. The R^2 values of symbolic expressions with decay state coefficients in the estimation of starboard propeller torque are slightly lower when compared to the symbolic expressions without decay state coefficients while the MAE values are higher in symbolic expressions with decay state coefficients when compared to the symbolic expressions obtained without decay state coefficients. The stopping criteria values in all six symbolic expressions are extremely low when compared to the fuel flow and ship speed analysis. Again, these values were never achieved so the GP execution was terminated after a maximum number of generations was reached. The values of the parsimony coefficient were low in all six symbolic expressions which generated very large symbolic expressions so the aforementioned coefficients were introduced to simplify their form. The other key factor that contributed to large symbolic expressions is the constants range which in all analyses is very low. Therefore, the GP algorithm had to replace the low constants range by increasing the size of symbolic expressions using mathematical functions. The estimation performance of starboard propeller torque with and without decay state coefficients compared to real data are shown in Figure 9.

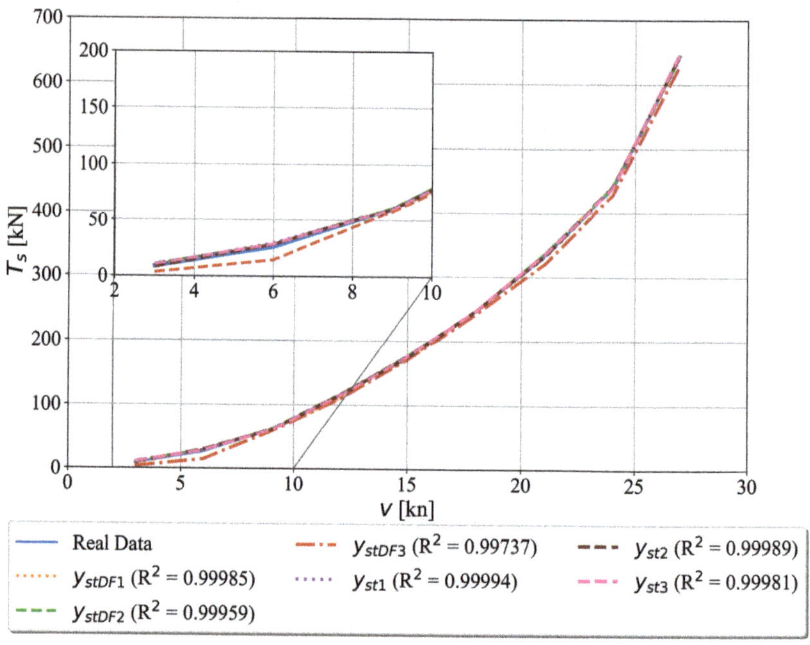

Figure 9. The Variation of Real and Estimated Starboard Propeller Torque Values versus Ship Speed.

In Figure 9, it can be noticed that all symbolic expressions have an accurate estimation of starboard propeller torque when compared to the values from the dataset. However, the third symbolic expressions with decay state coefficients have the lowest estimation accuracy when compared to the remaining five which can also be indicated with a lower R^2 value or higher MAE value, respectively.

3.1.4. The Symbolic Expressions for Port Propeller Torque Estimation with and without Decay State Coefficients

The procedure of obtaining symbolic expressions for estimation of port propeller torque with and without decay state coefficient is similar to the procedure of obtaining the symbolic expressions for starboard propeller torque. The starboard propeller torque was omitted as an input variable from the investigation due to the equal values as port propeller torque. Initial investigation of port propeller torque using GP algorithm with the inclusion of starboard propeller torque showed early termination of GP algorithm. In the case of symbolic expressions with decay state coefficient included there was a total of 15 input variables while in the case without decay state coefficients there was a total of 13 input variables, while the port propeller torque was output variable. The list of input and output variables is shown in Table 2. The equations are not based on previous knowledge or derived from other findings-but generated purely through the evolutionary process of GP described in the Methodology, which attempts to, in a heuristic manner, develop equations that provide a high fitness value for the used dataset. After multiple executions with the GP algorithm using the training dataset part the obtained symbolic expressions were evaluated on the testing dataset part to determine the R^2 and MAE value. Based on the highest R^2 value and lowest MAE values the three best symbolic expressions with and without decay state coefficients were chosen and shown in Tables 10 and 11.

Table 10. Three best symbolic expressions for port propeller torque estimation with decay state coefficients with corresponding R^2 and MAE score.

GP Parameters - Population, Generations, Selection Size, Tree Depth, Crossover Coef., Subtree Mutation Coef., Hoist Mutation Coef., Point Mutation Coef., Stopping Criteria, Samples, Constant Range, Parsimony Coef.	Symbolic Expression	R^2	MAE
[788, 470, 94, (6, 8), 0.93, 0.016, 0.017, 0.032, 6.00×10^{-9}, 0.93, $(-0.072, 0.083)$, 0.0044]	$y_{pptDF1} = (\log(\log(X_0)) + X_{12}) \log(X_9 - X_0) + X_{PPTDF11}$	0.99964	1.9885
[979, 263, 77, (6, 8), 0.91, 0.047, 0.012, 0.022, 6.86×10^{-9}, 0.91, $(-0.062, 0.0027)$, 0.0043]	$y_{pptDF2} = X_1 X_{10} + X_{PPTDF21}$	0.9996	2.61963
[986, 394, 53, (3, 12), 0.91, 0.018, 0.051, 0.013, 9.47×10^{-9}, 0.946, $(-0.02, 0.016)$, 0.0095]	$y_{pptDF3} = X_0 X_{PPTDF31}$	0.99427	14.0996

Table 11. Three best symbolic expressions for port propeller torque estimation without decay state coefficients with corresponding R^2 and MAE score.

GP Parameters - Population, Generations, Selection Size, Tree Depth, Crossover Coef., Subtree Mutation Coef., Hoist Mutation Coef., Point Mutation Coef., Stopping Criteria, Samples, Constant Range, Parsimony Coef.	Symbolic Expression	R^2	MAE
[709, 445, 70, (5, 11), 0.9, 0.023, 0.029, 0.041, 5.72×10^{-7}, 0.94, (−0.041, 0.01), 0.0031]	$y_{ppt_1} = \dfrac{X_0 X_{11} X_{PPT11}}{X_{11} + X_9}$	0.9994	3.35254
[986, 294, 74, (4, 12), 0.91, 0.042, 0.012, 0.025, 6.99×10^{-7}, 0.91, (−0.021, 0.081), 0.0061]	$y_{ppt_2} = X_{10}(\min(X_{13}, \log(|X_{PPT21}|)) + X_1 + 0.276)$	0.99922	4.06154
[769, 415, 69, (3, 11), 0.93, 0.014, 0.011, 0.028, 4.21×10^{-7}, 0.96, (−0.067, 0.035), 0.0046]	$y_{ppt_3} = (X_1 + X_{13})\left(\dfrac{X_{12}\sin(X_0)\sqrt{\sin^3\left(\sin\left(\sqrt{X_{12}}\right)\right)X_{PPT31}}}{X_{10}} + X_{PPT32}\right)^{\frac{1}{2}}$	0.99891	5.11714

Due to the large size of obtained symbolic expressions the coefficients $X_{PPTDF11}$, $X_{PPTDF21}$, $X_{PPTDF31}$, X_{PPT11}, X_{PPT21}, X_{PPT31}, and X_{PPT32}. The full form of these coefficients is given Appendices A.3 and A.4. Although the parsimony coefficient value for all symbolic expressions is low the bloat phenomenon did not occur. However, the large size of obtained symbolic expressions could be explained by the low range of constant values. Since this range is very low the GP algorithm used a large number of mathematical functions and input variables to achieve high estimation accuracy. Based on R^2 and MAE values the symbolic expressions with and without decay state coefficients have almost similar performance except for the third symbolic expression which has the lowest R^2 value and highest MAE value. The estimation performance of these six symbolic expressions are compared to the real data and shown in Figure 10.

The estimation performance of all six symbolic expressions is very high when compared to the real data except for the third symbolic expression with decay state coefficient which performed poorly when compared to the other symbolic expressions.

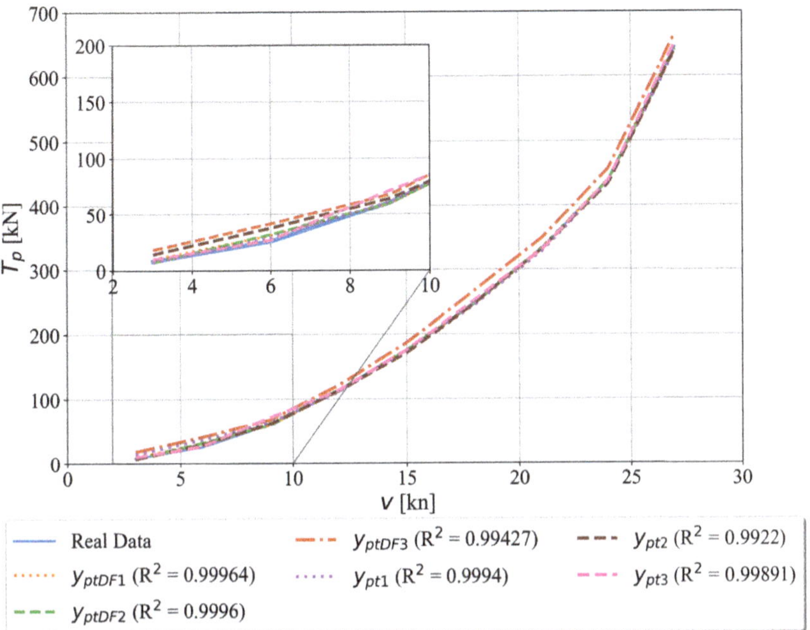

Figure 10. The variation of real and estimated port propeller torque versus the ship speed.

3.1.5. The Symbolic Expressions for Total Propeller Torque Estimation with and without Decay State Coefficients

To obtain symbolic expressions for total propeller torque estimation the starboard and port propeller torque were added together. This variable was used as the output variable in the training and testing portion of the dataset. The starboard and port propeller torque as input variables were omitted from the analysis so the total number of variables was 15 in the case where decay state coefficients were used and 13 in the case without decay state coefficients. After multiple GP executions using the training portion of the dataset the obtained symbolic expressions were evaluated on the testing portion of the dataset to determine R^2 and MAE value. Based on the highest R^2 and lowest MAE value the best symbolic expressions with and without decay state coefficients are chosen. The symbolic expressions with and without decay state coefficients are shown in Tables 12 and 13.

In Table 12 each symbolic expression has at least one decay state coefficient since the GP algorithm could not obtain the symbolic expression for estimation of total torque with both decay state coefficients. To simplify presentation of symbolic expressions in Tables 12 and 13 coefficients X_{TTDF11}, X_{TT11}, X_{TT12}, X_{TT21}, X_{TT31}, and X_{TT32} were introduced. The full form of these coefficient is given in Appendices A.5 and A.6. The R^2 values of symbolic expressions with decay state coefficients are lower while MAE values are higher than those values obtained using symbolic expressions without decay state coefficients. The graphical representation and estimation performance of six symbolic expressions from Tables 12 and 13 are shown in Figure 11.

Table 12. Three best symbolic expressions for total propeller torque estimation with decay state coefficients with corresponding R^2 and MAE score.

GP Parameters Population, Generations, Selection Size, Tree Depth, Crossover Coef., Subtree Mutation Coef., Hoist Mutation Coef., Point Mutation Coef., Stopping Criteria, Samples, Constant Range, Parsimony Coef.	Symbolic Expression	R^2	MAE
$[910, 285, 69, (5, 9),$ $0.91, 0.038, 0.016, 0.015,$ $8.42 \times 10^{-7}, 0.94,$ $(-0.029, 0.09), 0.0096]$	$y_{ttDF1} = \lvert X_{TTDF11} \rvert + X_{12}$ $\sin(\sin(\sin(\log(\tan(\sin(\sqrt{X_0})) - X_0)))) $	0.99848	11.697387
$[664, 116, 75, (3, 10),$ $0.9, 0.058, 0.015, 0.017,$ $3.4e \times 10^{-7}, 0.93,$ $(-0.048, 0.015), 0.0096]$	$y_{ttDF2} = \min(X_{13}, X_{14}) \max(X_5, X_{13}X_7)$ $- \sqrt{\max(X_5, X_{13}^2 X_7)} - 2\tan(\sqrt{X_3})$	0.991606	26.33334
$[790, 112, 79, (3, 12),$ $0.91, 0.012, 0.031, 0.021,$ $7.67 \times 10^{-7}, 0.9,$ $(-0.02, 0.054), 0.007]$	$y_{ttDF3} = X_{13}X_{15}\min(X_5, X_7)$	0.97971	49.89208

Table 13. Three best symbolic expressions for total propeller torque estimation without decay state coefficients with corresponding R^2 and MAE score.

GP Parameters Population, Generations, Selection Size, Tree Depth, Crossover Coef., Subtree Mutation Coef., Hoist Mutation Coef., Point Mutation Coef., Stopping Criteria, Samples, Constant Range, Parsimony Coef.	Symbolic Expression	R^2	MAE
$[682, 172, 56, (4, 7),$ $0.9, 0.018, 0.025, 0.029,$ $4.09 \times 10^{-7}, 0.93,$ $(-0.012, 0.065), 0.0026]$	$y_{tt1} = \lvert X_{12} - X_{TT11} \rvert$ $- X_{TT12} - \sqrt{\frac{X_3}{X_8}} + X_6 X_8$	0.99808	9.2407
$[798, 103, 77, (4, 11),$ $0.9, 0.01, 0.061, 0.021,$ $6.14 \times 10^{-7}, 0.92,$ $(-0.039, 0.046), 0.0069]$	$y_{tt2} = \frac{X_{12}X_8 + \sqrt{X_2}}{X_{TT21}}$	0.99806	13.25
$[883, 209, 64, (6, 9),$ $0.93, 0.013, 0.023, 0.028,$ $8.75 \times 10^{-7}, 0.96,$ $(-0.057, 0.057), 0.0099]$	$y_{tt3} = \max\left(\frac{X_{TT31}}{X_{TT32}} + \sqrt{X_3},\right.$ $\left.\log(X_2) - X_{12}\right) + X_{12}$	0.9976	13.6284

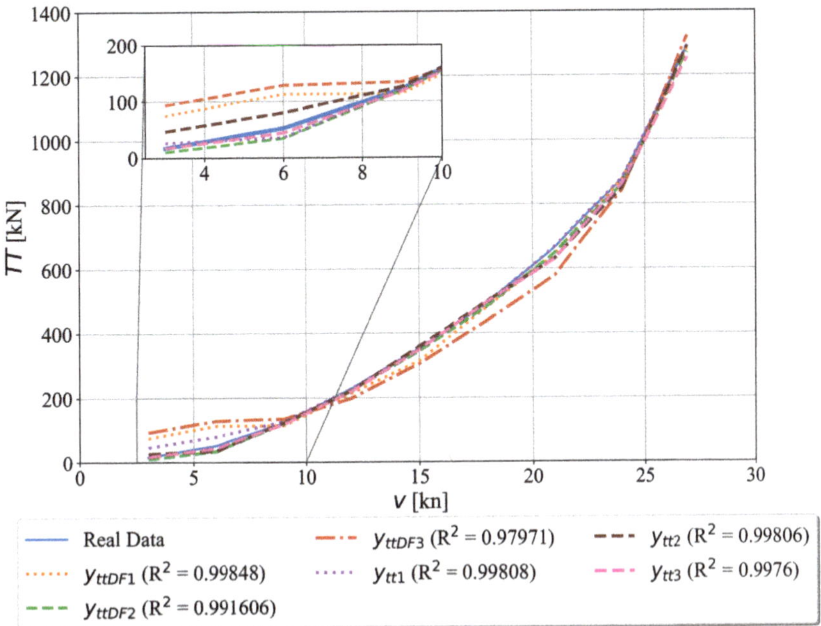

Figure 11. The variation of real and estimated total torque versus ship speed.

As seen in Figure 11 the y_{ttDF1}, y_{ttDF3} and y_{tt1} have some deviation from the real data at lower ship speeds. However, the highest deviation from the real data through entire ship speed range is produced by y_{ttDF3}.

3.2. Discussion

From conducted investigation, it can be noted that two correlation analyses showed that 14 out of 18 dataset variables (without decay state coefficients, T1, and P1) have positive correlation values with remaining variables in the range from 0.8791 up to 1.0. The T1 and P1 showed no correlation with any other variable except with itself. The reason why these two variables do not correlate is that they are constant values through the entire dataset as seen from Table 1. As already stated these two variables represent ambient temperature and pressure which were constant during the simulation of the CODLAG propulsion system. The turbo compressor and turbine decay state coefficients have positive, negative, or no correlation with other variables in the dataset. The analysis showed that with ship speed two decay state coefficients do not have any correlation at all since the correlation values are equal to zero. The Pearson's correlation analysis showed that two decay state coefficients have a small positive correlation (0.0008, 0.0001) with starboard and port propeller torque while Spearman's correlation analysis showed that two decay state coefficients have a negative and positive correlation (−0.0295, 0.0174) with starboard and port propeller torque. It should be noted that the correlation with fuel flow and decay state coefficients is negative in Pearson's and Spearman's correlation analysis.

Regardless of the results from two correlation analysis, the idea was to investigate the possibility of using the GP algorithm to obtain symbolic expressions for estimation of fuel flow, ship-speed, starboard, port, and total propeller torque with and without decay state coefficients since those two coefficients are possible indicators of GT system parts degradation. The total propeller torque was generated by adding together values of starboard and port propeller torque. All symbolic expressions were obtained on the training portion of the dataset with the proper definition of input and output dataset values as indicated in Table 2 and with a random selection of GP, parameters range shown in Table 3 in each GP algorithm execution. It should be noted that in the entire investigation using the GP algorithm the crossover operation was the dominant genetic operation and that predefined (randomly selected) stopping criteria value was never achieved by any of the population members. Therefore, each execution of the GP algorithm was terminated after the maximum number of generations was reached. After the symbolic expressions were obtained they were tested on the test part of the dataset to obtain R^2 and MAE values. The three best symbolic expressions in each case with and without decay state coefficients were chosen based on their highest R^2 value and the lowest MAE values. Another interesting thing is that all these symbolic expressions were obtained with a minimum range of constants which means that in the majority of cases the symbolic expressions consist of mathematical expressions and input variables. Some symbolic expressions grew in size to achieve low estimation error between calculated output and desired output. However, the parsimony coefficient range was low but the bloat phenomenon did not occur.

In the case of symbolic expressions for fuel flow estimation the symbolic expressions with decay state coefficients have slightly lower R^2 values (0.99398, 0.993, 0.95526) and slightly higher MAE (0.02664, 0.03695, 0.08184) values when compared to R^2 (0.9964, 0.99591, 0.99578) and MAE (0.02276, 0.02341, 0.023027) values obtained using symbolic expressions without decay state coefficients. The best symbolic expression with decay state coefficients has almost similar estimation performance of fuel flow when compared to the symbolic expressions obtained without decay state coefficients. Therefore, including those two decay state coefficients resulted in slightly lower performance of obtained symbolic expressions. However, these three symbolic expressions with decay state coefficients are highly valuable since they could indicate potential degradation of the GT propulsion system in terms of higher fuel consumption without noticeable improvement in propeller torque or ship speed.

In the case of ship speed estimation the three obtained symbolic expressions with decay state coefficients have achieved lower R^2 (0.99843, 0.99788, and 0.99593) and higher MAE (0.2858, 0.32584, and 0.41067) values when compared to R^2 (0.9998925, 0.999825, and 0.999541) and MAE (0.06729, 0.08665, 0.11797) values achieved with symbolic expressions obtained without decay state coefficients. Interestingly, those two decay state coefficients do not influence ship speed since Pearson's and Spearman's correlation analysis showed that these two coefficients do not have any correlation with ship speed. Therefore, in the case of those three symbolic expressions obtained with decay state coefficients, the other input variables are X_0, X_4, X_{10}, and X_{14} which are lever position, starboard propeller torque, turbo compressor inlet air pressure (P1), GT exhaust gas pressure, and fuel flow, respectively.

The estimation performance of starboard propeller torque with decay state coefficient is lower when R^2 (0.99985, 0.99959, and 0.99737) and MAE (1.98477, 3.16776, and 7.9579) values are compared to R^2 (0.99994, 0.99989, and 0.99981) and MAE (1.0697, 1.3387, and 1.8535) values of three symbolic expressions obtained without decay state coefficients. It should be noted that in these symbolic expressions the additional coefficients were introduced to simplify their presentation in Tables 8 and 9 while the full form of these coefficients is shown in Appendices A.1 and A.2. The correlation analysis showed that starboard propeller torque has a positive Pearsons correlation with both decay state coefficients while negative correlation coefficient with GCDSC and positive correlation with GTDSC. The symbolic expressions for estimation of port propeller torque showed similar behavior as in the case of starboard propeller torque. The values of Pearson's and Spearman correlation values of port propeller torque and decay state coefficients are the same as in the case of starboard propeller torque.

In the case of total propeller torque, the symbolic expressions with decay state coefficients achieved higher MAE values than those obtained without decay state coefficients which means that the decay state coefficient contributed to higher error rates. In terms of R^2 values, the first symbolic expression in Table 12 achieved a similar value as those symbolic expressions obtained without decay state coefficients which are shown in Table 13.

With the use of the GP algorithm, none of the obtained symbolic expressions with decay state coefficients, including the best three symbolic expressions shown in Table 12 did not include both of the decay state coefficients. The estimation performance is lower at low ship speeds and they are increasing as the ship speed also increases. Generally, lower estimation performance can be noticed for symbolic expressions with decay state coefficients when compared to those obtained without decay state coefficients. In comparison to the previous work in the field, refs [10–12] it can be seen that GP implementation in this paper achieves comparable results to other works using it. The same can be said for other researchers with similar goals, such as [36] in which the used methods achieve results that are comparable to the ones achieved by GP. In comparison to the performance of the existing work in AI-based CODLAG system modeling, which used other ML algorithms it is seen that results achieved by GP are comparable or better, with the benefit of clearer models. The clearer models in question make it possible to see which of the inputs (such as decay coefficients) ended up not being included in the best performing models signifying their low influence in the final model.

4. Conclusions

In this paper, the publicly available dataset of the CODLAG propulsion system was used in the GP algorithm to obtain the symbolic expressions for fuel flow, ship speed, starboard propeller torque, port propeller torque, and total propeller torque estimation with and without decay state coefficients. From the extensively conducted investigations, the following conclusions can be drawn:

- the Pearson's and Spearman's correlation analysis showed that from a total of 18 variables in the dataset 14 of them (without decay state coefficient, T1, and P1) have positive correlation values. The turbo compressor decay state coefficient and turbine decay state coefficient do not correlate with ship speed, have positive Pearsons correlation with starboard and port propeller torque, have positive and negative Spearman's correlation with starboard and port propeller torque, and negative correlation with fuel flow. The T1 and P1 represent ambient temperature and pressure so they are constant values throughout the entire dataset. Hence there are not any correlation values with other parameters in the dataset.
- the GP algorithm can be used to obtain symbolic expressions for estimation of fuel flow, ship speed, starboard propeller torque, port propeller torque, and total propeller torque with and without decay state coefficients for the observed CODLAG propulsion system,
- the symbolic expressions for estimation of fuel flow, ship speed, starboard propeller, port propeller and total propeller torque with decay state coefficients generally have slightly lower R^2 and slightly higher MAE values when compared to those symbolic expressions obtained without decay state coefficients. However, those symbolic expressions with decay state coefficients are more valuable from the CBM perspective which mean that they could be used to estimate or potentially predict possible degradation system states and schedule the system maintenance,
- the symbolic expressions for estimation of starboard propeller, port propeller, and total propeller torque with and without decay state coefficients showed slightly lower estimation performance for lower ship speeds.

Based on the conducted investigation, it can be concluded that the GP algorithm can be used for the estimation of CODLAG propulsion system-specific variables. The use of decay state coefficients in symbolic expressions can produce more realistic symbolic expressions which potentially could be used to predict possible performance degradation of the CODLAG propulsion system. The findings of the paper demonstrate the ability of the application of GP for the regression of the CODLAG system parameters. Academical applications are the possibility for the use of the determined equations for a precise determination of the regressed system parameters. Such an approach can greatly decrease the time necessary for the modeling of the system at various operating points. The use of GP as opposed to different AI-based modeling techniques is the shape of the generated models, which are mathematical equations, that can be easily and more simply implemented within existing or newly developed systems as they are not limited to an individual programming language or a specific library as is commonly the case. While only the CODLAG system, in particular, is modeled, the approach may be applied to different propulsion systems for which the data can be collected in future work.

Author Contributions: Conceptualization, N.A., I.P., V.M. and Z.C.; methodology, N.A., S.B.Š., I.L. and V.M.; software, N.A., S.B.Š. and I.L.; validation, I.P., V.M. and Z.C.; formal analysis, S.B.Š., I.L. and I.P.; investigation, N.A. and I.L.; resources, N.A., S.B.Š. and I.L.; data curation, S.B.Š., I.L., I.P. and V.M.; writing—original draft preparation, N.A., S.B.Š. and I.L.; writing—review and editing, I.P., V.M. and Z.C.; visualization, N.A. and V.M.; supervision, I.P. and Z.C.; project administration, V.M. and Z.C.; funding acquisition, V.M. and Z.C. All authors have read and agreed to the submitted version of the manuscript.

Funding: This research received no external funding.

Data Availability Statement: The study used a publicly available dataset obtainable at: https://archive.ics.uci.edu/ml/datasets/Condition+Based+Maintenance+of+Naval+Propulsion+Plants (accessed on 25 April 2021).

Acknowledgments: This research has been supported by the Croatian Science Foundation under the project IP- 2018-01-3739, CEEPUS network CIII-HR-0108, European Regional Development Fund under the grant KK.01.1.1.01.0009 (DATACROSS), project CEKOM under the grant KK.01.2.2.03.0004, CEI project "COVIDAi" (305.6019-20), University of Rijeka scientific grants: uniri-tehnic-18-275-1447, uniri-tehnic-18-18-1146 and uniri-tehnic-18-14.

Conflicts of Interest: The authors declare no conflict of interest.

Appendix A. Coefficients in Symbolic Expressions

The coefficient of symbolic expressions that are defined for estimation of starboard propeller torque, port propeller torque, and total propeller torque with and without decay state coefficients are given. It should be noted that the GP differently treats division, natural logarithm, and square root function during its execution to avoid infinite values and complex numbers. The division function:

$$y_{DIV}(x_1, x_2) = \begin{cases} \frac{x_1}{x_2} \text{ if } |x_2| > 0.001 \\ \frac{x_1}{x_2} = 1 \text{ if } x_2 = 0 \end{cases}. \tag{A1}$$

The natural logarithm function:

$$y_{LOG}(x_1) = \begin{cases} \log(|x_1|) \text{ if } |x_1| > 0.001 \\ \log(x_1) = 0 \text{ otherwise} \end{cases}. \tag{A2}$$

The square root function:

$$y_{SQRT}(x_1) = \sqrt{|x_1|}, \tag{A3}$$

The variables x_1 and x_2 do not have any connections with input variables that were used in symbolic expressions since they are general variable names used as arguments in previously defined functions.

Appendix A.1. Coefficients in Symbolic Expressions for Starboard Propeller Torque Estimation with Decay State Coefficients

$$\begin{aligned} X_{SPTDF11} = & \min\left(X_{13} - 3X_9, X_0\left(X_{15} - \sqrt{X_8}\right)\tan(\tan(-\tan(X_0) + X_{13} - 2X_{15}))\right) + \\ & \min(X_{13} - X_9, \tan(X_0)) + \min(X_0, \min(\cos(X_0) - 2\tan(X_0) - X_9, \\ & \tan(X_0)) - \tan(\tan(X_0)) - \tan(X_{13} - 2X_9) - X_9) \end{aligned} \tag{A4}$$

$$X_{SPTDF12} = X_0 \cos(X_{10})\left(X_9 - \sqrt{X_8}\right) + \left(X_0 - \sqrt{X_8}\right)\cos(X_{10})\left(X_9 - \sqrt{X_8}\right) \qquad \text{(A5)}$$

$$
\begin{aligned}
X_{SPTDF21} = {} & \sin(\sin((\min(X_0, X_{12}) + \sin(\min(X_0, X_{12}) + \sin(\sin(\sin(\min(X_0, X_{12}) + \\
& \sin(\min(X_0, X_{12}) + X_0) + \sin(\sin(\min(X_0, X_{12}) + \sin(X_0))) + X_0)))) + \\
& |\cos(X_{10})| + X_0 + \sin(\sin(X_0)))^{\frac{1}{2}} + \min(X_0, X_{12}) + \sin(\sin(2X_0)))) + \\
& \min(X_0, X_{12}, X_8, \tan(X_{15}))
\end{aligned}
\qquad \text{(A6)}
$$

$$
\begin{aligned}
X_{SPTDF31} = {} & \min(X_1 \cos(X_9), X_1 \cos(X_9) \log(X_1 \cos(X_6) \log(X_1 \cos^2(X_9))) \\
& \log(X_1 \cos(X_9) \cos(\sqrt{\min(X_{11}, X_{14})})), \log(X_6 \log(X_6 \cos(\log(X_6 \\
& \log(X_6 \log(\cos(X_8 X_9))) \log(\cos(X_6) \cos(X_9) \sin(\tfrac{X_7}{X_6}))))))))
\end{aligned}
\qquad \text{(A7)}
$$

Appendix A.2. Coefficients in Symbolic Expression for Starboard Propeller Torque Estimation without Decay State

$$
\begin{aligned}
X_{SPT11} = {} & \Bigg(\max(\log(X_3 - \frac{\min(X_{12}, X_2^4 \tan(X_0))}{X_{12}}), -|\sin(X_8)| \\
& - \tan(X_0)\sin(\sqrt{\tan(X_0) - X_1}) + X_1 + \sin(X_1) + X_{12} - X_9) \Bigg)^{\frac{1}{2}}
\end{aligned}
\qquad \text{(A8)}
$$

$$
\begin{aligned}
X_{SPT12} = {} & \min(X_5 \cos(\log(X_1)), \min(X_{12}, X_6) - \Bigg(\min(\sqrt{X_8} \cos(\log(X_1)), \frac{\log(X_4)}{X_2}) + \\
& \left(\frac{\log(X_4)}{X_2}\right)^{\frac{1}{2}} | \frac{X_2 X_8}{\cos(X_{11}) - \log(X_3 + X_2 X_8 - \frac{\min(X_2^2 X_5 \sin(X_1), X_5 \sin(\sin(\sin(\sin(X_1)))))}{\min(X_{12}, \frac{\log(X_4)}{X_2})})} |)
\end{aligned}
\qquad \text{(A9)}
$$

$$
X_{SPT21} = \frac{\max(X_0 X_8, \frac{X_1}{X_9^{11}} + X_{13} + \sin(X_3) + 2\sin(\log(X_3)))}{X_9^3}
\qquad \text{(A10)}
$$

$$
\begin{aligned}
X_{SPT31} = {} & \sin\Bigg(\Bigg(\max(-0.057\sin(\sqrt{X_1 X_{10}})\csc(\sin(\sin(\sin(\sqrt{X_2})))))(\log(\min(X_{12}, \\
& \sqrt{\min(X_7, X_{11} + \tan(\tfrac{X_3}{X_7}))}))) + \sin(\frac{X_{10}|X_1 X_{10} \sec(X_1)|}{\sqrt{X_2}}) + \sin(\frac{|X_1 X_{10} \sec(X_1)|}{X_{10}}) + \\
& 2\sin(\frac{|\frac{X_1 X_{10}}{\log(\tan(X_1))}|}{X_{10}}) + \sqrt{X_2}\cot(\tan(X_9))\sin(\cos(\frac{\tan(\tan(X_9))}{X_1 X_{10}})) + X_1 X_{10} \cot(\tan(X_9)) \\
& \sin(\sec(X_1)\sin(\frac{\sqrt{X_2}}{X_{10}})) + X_1 X_{10} \sin(X_1)\cot(\tan(X_9)) + X_1 X_{10} + \sin(X_1 \cot(X_9)) + \\
& 2\sin(X_1) + 4\sin(\frac{\sqrt{X_2}}{X_{10}}) + \log(X_{13}) + \sin(\sqrt{X_2})), \tan(X_9)) \Bigg)^{\frac{1}{2}}\Bigg)
\end{aligned}
\qquad \text{(A11)}
$$

$$\begin{aligned}
X_{SPT32} =\; & \log\Big(\min\big(X_{12}, \sqrt{\min(X_7, X_{11} + \tan(\tfrac{X_3}{X_7}))}\big)\Big) + \sin\Big(\frac{X_{10}|X_1 X_{10}\sec(X_1)|}{\sqrt{X_2}}\Big) + \\
& \sin\Big(\frac{|X_1 X_{10}\sec(X_1)|}{X_{10}}\Big) + 2\sin\Big(\frac{|\frac{X_1 X_{10}}{\log(\tan(X_1))}|}{X_{10}}\Big) + \sqrt{X_2}\cot(\tan(X_9)) \\
& \sin\Big(\cos\big(\frac{\tan(\tan(X_9))}{X_1 X_{10}}\big)\Big) + X_1 X_{10}\cot(\tan(X_9))\sin\big(\sec(X_1)\sin(\frac{\sqrt{X_2}}{X_{10}})\big) + \\
& X_1 X_{10}\sin(X_1)\cot(\tan(X_9)) + \sin(X_1\cot(X_9)) + \cot(X_1)\tan(X_9) + \\
& 2\sin(X_1) + 3\sin\big(\frac{\sqrt{X_2}}{X_{10}}\big) + \sin\big(\sin(\frac{\sqrt{X_2}}{X_{10}})\big) + \sin(\sqrt{X_2})
\end{aligned} \tag{A12}$$

Appendix A.3. Coefficients in Symbolic Expressions for Port Propeller Torque Estimation with Decay State Coefficients

$$\begin{aligned}
X_{PPTDF11} =\; & \max\big(X_{10} + X_{12}, X_{15}(\max(X_9(\min(\log(\log(\tan(\log(\tan(X_1)))))), \\
& \min(\tan(X_1), 2X_{10}\log(\log(X_0))|X_{12} + \tan(\log(\tan(\tan(X_1))))|)) + \\
& \log(X_0) + X_{15}(X_{14}(\tan(X_1) + X_{13}X_6) + \tan(X_1) + X_9) + \tan(X_1)) + \\
& \min(\log(\log(\tan(\log(\tan(\tan(X_1)))))), \tan(X_1)) + 2\log(\log(\tan(\log(X_{15} - X_0)))) + \\
& \log(X_9 - X_0) - \sqrt{-\sin(X_0 - X_9)} + 3\log(X_0) + 3\tan(X_1) + \\
& 5\log(\tan(\tan(X_1))) - X_{15}, \log(\tan(X_1)) + 3\tan(\log(X_1)) + 2X_{12}) + \\
& \log(\tan(\log(\sqrt{-\sin(X_0 - X_{15})} - X_0)))) + \tan(X_1))
\end{aligned} \tag{A13}$$

$$\begin{aligned}
X_{PPTDF21} =\; & (-X_{14} - \cos(X_{14}) + X_8)\min((X_8 - X_{14})^2\tan(\log(\tan(X_0)(X_8 - X_{14})^3 \\
& \tan(\log((X_8 - 0.999352)(X_8 - X_{14})(X_8 - |X_{15}|)))))), X_0) + \\
& \log((X_8 - X_{14})^3\log(X_0(X_8 - X_{14}))\tan(\log(X_1 X_{10}))\tan(\log(X_0(X_8 - X_{14})))) + \\
& \log(X_0 X_8(X_8 - X_{14})^3(\cos(X_0) - X_{14} + X_8)\tan(\log(X_1 X_{10}))\tan(\log(X_0(X_8 - X_{14}))))
\end{aligned} \tag{A14}$$

$$\begin{aligned}
X_{PPTDF31} =\; & \frac{X_{14}}{X_0(-0.181111|\sin(X_1)| - 0.181111|\sin(\sin(X_1))| + X_{13})} + \\
& \frac{X_{13}^4 X_{15}^{15}}{X_9^{13}|\sin(X_1)|^2} + \frac{X_{13}^2 X_{15}}{\sin(X_{10}) + \log(X_8)} + X_{10} - \frac{4.19971 X_9}{X_{13}} + 0.004 X_4
\end{aligned} \tag{A15}$$

Appendix A.4. Coefficients in Symbolic Expressions for Port Propeller Torque Estimation without Decay State Coefficients

$$\begin{aligned}
X_{PPT11} =\; & \Bigg(\max(X_{10}, X_{12}) + \Bigg(X_3\cos(X_1(X_9 - 0.002) + \sqrt{X_1})\cos\Big(\sqrt{X_1\cos(X_1 + X_{11})} + \\
& X_1 X_9)\cos\Bigg(\frac{\cos\Big(\sqrt{X_{12}(X_1(X_9 - 0.002)X_9 + \sqrt{X_1})\cos(\frac{\cos(X_1 + X_{11})}{X_{11} + X_9})} + X_{11} + X_9\Big)}{\cos(\frac{X_{11}}{X_{11} + X_9}) + X_9}\Bigg)\Bigg)\Bigg)^{\frac{1}{2}}\Bigg)
\end{aligned} \tag{A16}$$

$$X_{PPT21} = \Big(\log\Big(\log\Big(\frac{\log(\log(\min(\frac{X_2}{X_5^2}, \log(\log(\frac{X_0}{X_5}) + \csc(X_0)\log(X_7)))) + \csc(X_0)\log(X_7))}{X_5}\Big) + \csc(X_0)\log(X_7)\Big)\Big) \tag{A17}$$

$$X_{PPT31} = \Bigg(\frac{X_{12}^2\sin(\sin(X_0))\log\Big(\sqrt{X_{12}}\sin(X_{12})\big(\frac{X_{12}^{5/2}\sin(X_{12})}{X_1} + X_{13}X_6\big)\Big)}{X_1^2} + \sqrt{X_6}\Bigg) \tag{A18}$$

$$X_{PPT32} = \tan(\cos(X_{10})) + 2X_{11} - X_{13}X_6 + 2\tan(X_{13}) + \sqrt{X_6} \tag{A19}$$

Appendix A.5. Coefficients in Symbolic Expressions for Total Propeller Torque Estimation with Decay State Coefficients

$$
\begin{aligned}
X_{TTDF1} = &\left(\min\left(X_0, \log\left(\sqrt{X_2}\right), X_{12}\sin(\log(\log(X_{12}))) \right) + \log(X_3) \right) \\
&\left(X_{12}\sin\left(\log\left(\min\left(X_0, \log\left(\sqrt{X_2}\right), \log(X_3)\sin(\log(X_0)) \right) \right) \right) \right) \\
&- \sqrt{\log(X_1)}\left(\sqrt{-\sin(X_0 - X_{13})} - \tan(X_{15}) \right) \\
&\tan\left(\min\left(X_{15}, \log\left(\sqrt{\sqrt[4]{\log(X_1)}\sqrt{\log(X_3)}} - X_0 \right) \right. \right. \\
&\left. \left. \sqrt{\tan\left(\min\left(X_0, \log\left(\sqrt{X_2}\right) \right) \right)} \right) \right)
\end{aligned}
\tag{A20}
$$

Appendix A.6. Coefficients in Symbolic Expressions for Total Propeller Torque Estimation without Decay State Coefficients

$$
\begin{aligned}
X_{TT11} = &\left(-\sqrt{X_{10}} - \tan\left(\log(X_{10}) + \tan\left(\sqrt{X_{10}}\right) \right) + \right. \\
&\left. X_{12} - \sqrt{\frac{X_3}{X_8}} - \tan(\tan(X_8)) - X_9 \right)
\end{aligned}
\tag{A21}
$$

$$
\begin{aligned}
X_{TT12} = \max\left(\right. &-\left| \left|\left| X_{12} - \tan(\log(X_{10}) + \sqrt{X_{10}}) \right| - \tan(X_8) \right| - \tan(\sqrt{X_{10}}) \right. \\
&- \sqrt{X_{12} - X_9 - \tan(\log(X_{10}) + \tan(\sqrt{X_{10}})) - \sqrt{X_{10}} - \sqrt{\frac{X_3}{X_8}}} \bigg| \\
&- |X_{12} - 2\tan(\sqrt{X_{10}}) - \tan(X_8) - \tan(X_9 + \sqrt{X_{10}})| + |X_6|, \\
&\left. X_{12} - \sqrt{\frac{X_3}{\tan(X_8)}} \right)
\end{aligned}
\tag{A22}
$$

$$
\begin{aligned}
X_{TT21} = \max(&| \max(| \cos(\sqrt{\frac{-|X_5| - X_{12}|X_8| + \frac{\sqrt{X_2} - X_{12}(X_{12}|X_8| + \sqrt{X_2})}{\sqrt{X_2}} - \sqrt{X_2}}{X_{12}|X_8| + \sqrt{X_2}}}) |, \cos(X_6))|, \\
&\log(X_{10}) - | \max(\cos(X_6), \cos(\sqrt{\frac{-X_{12}X_8 - |X_5|}{X_{12}}}))| \\
&(\frac{\log(X_{10}) - X_{12}(\log(X_{10})\min(X_{12}, X_3) + \sqrt{X_2})}{\sqrt{X_2}} + X_{12}|X_8|))
\end{aligned}
\tag{A23}
$$

$$
\begin{aligned}
X_{TT31} = \min\left(\log(X_5), \left(\min(X_{12}, -\min(X_1 - 2X_{10} + 2\sqrt{X_{12}}, \sqrt{\cos(\sqrt{X_3})}) + \right. \right. \\
\left. \left. \sqrt{\min(X_1 + \sqrt{X_1} - X_{10}, \sqrt{X_{12}})} + X_1 - X_{10} \right)^{\frac{1}{2}} + X_1 - X_{10} \right)
\end{aligned}
\tag{A24}
$$

$$X_{TT32} = \left| \sin\left(\sin\left(\sin\left(\sin\left(\sin\left(\sin\left(\sin\left(\sin\left(\sin\left(\sin\left(\sin\left(\sin\right.\right.\right.\right.\right.\right.\right.\right.\right.\right.\right.\right.$$

$$\left(\sin\left(\sin\left(\sin\left(\sin\left(\sin\left(\sin\left(\sin\left(\sin\left(\sin\left(\sin\left(\sin\left(\sin\right.\right.\right.\right.\right.\right.\right.\right.\right.\right.\right.\right. \tag{A25}$$

$$\left.\left.\left.\left.\left(\sin\left(\sin\left(\sin\left(\sin\left(\frac{\cos X_9}{X_{12}}\right) \right|$$

References

1. Fernández, I.A.; Gómez, M.R.; Gómez, J.R.; Insua, Á.B. Review of propulsion systems on LNG carriers. *Renew. Sustain. Energy Rev.* **2017**, *67*, 1395–1411. [CrossRef]
2. Mrzljak, V.; Poljak, I.; Medica-Viola, V. Dual fuel consumption and efficiency of marine steam generators for the propulsion of LNG carrier. *Appl. Therm. Eng.* **2017**, *119*, 331–346. [CrossRef]
3. Armellini, A.; Daniotti, S.; Pinamonti, P. Gas Turbines for power generation on board of cruise ships: a possible solution to meet the new IMO regulations? *Energy Procedia* **2015**, *81*, 540–547. [CrossRef]
4. Kothamasu, R.; Huang, S.H. Adaptive Mamdani fuzzy model for condition-based maintenance. *Fuzzy Sets Syst.* **2007**, *158*, 2715–2733. [CrossRef]
5. Widodo, A.; Yang, B.S. Support vector machine in machine condition monitoring and fault diagnosis. *Mech. Syst. Signal Process.* **2007**, *21*, 2560–2574. [CrossRef]
6. Altosole, M.; Dubbioso, G.; Figari, M.; Viviani, M.; Michetti, S.; Millerani Trapani, A. Simulation of the dynamic behaviour of a CODLAG propulsion plant. In Proceedings of the WARSHIP 2010 Conference: Advanced Technologies in Naval Design and Construction, Beijing, China, 23–25 November 2010; pp. 9–10.
7. Altosole, M.; Campora, U.; Martelli, M.; Figari, M. Performance Decay Analysis of a Marine Gas Turbine Propulsion System. *J. Ship Res.* **2014**, *58*, 117–129. [CrossRef]
8. Martelli, M.; Figari, M. Real-Time model-based design for CODLAG propulsion control strategies. *Ocean. Eng.* **2017**, *141*, 265–276. [CrossRef]
9. Coraddu, A.; Oneto, L.; Ghio, A.; Savio, S.; Anguita, D.; Figari, M. Machine Learning Approaches for Improving Condition?Based Maintenance of Naval Propulsion Plants. *J. Eng. Marit. Environ.* **2014**, *230*, 136–153. [CrossRef]
10. Lorencin, I.; Anđelić, N.; Mrzljak, V.; Car, Z. Multilayer perceptron approach to condition-based maintenance of marine CODLAG propulsion system components. *Pomorstvo* **2019**, *33*, 181–190. [CrossRef]
11. Baressi Šegota, S.; Lorencin, I.; Musulin, J.; Štifanić, D.; Car, Z. Frigate Speed Estimation Using CODLAG Propulsion System Parameters and Multilayer Perceptron. *NAŠE MORE Znan. Časopis More Pomor.* **2020**, *67*, 117–125. [CrossRef]
12. Anđelić, N.; Baressi Šegota, S.; Lorencin, I.; Car, Z. Estimation of gas turbine shaft torque and fuel flow of a CODLAG propulsion system using genetic programming algorithm. *Pomorstvo* **2020**, *34*, 323–337. [CrossRef]
13. Cheliotis, M.; Lazakis, I.; Theotokatos, G. Machine learning and data-driven fault detection for ship systems operations. *Ocean. Eng.* **2020**, *216*, 107968. [CrossRef]
14. Uyanık, T.; Karatuğ, Ç.; Arslanoğlu, Y. Machine learning approach to ship fuel consumption: A case of container vessel. *Transp. Res. Part Transp. Environ.* **2020**, *84*, 102389. [CrossRef]
15. Berghout, T.; Mouss, L.H.; Bentrcia, T.; Elbouchikhi, E.; Benbouzid, M. A deep supervised learning approach for condition-based maintenance of naval propulsion systems. *Ocean. Eng.* **2021**, *221*, 108525. [CrossRef]
16. Tsaganos, G.; Nikitakos, N.; Dalaklis, D.; Ölcer, A.; Papachristos, D. Machine learning algorithms in shipping: improving engine fault detection and diagnosis via ensemble methods. *WMU J. Marit. Aff.* **2020**, *19*, 1–22. [CrossRef]
17. Bachmayer, R.; Kampmann, P.; Pleteit, H.; Busse, M.; Kirchner, F. Intelligent Propulsion. In *AI Technology for Underwater Robots*; Springer: Berlin/Heidelberg, Germany, 2020; pp. 71–82.
18. Turing, A.M. Computing machinery and intelligence. In *Parsing the Turing Test*; Springer: Berlin/Heidelberg, Germany, 2009; pp. 23–65.
19. Forsyth, R. BEAGLE—A Darwinian approach to pattern recognition. *Kybernetes* **1981**, *10*, 159–166. [CrossRef]
20. Koza, J.R. Genetic programming for economic modeling. *Stat. Comput.* **1994**, *4*, 187–197.
21. Loginov, A.; Heywood, M.; Wilson, G. Stock selection heuristics for performing frequent intraday trading with genetic programming. *Genet. Program. Evolvable Mach.* **2021**, *22*, 35–72. [CrossRef]
22. Anđelić, N.; Baressi Šegota, S.; Lorencin, I.; Jurilj, Z.; Šušteršič, T.; Blagojević, A.; Protić, A.; Ćabov, T.; Filipović, N.; Car, Z. Estimation of covid-19 epidemiology curve of the united states using genetic programming algorithm. *Int. J. Environ. Res. Public Health* **2021**, *18*, 959. [CrossRef]
23. Salgotra, R.; Gandomi, M.; Gandomi, A.H. Time series analysis and forecast of the COVID-19 pandemic in India using genetic programming. *Chaos Solitons Fractals* **2020**, *138*, 109945. [CrossRef] [PubMed]
24. Benesty, J.; Chen, J.; Huang, Y.; Cohen, I. Pearson correlation coefficient. In *Noise Reduction in Speech Processing*; Springer: Berlin/Heidelberg, Germany, 2009; pp. 1–4.

25. Schirinzi, A.; Cazzolla, A.P.; Lovero, R.; Lo Muzio, L.; Testa, N.F.; Ciavarella, D.; Palmieri, G.; Pozzessere, P.; Procacci, V.; Di Serio, F.; et al. New insights in laboratory testing for COVID-19 patients: looking for the role and predictive value of Human epididymis secretory protein 4 (HE4) and the innate immunity of the oral cavity and respiratory tract. *Microorganisms* **2020**, *8*, 1718. [CrossRef] [PubMed]
26. Lupton, R. 11. Least Squares Fitting for Linear Models. In *Statistics in Theory and Practice*; Princeton University Press: Princeton, NJ, USA, 2020; pp. 81–97.
27. Car, Z.; Baressi Šegota, S.; Anđelić, N.; Lorencin, I.; Mrzljak, V. Modeling the spread of COVID-19 infection using a multilayer perceptron. *Comput. Math. Methods Med.* **2020**, *2020*, 5714714. [CrossRef]
28. Liu, Y.; Mu, Y.; Chen, K.; Li, Y.; Guo, J. Daily activity feature selection in smart homes based on pearson correlation coefficient. *Neural Process. Lett.* **2020**, *51*, 1–17. [CrossRef]
29. Sedgwick, P. Spearman's rank correlation coefficient. *BMJ* **2014**, *349*, g7327. [CrossRef]
30. Li, C.; Yang, H.; Bao, B.; Guo, H.; Jiang, Y.; Zhang, J. Spearman Correlation Coefficient Abnormal Behavior Monitoring Technology Based on RNN in 5G Network for Smart City. In Proceedings of the 2020 International Wireless Communications and Mobile Computing (IWCMC), Limassol, Cyprus, 15–19 June 2020; pp. 1440–1442.
31. Okpala, B. A Measure of the Impact of Employee Motivation on Multicultural Team Performance Using the Spearman Rank Correlation Coefficient. *SSRN 3702059* **2020**, *2020*, 1–12. [CrossRef]
32. Poli, R.; Langdon, W.B.; McPhee, N.F.; Koza, J.R. *A Field Guide to Genetic Programming*; Lulu. com, Lulupress: Morrisville, NC, USA, 2008.
33. Strušnik, D.; Avsec, J. Artificial neural networking and fuzzy logic exergy controlling model of combined heat and power system in thermal power plant. *Energy* **2015**, *80*, 318–330. [CrossRef]
34. Fang, Y.; Li, J. A review of tournament selection in genetic programming. In *International Symposium on Intelligence Computation and Applications*; Springer: Berlin/Heidelberg, Germany, 2010; pp. 181–192.
35. Poli, R.; McPhee, N.F. Parsimony pressure made easy. In Proceedings of the 10th Annual Conference on Genetic and Evolutionary Computation, Atlanta, Georgia, 12–16 July 2008; pp. 1267–1274.
36. Agrež, M.; Avsec, J.; Strušnik, D. Entropy and exergy analysis of steam passing through an inlet steam turbine control valve assembly using artificial neural networks. *Int. J. Heat Mass Transf.* **2020**, *156*, 119897. [CrossRef]

Journal of
Marine Science and Engineering

Article

Research on Improving the Working Efficiency of Hydraulic Jet Submarine Cable Laying Machine

Zhifei Lu [1], Chen Cao [2], Yongqiang Ge [2], Jiamin He [2], Zhou Yu [2], Jiawang Chen [2,*] and Xinlong Zheng [1]

1   State Grid Zhoushan Electric Power Company, Zhoushan 316021, China; xnjdlzf@126.com (Z.L.);
    zhengxin3942607@163.com (X.Z.)
2   Ocean College, Zhejiang University, Zhoushan 316021, China; cc666@zju.edu.cn (C.C.);
    ge_yongqiang@zju.edu.cn (Y.G.); 21934190@zju.edu.cn (J.H.); oceanman@zju.edu.cn (Z.Y.)
*   Correspondence: arwang@zju.edu.cn; Tel.: +86-1866-717-1179

Abstract: The anchoring and hooking of ships, bedrock friction and biological corrosion threaten the safety and stability of submarine cables. A hydraulic jet submarine cable laying machine manages to bury the submarine cables deep into the seabed, and effectively reduces the occurrence of external damage to the submarine cables. This machine uses a hydraulic jet system to realize trenching on the seabed. However, the hydraulic jet submarine cable laying machine has complicated operation and high power consumption with high requirements on the mother ship, and it is not yet the mainstream trenching method. In this paper, a mathematical model for the hydraulic jet nozzle of the submarine cable laying machine is established, and parameters that affect the trenching efficiency are studied. The effects of jet target distance, flow, angle and nozzle spacing on the working efficiency of the burying machine are analyzed by setting up a double-nozzle model. The results of the theory, numerical simulation and experiment show that the operational efficiency of the hydraulic jet submarine cable laying machine can be distinctly improved by setting proper jet conditions and parameters.

Keywords: submarine cable; hydraulic jet; jet parameter; operation efficiency

Citation: Lu, Z.; Cao, C.; Ge, Y.; He, J.; Yu, Z.; Chen, J.; Zheng, X. Research on Improving the Working Efficiency of Hydraulic Jet Submarine Cable Laying Machine. *J. Mar. Sci. Eng.* **2021**, *9*, 745. https://doi.org/10.3390/jmse9070745

Academic Editor: Igor Poljak

Received: 1 June 2021
Accepted: 1 July 2021
Published: 5 July 2021

Publisher's Note: MDPI stays neutral with regard to jurisdictional claims in published maps and institutional affiliations.

1. Introduction

Submarine cables directly laid on the seabed are vulnerable to damage caused by the anchoring and hooking of ships, bedrock friction and biological corrosion. Among them, defects caused by the ship anchoring and hooking process account for around 95%, indicating the highest risk [1–3]. Therefore, burying the submarine cables into the seabed can effectively reduce the occurrence of external damage, making it necessary to develop a submarine cable laying machine. There are mainly two types of laying machine, namely self-propelled and towed, depending on the embodiment of trenching, while towed submarine cable laying machines can be further divided into the water jet, the plow chain wheel and the Plough type [4–6]. Compared with the other two towed submarine cable laying machines, the hydraulic jet one has a large load requirement on the mother ship, while the related equipment is complicated to operate. However, its trench depth can be adjusted, making the protection of the cable more direct and effective [7,8]. Therefore, further improvement of its operational efficiency has become a research hotspot.

Scholars at home and abroad have paid less attention to submarine cable laying machines, but research on underwater operation systems is more extensive. Mai The Vu et al. conducted analyses on the design and mechanics of a developing UTV (underwater tracked vehicle) with a rotating RC (radial component) tool for rock excavation. They analyzed the parameters that affect the performance, including the cutting forces, torque, and power requirements of the UTV with the RC tool in rock conditions for designing [9]. RC is an effective tool for trenching but will require more energy when used in a submarine cable laying machine. Simultaneously, Mai The Vu et al. conducted physical analysis of the

181

design and mechanics of a UTV with an LT (ladder trencher). They studied the factors that affect the feasibility of the UTV with LT in soft soil conditions and sought to understand the factors that affect the cutting performance to provide an improved trencher performance prediction model [10]. Compared with RC tools, LT is a more effective trenching tool in soft mud conditions. However, LT has higher requirements for installation and operability, and it is more suitable for a UTV than a submarine cable laying machine as it moves mainly through the drag of the ship. Mai The Vu et al. also described how the analytical model is derived and implemented for the design and analysis of the mechanics of a UTV with a rotating LT for cutting underwater soil by considering all target specifications [11]. The rotating LT is obviously more effective, but the limitations of its application in a submarine cable laying machine are the same as those of the ordinary LT as described above. In addition, Mai The Vu et al. showed how an analytical trenching machine model is derived and they designed and analyzed the trenching machine operation in the up-cutting operation mode. To obtain improved trenching performance modeling, the factors that affect the cutting performance of the UTV with the CB in soft soil conditions regarding the cutting-mode operation were analyzed [12]. CB is an overly complicated trenching tool that is very expensive to develop and use, while it exceeds the trenching requirements of the CD submarine cable laying machine. In summary, trenching methods are complex and diverse, but they do not meet the actual needs of submarine cable engineering with high operating costs. In applications, as a practical tool, further improving the operational efficiency of the hydraulic jet is more important than using other complex methods.

The State Grid Zhoushan Power Supply Company has a professional construction team for submarine power cable laying in China, equipped with the most advanced and dedicated submarine cable laying construction ship, Qifan No. 9. This workboat adopts a self-developed hydraulic jet submarine cable laying machine to dig trenches. The submarine cable laying construction ship Qifan No. 9 and the hydraulic jet submarine power cable laying machine are shown in Figures 1 and 2, respectively.

Figure 1. Qifan No. 9 with a cable capacity of 5000 tons.

Figure 2. The hydraulic jet submarine cable laying machine.

In this paper, the influences of nozzle standoff distance, jet flow rate, jet angle and nozzle spacing on the trench depth and width of the laying machine are numerically and theoretically analyzed, based on which factors that affect the operational efficiency of the laying machine are determined. The theoretical and numerical simulation analysis are then verified by experiments, and from the above investigation, the design parameters of the hydraulic jet nozzle are given.

2. Mechanism Analysis

The schematic diagram of the soil-breaking of a hydraulic impinging jet is shown in Figure 3. With the impinging of the hydraulic jet, the disturbing of the soil depends on the characteristics of both the jet and soil. One of the most important parameters considered here is the resistance in the process of soil-breaking, which is called the critical failure pressure of soil. The critical failure pressure of soil under the action of jet flow is related to the soil particle size, permeability, density and other parameters, expressed as follows [13–16]:

$$F_{cr} = \beta \tau_f^2 \left(\frac{d_{60}}{k}\right)^{-2} \gamma_d^{-1} \tag{1}$$

where F_{cr} is the critical jet pressure on the failure surface, τ_f is the shear strength of soil, d_{60} is the soil particles' limited size, γ_d is the dry unit weight of soil, k is the soil permeability coefficient, d_{60}/k is the erosion resistance of soil, and N and β are correction factors. It was experimentally determined that $\beta = 1.8 \times 10^{13}$ [17].

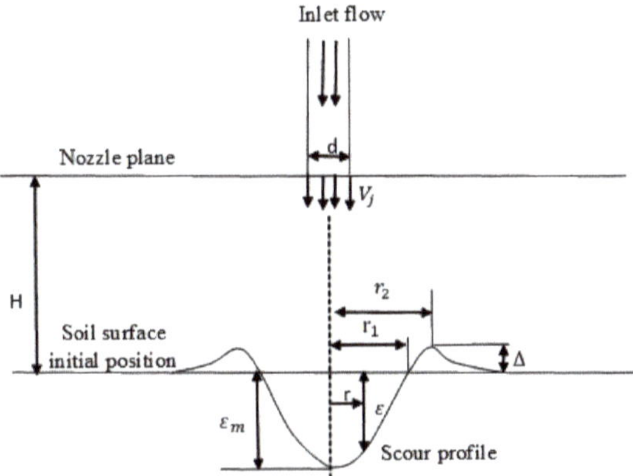

Figure 3. The schematic diagram of soil-breaking of vertical impinging jet.

Equation (1) is an empirical model obtained from experimental research and is only related to the properties of soil. The condition of hydraulic flow rate should also be taken into account considering that the jet flow at the nozzle tip is perpendicular to the soil surface. The total pressure in the half width range is:

$$F_b = \int_0^{b_{1/2}} p(y) \cdot 2\pi y\pi dy = \frac{5}{7}\pi b_{1/2}^2 P_m \tag{2}$$

where $b_{1/2}$ is the half width and thickness of the jet, and P_m is the dynamic pressure at the center of the jet stream, which can be obtained as:

$$P_m = \frac{1}{2}\rho u_{max}^2 \tag{3}$$

The average stress within the half width can be calculated as:

$$\overline{P_b} = \frac{F_b}{S} = \frac{5}{7}P_m = \frac{45\rho Q^2 u^2}{896\pi^2 v^2 l^2} \tag{4}$$

where Q is the flow rate, u is the nozzle exit velocity, v is the hydrodynamic viscosity and l is the distance from the nozzle to the jet surface.

In reality, there is a certain inclination angle between the scouring jet and the stressed soil surface. The diagram is shown in Figure 4.

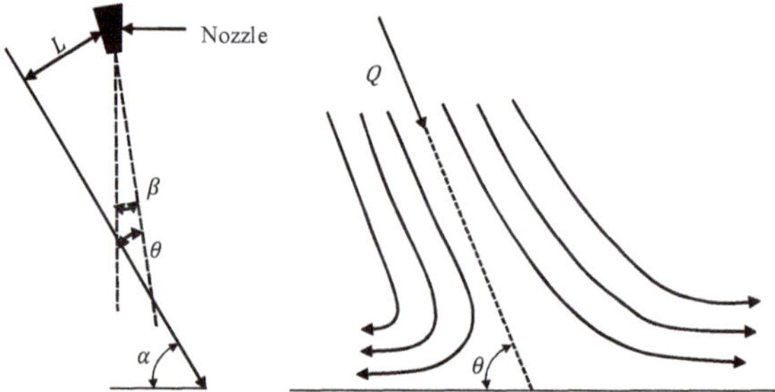

Figure 4. The schematic diagram of soil-breaking of oblique impinging jet.

The curved surface equation of the scour surface is:

$$(x \tan q + z - L \sin q)^2 - [x^2 + y^2 + (z - L \sin q)] \sin a (\tan^2 q + 1) = 0 \tag{5}$$

The equation of the scour surface contour curve is:

$$\left(\sqrt{x^2 + y^2 + (L \sin \alpha)^2} - L \cos \alpha\right)^2 + (L \sin \alpha)^2 = (x - L \cos \alpha)^2 + y^2 \tag{6}$$

where θ is the supplementary angle of the jet scour angle, L is the distance between the jet pole and the jet hitting the surface along the direction of the jet, α is the angle of jet expansion and $\{x, y, z\}$ is the coordinate of any point on the surface.

The average force acting on the jet plane during the tilting scour is [18,19]:

$$\overline{P_b} = \frac{\rho Q u}{f(L, \theta)} \tag{7}$$

where $f(L,\theta)$ is jet area. Research shows that the sediment settlement is faster when the scour angle of the nozzle is increased, so the post-spray should be considered in the actual scour to wash off the suspended sediment.

The formula of the bed-load transport rate is as follows [20,21]:

$$q_b = \frac{\pi}{6} \rho_s d u_b p \tag{8}$$

where ρ_s is the sediment density, u_b is the bottom critical average velocity of sediment entrainment, and p is the probability of sediment entrainment, $p = nd^3$. It can be inferred that the sediment transport rate is mainly related to the flow velocity on the surface of the sand bed, so increasing the flow velocity can enhance the sediment transport volume.

The essential condition for the destruction of the upper body under the jet impinging is that the jet impact force of the upper body is greater than the critical failure pressure. In other words, the average force within the half width range is larger than the critical failure pressure of the soil.

3. Numerical Simulation Analysis

3.1. Finite Element Method

The process of jet trenching is a complex solid–liquid two-phase flow problem. In this paper, the Euler multiphase flow model is adopted [22–24]. Its continuity equation is expressed as:

$$\frac{\partial}{\partial t}\int \alpha_i\rho_i\chi dV + \oint_A \alpha_i\rho_i\chi(v_i - v_g)\cdot da = \int_V \sum_{j\neq i}(m_{ij} - m_{ji})\chi dV + \int_V S_i^a dV \tag{9}$$

where α_i is the volume fraction of phase i, ρ_i is the density of phase i, χ is the cavitation rate, v_i is the rate of phase i, v_g is the grid velocity, m_{ij} is the mass transfer from phase j to phase i, m_{ji} is the mass transfer from phase i to phase j, and S_i^α is the quality source term. In addition, the volume fraction satisfies: $\Sigma_i\alpha_i = 1$. The momentum equation of multiphase separation flow is:

$$\begin{aligned}
\frac{\partial}{\partial t}\int \alpha_i\rho_i\chi dV + \oint_A \alpha_i\rho_i\chi(v_i - v_g)\cdot da = \\
-\int_V \alpha_i\chi\nabla\rho dV + \int_V \alpha_i\rho_i\chi g dV + \oint_A [\alpha_i(\tau_i + \tau_i^t)]\chi\cdot da \\
+\int_V M\chi dV + \int_V \iint_V S_i^a dV \int_V \Sigma\Sigma(m_{ij}v_j - m_{ji}v_i)\chi dV
\end{aligned} \tag{10}$$

where p is pressure, assuming that it is equal in the two phases; g is the acceleration vector; τ_i is molecular stress; τ_i^t is turbulent stress; M_i is the interphase momentum transfer per unit volume; $(F_{int})_i$ is the internal force; S_i^v is the phase quality source term; m_{ij} is the mass transfer rate from phase j to phase i, and m_{ji} is the mass transfer rate from phase i to phase j. The interphase momentum transfer represents all the forces acting from phase to phase and satisfies the following equation:

$$\sum_i M_i = 0 \tag{11}$$

To simplify the simulation computation, the following assumptions are made when establishing the numerical simulation model, on the premise of meeting the simulation requirements: (1) soil is an isotropic medium; (2) the fluid is incompressible; (3) the influence of the ocean current on the soil-breaking of the hydraulic jet is ignored. (Based on the above three assumptions and formulas, we can use the multiphase separation flow model to simulate the trenching process through a hydraulic jet.) The ocean sediment in the case of soil-breaking is mainly silt and sand, so we select clayey sand powder as the simulation object. The soil parameters are shown in Table 1.

Table 1. The clayey sand powder soil parameters.

Soil Parameters	Value
shear strength τ_f	54 kPa
critical pressure of failure surface F_{cr}	0.23 MPa
soil particles limited size d_{60}	1.2 mm
density ρ	2560 kg/m^3
turbulent Prandtl number P_r	0.9
particle distribution size (Sauter average diameter)	1 mm

The dual-nozzle numerical simulation model is shown in Figure 5. The three-dimensional model simulates the soil-breaking of the hydraulic jet in still water, including the nozzle, cement interface and bottom mud [25,26]. During the simulation, some parameters, such as standoff distance, flow rate, jet angle and nozzle spacing, are adjusted according to the change in the study object [27].

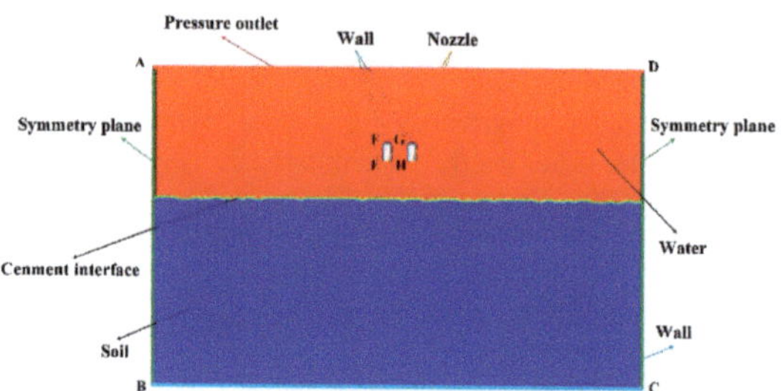

Figure 5. The dual-nozzle numerical simulation model.

We use STAR-CCM + [28,29] to simulate and analyze the effects of jet target distance, jet flow, jet angle and nozzle spacing on the jet trenching effect in a 2D plane, which is mainly judged based on the depth and width of the trenching.

3.2. Simulation Results

3.2.1. Influence of Standoff Distance on Jet Flow Effect

We set the nozzle angle $\theta = 90°$, nozzle diameter d = 60 mm, nozzle spacing as 300 mm and jet flow rate as 1.187 m^3/min and analyze the scour depth and width at different jet standoff distances. The analysis results are shown in Figures 6 and 7.

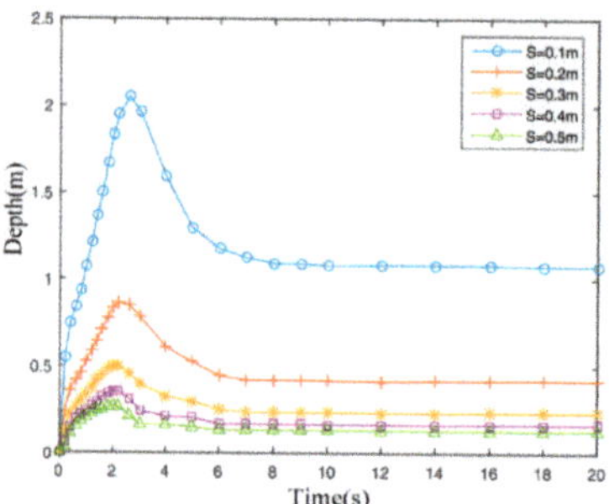

Figure 6. The relationship between scour depth and time at different standoff distances.

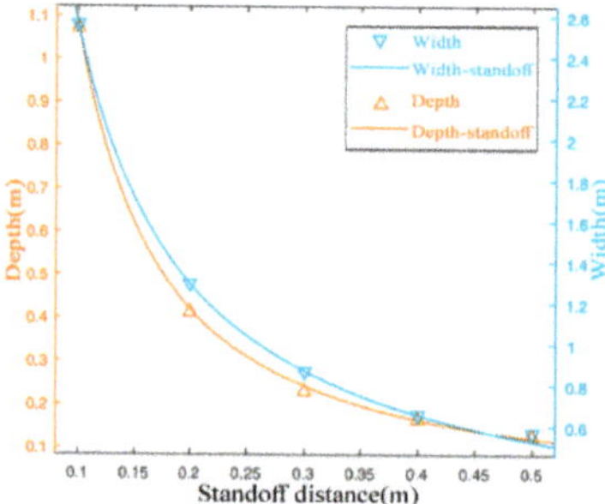

Figure 7. The correlation curve of scouring performance and standoff distance.

It is not difficult to see from the figure that the scouring depth reaches the maximum at around 3 s under different target distances, and then the scouring depth decreases and stabilizes with the siltation of the soil. The increase in the standoff distance will reduce the depth and width of the scouring. When the target distance is 0.1 m, the scouring depth and scouring width reach the maximum.

3.2.2. Influence of Jet Flow Rate on Jet Flow Effect

We set the nozzle angle θ = 90°, nozzle diameter d = 60 mm, nozzle spacing as 300 mm, standoff distance as 300 mm and jet flow rate ranging from 3 m/s to 20 m/s—that is, jet flow rate changing from 0.509 m^3/min to 3.393 m^3/min. We analyze the scour depth and width at different jet flow rates and Figures 8 and 9 show the numerical simulation results.

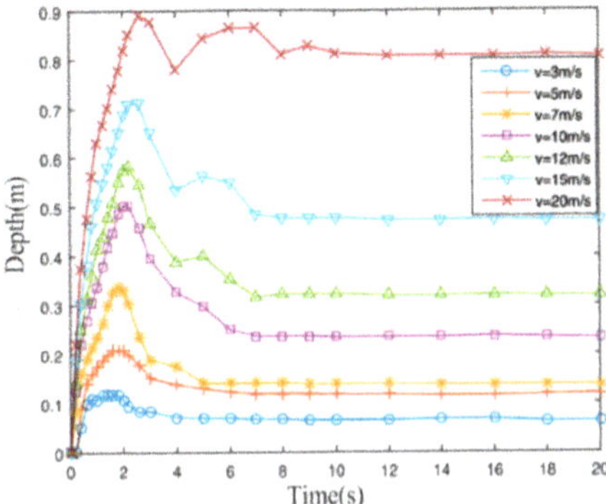

Figure 8. The relationship between scour depth and time at different flow rates.

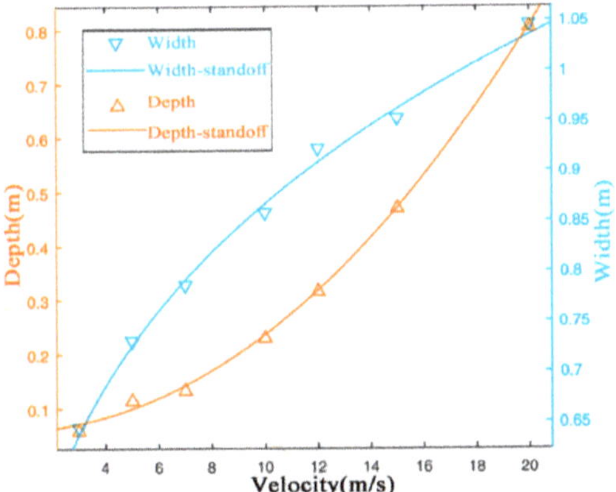

Figure 9. The correlation curve of scouring performance and flow rates.

It can be seen from the figure that as the flow rate increases, the scouring depth and width will increase, while the time for the scouring depth to stabilize will become longer. Once the flow rate is greater than 10 m/s, there will be two scouring effects—that is, the scouring depth will increase again, which is unfavorable. Therefore, the flow velocity is selected as 7 m/s in the subsequent experiments—that is, the flow rate is 1.187 m^3/min.

3.2.3. Influence of Jet Angle on Jet Flow Effect

We set the nozzle diameter d = 60 mm, nozzle spacing as 300 mm, standoff distance as 300 mm and jet flow rate as 1.187 m^3/min. We analyze the scour depth and width at different jet angles and Figures 10–13 show the numerical simulation results.

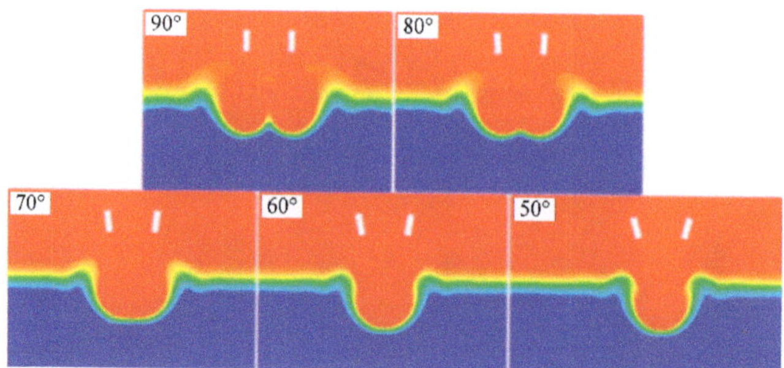

Figure 10. Numerical simulation results of different jet angles.

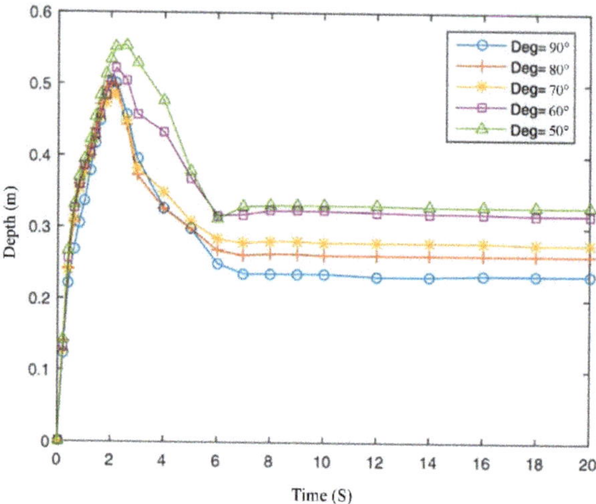

Figure 11. The relationship between scour depth and time at different jet angles.

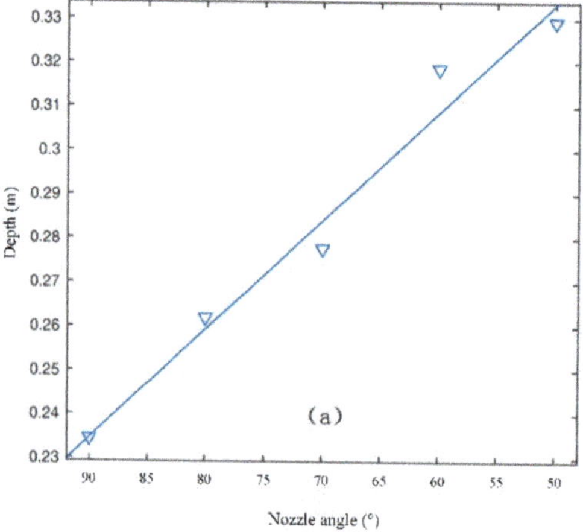

Figure 12. The correlation curve of scour depth and jet angle.

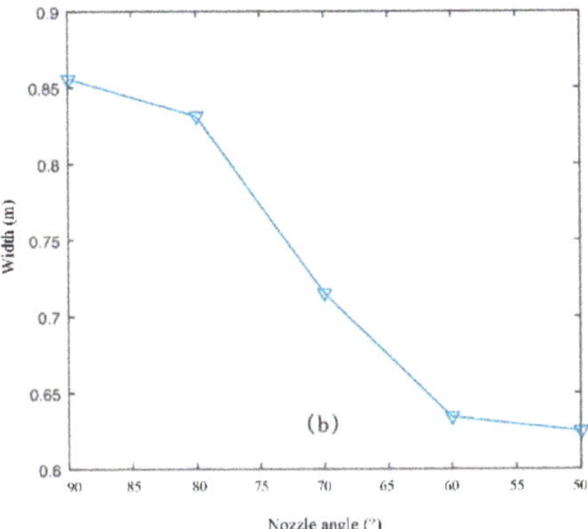

Figure 13. The correlation curve of scour width and jet angle.

Obviously, as the nozzle angle decreases, the scour depth increases, but the width decreases. Thus, the selection of the nozzle angle is analyzed in detail.

3.2.4. Influence of Jet Spacing on Jet Flow Effect

We set the nozzle angle $\theta = 90°$, nozzle diameter d = 60 mm, standoff distance as 300 mm and jet flow rate as 1.187 m^3/min. We analyze the scour depth and width at different jet spacing values and Figures 14–16 show the numerical simulation results.

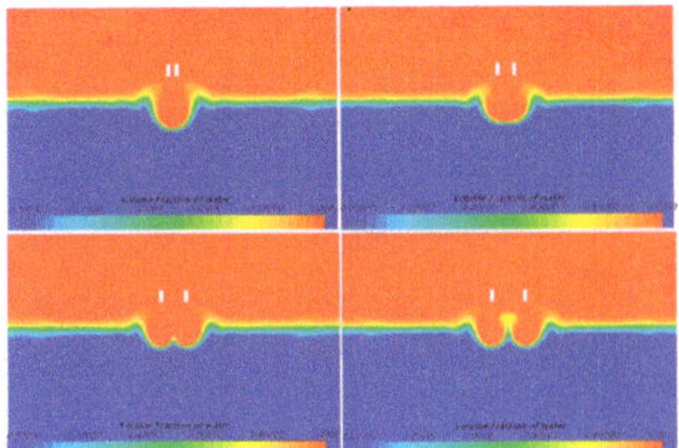

Figure 14. The correlation curve of scour width and jet angle.

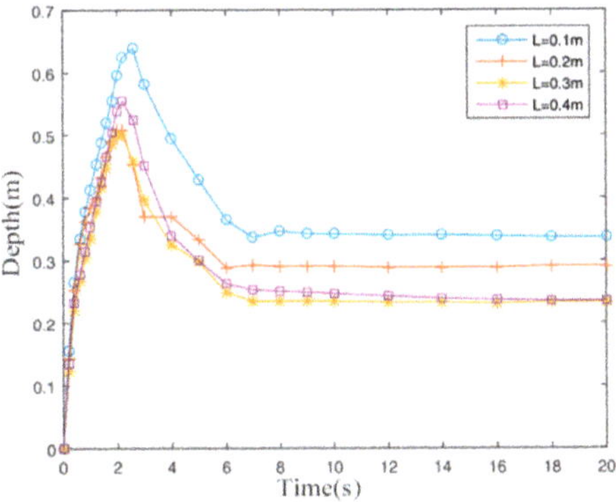

Figure 15. The relationship between scour depth and time at different jet spacing values.

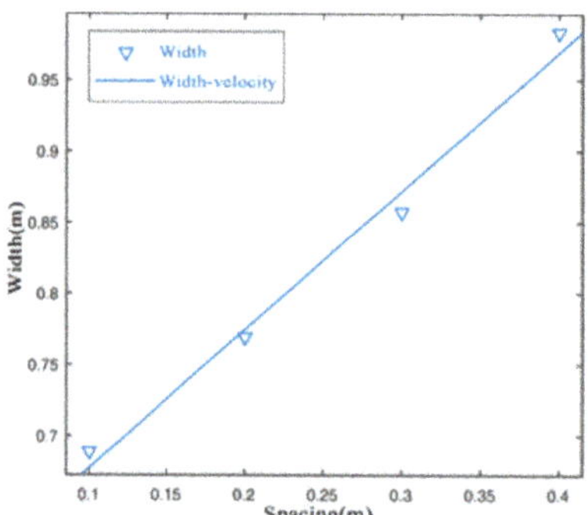

Figure 16. The correlation curve of scour width and jet spacing.

It can be obtained from the figure that the scour depth decreases as the nozzle spacing increases. When the distance is greater than 0.3 m, the scouring depth will no longer change. Correspondingly, the scouring width will increase accordingly, but when the spacing is greater than 0.3 m, there will be siltation of unscoured soil in the middle of the trench, which is obviously undesirable.

4. Experiment Analysis

The jet parameters of the numerical simulation analysis are set based on actual working conditions, which are difficult to establish under experimental conditions. Therefore, a model experiment is conducted that follows the Froude similarity criterion [30,31]. In hydrodynamics, the Froude number is expressed as the ratio of the inertial force and gravity of the fluid. Therefore, the prototype is scaled down according to the similarity principle, the similarity model is observed and analyzed, and then the results of the model experiment are converted to the engineering laying machine, thus obtaining the analytical results of the engineering machine. The experimental parameters are shown in Table 2.

Table 2. Experimental parameters and numerical simulation parameters.

Related Parameters	Experimental Prototype	Simulation Parameters (Engineering Embedding Machine)
scale factor	1	10
nozzle diameter	6 mm	60 mm
jet flow rate	$(1.61–10.73) \times 10^{-3}$ m^3/min	$(0.509–3.393)$ m^3/min
jet standoff distance	0–140 mm	0–1400 mm
jet angle	0–90°	0–90°
shear strength of soil	5.4 kPa	54 kPa

4.1. Design of Experiment Platform

As shown in Figures 17 and 18, the influences of jet flow rate, jet standoff distance and jet angle on trenching morphology and trenching depth [32] are studied using the experimental platform for the soil-breaking of the hydraulic jet. The experimental platform is composed of the experimental substrate, water tank, sandbox, bracket, pump and its auxiliary facilities, the driving and debugging system, as well as the observation and measurement system. Among them, the experimental soil samples are prepared in batches according to the unified production standard through a certain proportion of kaolin and water. Moreover, the jet flow of the nozzle is controlled by a water pump and speed-regulating valve, the standoff distance and angle of the jet are adjusted by swinging the support, and the depth and width of the jet are recorded by an HD camera.

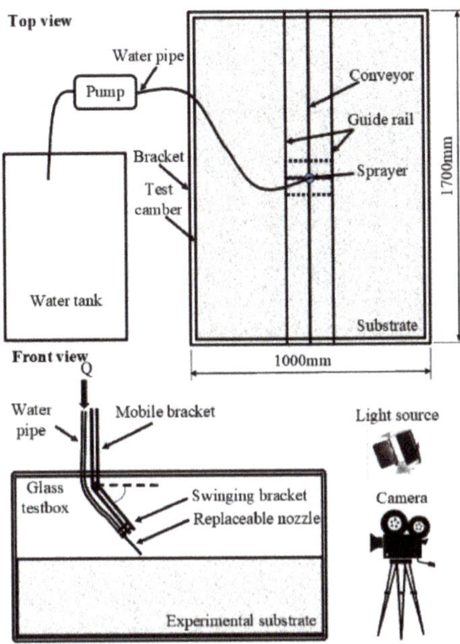

Figure 17. Schematic diagram of experimental analysis platform and observation platform.

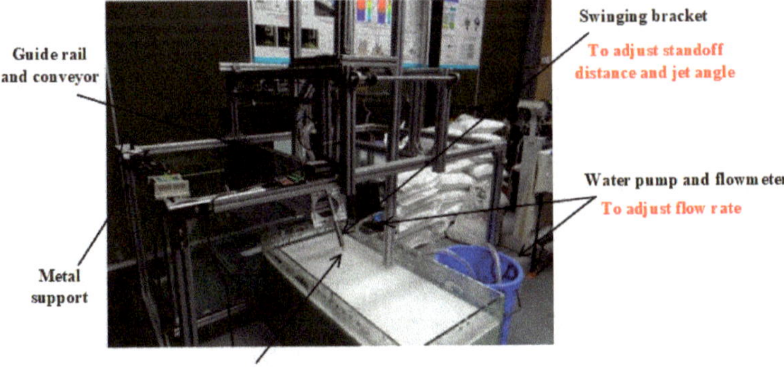

Figure 18. The experimental platform for soil-breaking of hydraulic jet.

Experiments include a static scouring experiment and dynamic moving scouring experiment. In the static scouring experiment, the influence of different scour angles on the scour depth and width, as well as the influence of the jet standoff distance and jet flow rate on the scour effect, are investigated. In the dynamic scouring experiment, the influence of different scour angles on the scour depth and width, and the influence of the jet standoff distance and jet flow rate, are studied when the nozzle is moving horizontally.

4.2. Experimental Results

4.2.1. Static Scouring Experiment

Firstly, the simulation and experimental results are compared from two aspects of the scour pit shape and depth to verify the reliability of the simulation results. According to the numerical simulation results, experimental restriction conditions are set as shown in Table 3. Table 4 compares the experimental and simulation results of the scour pit depth and width, and the simulation data refer to the scaling criterion to scale the original simulation results.

Table 3. Experimental restriction conditions.

Parameters	Value
scour angle	90°
jet standoff distance	30 mm
scour flow rate	3.75×10^{-3} m^3/min
nozzle diameter	6 mm

Table 4. Comparison between numerical simulation results and experimental results.

Nozzle Diameters	Scour Depth (mm)		Scour Width (mm)	
	Simulation	Experiment	Simulation	Experiment
60 mm/6 mm	193.23	198	84.24	78

The development trend of the scour pit depth and width of the simulation results is close to the results from the experiment. However, the scour depth from the simulation is relatively small compared with the experimental value, which is mainly caused by the wall effect of the glass tank [14,15]. On the other hand, the simulated scouring width is larger than the experimental result, which is due to the certain deviation in the smoothness of the sand surface and the viscosity of the glass wall to the fine sand, causing the width of the scouring pit on the upper part to narrow during the scouring.

(a) Influence of jet standoff distance on scour effect

According to the curve trend in Figure 7 of the single-nozzle simulation, once the standoff distance exceeds 600 mm, the scouring depth will further decrease, even not exceeding 200 mm, which is not in line with our ideal situation. Therefore, in the experiment, 8 standoff distances are set to verify the changes in scour pit depth and width at different standoff distances of 0–70 mm, where the experiment is repeated twice for each scour condition.

As shown in Table 5 and Figure 19, the scour pit depth decreases with the increase in the jet standoff distance, while the scour pit width increases. Here, the jet is submerged, leading to an entrainment flow during the spraying process. Currently, the flow rate increases while the average velocity decreases. With the increase in the jet standoff distance, the hydraulic jet flow expands along the direction of the jet. Although the shear generated by the flow velocity decreases, the scour flow rate and effective area increase, resulting in a decrease in the scour pit depth and increase in the scour width [30,33,34].

Table 5. Scouring effect at different standoff distances of nozzle.

Working Condition	Flow Rate ($\times 10^{-3}$ m^3/min)	Nozzle Diameter (mm)	Standoff Distance (mm)	Angle (°)	Scouring Width (mm)	Scouring Depth (mm)
1	3.75	6	0	90	63.3	229.5
2	3.75	6	10	90	68.1	221.7
3	3.75	6	20	90	73.5	210.3
4	3.75	6	30	90	89.5	196.8
5	3.75	6	40	90	111.3	177.1
6	3.75	6	50	90	129.3	147.4
7	3.75	6	60	90	165.2	116.3
8	3.75	6	70	90	209.4	77.3

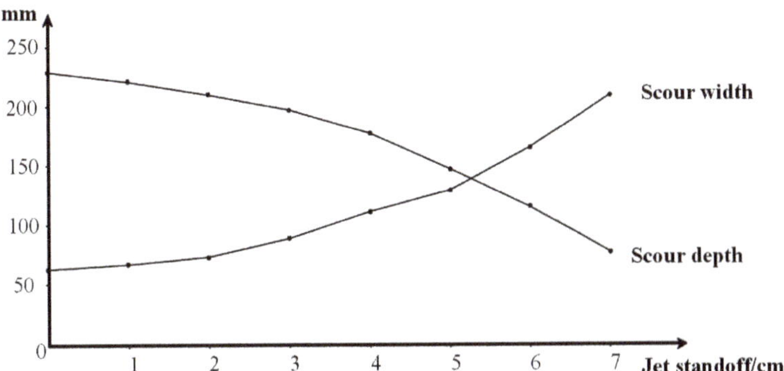

Figure 19. Scour depth and width at different standoff distances of the nozzle.

(b) Influence of jet angle on scouring effect

According to the results of the single-nozzle simulation, when the jet inclination angle is 60–90°, the scour depth is relatively deep, with little influence on the side wall. When the jet inclination angle is less than 60°, the jet depth decreases sharply. Therefore, 4 scour conditions are set in this experiment, where the jet angle ranges from 60° to 90°. The experiment is repeated twice for each scour condition.

As shown in Table 6 and Figure 20, the maximum scour depth can be achieved at the jet angle of 90°. The scour pit width is the largest when the scour angle is 60°, while the scour pit widths are similar at other different scour angles.

Table 6. Scouring effect at different jet angles.

Working Condition	Flow Rate ($\times 10^{-3}$ m³/min)	Nozzle Diameter (mm)	Standoff Distance (mm)	Angle (°)	Scouring Width (mm)	Scouring Depth (mm)
1	3.75	6	30	90	159.3	147.6
2	3.75	6	30	80	180.5	108.4
3	3.75	6	30	70	187.0	93.0
4	3.75	6	30	60	197.7	89.3

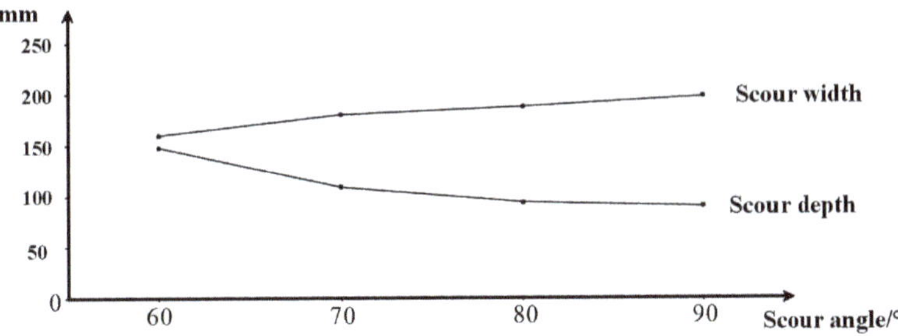

Figure 20. Scour depth and width at different jet inclination angles.

(c) Influence of jet flow rate on scour effect

In order to verify the changes in the scour pit depth and width at different jet flow rates, 6 scour conditions are set, and the experiment is repeated twice for each scour condition. Table 7 and Figure 21 show the scour pit depths and widths under different jet flow rate conditions.

Table 7. Scouring effect at different jet flow rate.

Working Condition	Flow Rate (×10⁻³ m³/min)	Nozzle Diameter (mm)	Standoff Distance (mm)	Angle (°)	Scouring Width (mm)	Scouring Depth (mm)
1	1.61	6	30	90	108.5	74.9
2	2.682	6	30	90	153.1	83.7
3	3.754	6	30	90	198.0	89.8
4	5.363	6	30	90	240.1	108.0
5	6.438	6	30	90	281.7	134.6
6	8.048	6	30	90	291.1	174.1
7	10.73	6	30	90	290.4	218.9

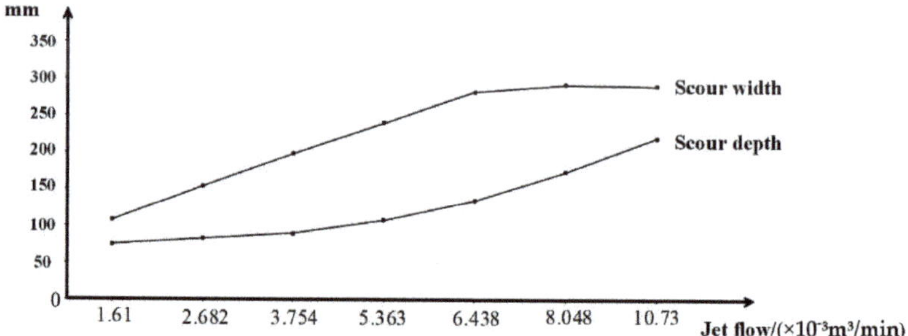

Figure 21. Scour depth and width at different jet flows.

The scouring depth does not increase with the increasing flow rate within the experimental range. In working conditions 6 and 7, the scour pit depth is basically the same, but the scour pit width in working condition 7 is larger. Moreover, compared with conditions 5 and 6, the scour pit depth changes slowly, but the scour pit width changes at a higher rate. This is because at a low flow rate, the scour is mainly achieved by the friction between the flow and the sediment surface, so, within this range, the greater the flow rate, the greater the scour pit depth. Once the velocity exceeds a certain value, the intensity of the hydraulic jet flow penetrating the water is enough to generate a turbulence vortex [25,35], thus decreasing the scour pit depth.

4.2.2. Dynamic Scouring Experiment

The dynamic scouring experiment mainly analyzes the influence of different scour angles and moving speeds of the scour platform on the scour effect. The maximum working speed of the laying machine is set as 150 m/h—that is, the maximum moving speed is 41.7 mm/s. Thus, 4 scour conditions are set in the experiment, as shown in Table 8, where the experiment is repeated twice for each group.

Table 8. Scouring effect of dynamic scour.

Working Condition	Flow Rate (×10⁻³ m³/min)	Nozzle Diameter (mm)	Standoff Distance (mm)	Moving Speed (mm/s)	Angle (°)
1	3.75	6	30	10	90
2	3.75	6	30	20	90
3	3.75	6	30	30	60
4	3.75	6	30	30	90
5	3.75	6	30	41.7	60

Figure 22 shows the change in the scour shape in the experiment at working condition 2. A deep scour pit is observed at the early stage of the experiment. Later, the overall scour depth decreases, and it is basically the same in the nozzle moving path. At the beginning of the scour, it can be approximated as static scour, forming a deeper scour pit. As the nozzle moves, the tilting nozzle flushes the sediment to the scour pits previously formed, resulting in a decrease in the scour depth [16].

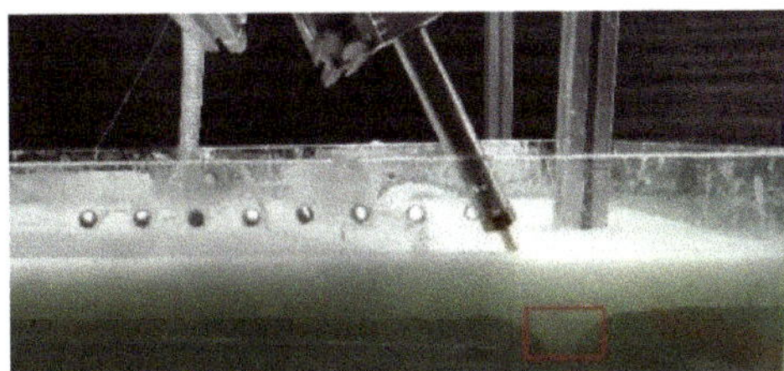

(a) Location 1

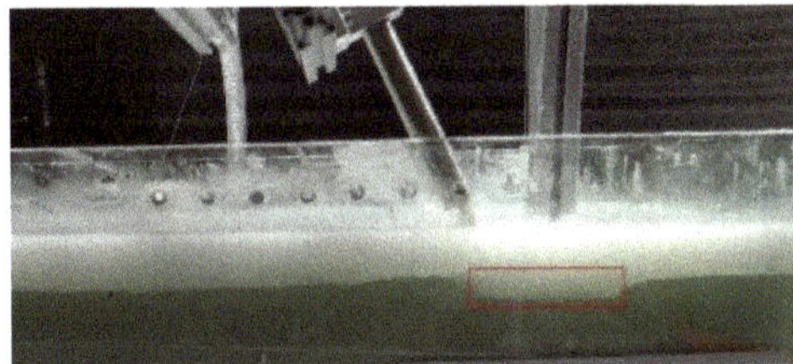

(b) Location 2

(c) Location 3

Figure 22. The scour contour of working condition 2.

Compared with the working condition 4 in the nozzle inclination variable experiment, the two working conditions only differ in the moving speed. It is obvious that the depth of the scour pit in the dynamic scouring condition is less than that in the static scouring condition. In the static scouring experiment, there is a lot of suspended sand in the scour pit during the scour process, and most of the suspended sand is settled during the mobile scour measurement, so the depth of the scour pit is smaller during dynamic scouring.

By comparing the results of working condition 3 and 4, it is found that the scour depth increases slightly when the scour angle increases. According to the results, the scour in all directions is equally difficult during the static scour, so the inclined scour has no advantage in the static scour experiment. On the contrary, it is easiest to scour the sediment backward for the dynamic scour experiment, as the inclined nozzle can flush the sediment into the scour pit formed by the previous period. Therefore, in the mobile scour experiment, increasing the jet inclination angle within a certain range can increase the scour sludge discharge effect.

Compared with the working conditions 1, 2, 3, and 5, the scour depth decreases slightly when the moving speed of the scour platform increases. In other words, for the substrate used in the experiment, the traveling speed change within a certain range has little influence on the trench depth and width.

5. Conclusions

The research shows that the depth of jet trenching first increases rapidly to a maximum value in a short period of time. As previously disturbed sediment is backfilled into the pit, the trench depth decreases to some extent. After some fluctuations, the trench depth finally settles at a certain value. The simulation result of sediment backfill is shown in Figure 23, and the backfilling effect is more obvious with the larger jet dip angle.

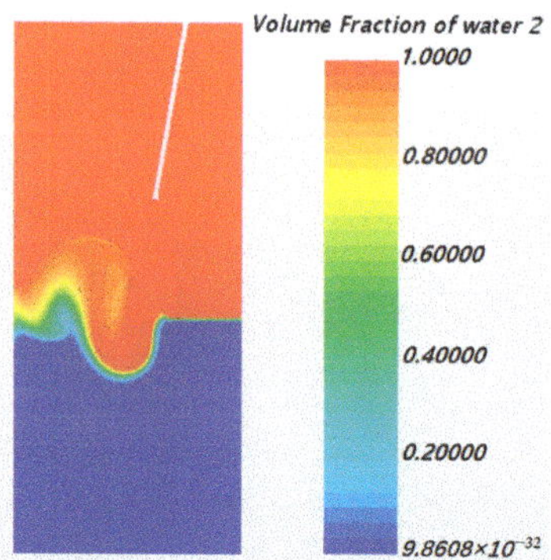

Figure 23. The numerical simulation result of sediment backfills.

The experimental results show that, under the same jet conditions, the greater the standoff distance from the nozzle to the sediment surface, the greater the depth and width of the jet trenching will be, and the relationship between scouring performance and the standoff distance can be matched in terms of two exponential functions. Moreover, under the same jet conditions, with the increase in the fluid flow velocity, the depth and width of the trench will increase, the change rate of the trench depth will gradually increase, and the change rate of the trench width will gradually decrease. Their relationships can be fit into two opposite exponential curves. The jet angle also has an influence on the trench effect, and the trench depth of the trenching increases with the increase in the angle within 0–40°. Moreover, the trench depth can be improved by changing the spacing between nozzles. When the spacing increases to a certain value, the double-nozzle jet system becomes two independent single-nozzle jet systems, and its influence on the jet trenching depth becomes very small. The width of the jet is linearly related to the spacing between nozzles, and they change in the same direction.

The fitting functions of the burying results (depth and width) and jet flow rate, jet standoff distance, jet spacing and jet angle are shown in Table 9. The trenching ability of the laying machine can be effectively enhanced by adjusting the jet flow velocity and the jet standoff distance. When the operating power of the laying machine is constant, the working efficiency of the laying machine can be effectively improved by changing the configuration of the spray arm nozzle, such as the nozzle angle and nozzle spacing.

Table 9. Relation fitting functions.

Curve	Fitting Function
depth–velocity	$f(x) = 0.0015x^{2.071} + 0.0572$
width–velocity	$f(x) = 0.4808x^{0.2559}$
depth–standoff distance	$f(x) = 0.047x^{-1.365}$
width–standoff distance	$f(x) = 0.2671x^{-0.9791}$
depth–nozzle spacing	$f(x) = 0.974x + 0.58$
depth–jet angle	$f(x) = 0.0025x + 0.235$

In the future, the authors will adjust the parameters (such as jet target distance, flow, angle, and nozzle spacing, etc.) of the hydraulic jet submarine cable laying machine on Qifan No. 9. On this basis, we will study whether the working efficiency of the machine has been significantly improved in actual applications.

Author Contributions: Methodology, Z.Y. and Z.L.; formal analysis, C.C., J.H. and Y.G.; investigation, J.H. and Y.G.; resources, Z.L.; data curation, Z.Y. and Y.G.; writing—original draft preparation, C.C.; writing—review and editing, C.C.; visualization, C.C.; project administration, J.C. and X.Z.; funding acquisition, J.C. and X.Z. All authors have read and agreed to the published version of the manuscript.

Funding: This research was supported by the Key Research and Development Project of Zhejiang Province (2019C03115).

Conflicts of Interest: The authors declare no conflict of interest.

References

1. Mei, X.; He, C.; Huang, X. Study on sea surface monitoring scheme of submarine cable routing in Hainan Networking Project. *China Water Transp.* **2011**, *11*, 86–87.
2. Szyrowski, T.; Sharma, S.; Sutton, R.; A Kennedy, G. Developments in subsea power and telecommunication cables detection: Part 2—Electromagnetic detection. *Underw. Technol.* **2013**, *31*, 133–143. [CrossRef]
3. Carter, L.; Gavey, R.; Talling, P.; Liu, J. Insights into Submarine Geohazards from Breaks in Subsea Telecommunication Cables. *Oceanography* **2014**, *27*, 58–67. [CrossRef]
4. Negishi, Y.; Ishihara, K.; Murakami, Y.; Yoshizawa, N. Design of Deep-Sea Submarine Optical Fiber Cable. *IEEE J. Sel. Areas Commun.* **1984**, *2*, 879–885. [CrossRef]
5. Zajac, E.E. Dynamics and Kinematics of the Laying and Recovery of Submarine Cable. *Bell Syst. Tech. J.* **1957**, *36*, 1129–1207. [CrossRef]
6. Choi, J.K.; Yokobiki, T.; Kawaguchi, K. Automated cable-laying system rov-based installation of donet2 ocean floor observatory. *Sea Technol.* **2015**, *56*, 43–55.
7. Kurashima, T.; Horiguchi, T.; Yoshizawa, N.; Tada, H.; Tateda, M. Measurement of distributed strain due to laying and recovery of submarine optical fiber cable. *Appl. Opt.* **1991**, *30*, 334–337. [CrossRef]
8. Prpi-Ori, J.; Nabergoj, R. Nonlinear dynamics of an elastic cable during laying operations in rough sea. *Appl. Ocean. Res.* **2005**, *27*, 255–264. [CrossRef]
9. Vu, M.T.; Choi, H.; Ji, D.H.; Jeong, S.-K.; Kim, J.Y. A study on an up-milling rock crushing tool operation of an underwater tracked vehicle. *Proc. Inst. Mech. Eng. Part M J. Eng. Marit. Environ.* **2019**, *233*, 283–300. [CrossRef]
10. Vu, M.T.; Choi, H.-S.; Kim, J.Y.; Tran, N.H. A study on an underwater tracked vehicle with a ladder trencher. *Ocean Eng.* **2016**, *127*, 90–102. [CrossRef]
11. Vu, M.T.; Jeong, S.-K.; Choi, H.-S.; Oh, J.-Y.; Ji, D.-H. Study on down-cutting ladder trencher of an underwater construction robot for seabed application. *Appl. Ocean Res.* **2018**, *71*, 90–104. [CrossRef]
12. Vu, M.T.; Choi, H.-S.; Nguyen, N.D.; Kim, S.-K. Analytical design of an underwater construction robot on the slope with an up-cutting mode operation of a cutter bar. *Appl. Ocean Res.* **2019**, *86*, 289–309. [CrossRef]
13. Aderibigbe, O.; Rajaratnam, N. Erosion of loose beds by submerged circular impinging vertical turbulent jets. *J. Hydraul. Res.* **1996**, *34*, 19–33. [CrossRef]
14. Beltao, S.; Rajaratnam, N. Impinging circular turbulent jets. *J. Hydraul. Div.* **1974**, *100*, 1313–1328. [CrossRef]
15. Beltao, S.; Rajaratnam, N. Impinging of axisymmentric developing jets. *J. Hydraul. Div.* **1977**, *15*, 311–326. [CrossRef]
16. Qi, M.; Fu, R.; Chen, Z. Study on equilibrium scour depth of impinging jet. *J. Hydrodyn.* **2005**, *20*, 368–372.
17. Li, F.; Du, J.; Shi, X.; Guo, W.; Wang, J. Research on the mechanism of hydraulic jet soil-breaking and its engineering application. *Fluid Machinery.* **1997**, *25*, 26–29.
18. Kochin, N.E.; Kibel, I.A.; Roze, N.V.; Boyanovitch, D.; Radok, J.R.M.; Talbot, L. Theoretical Hydromechanics. *Phys. Today* **1966**, *19*, 76. [CrossRef]

19. Fursikov, A.V.; Vishi, M.I. Mathematical problems of statistical hydromechanics. *Math. It's Appl.* **1988**, *39*, 620–622.
20. Einstein, H.A. *The Bed-Load Function for Sediment Transportation in Open Channel Flows*; US Department of Agriculture: Washington, DC, USA, 1950; pp. 12–35.
21. Engelund, F.; Fredsøe, J. A Sediment Transport Model for Straight Alluvial Channels. *Hydrol. Res.* **1976**, *7*, 293–306. [CrossRef]
22. Drew, D.A. Averaged Field Equations for Two-Phase Media. *Stud. Appl. Math.* **1971**, *50*, 133–166. [CrossRef]
23. Ishii, M. *Thermo-Fluid Dynamics Theory of Two-Phase Flow*; Eyrolles: Paris, France, 1975; pp. 107–110.
24. Enwald, H.; Peirano, E.; Almstedt, A.E. Eulerian two-phase flow theory applied to fluidization. *Int. J. Multiph. Flow* **1996**, *22*, 21–66. [CrossRef]
25. Huai, W.-X.; Xue, W.-Y.; Qian, Z. Numerical Simulation of Sediment-Laden Jets in Static Uniform Environment using Eulerian Model. *Eng. Appl. Comput. Fluid Mech.* **2012**, *6*, 504–513. [CrossRef]
26. Xue, W.-Y.; Huai, W.-X.; Li, Z.-W.; Zeng, Y.-H.; Qian, Z.; Yang, Z.-H.; Zhi-Wei, L.; Yu-Hong, Z.; Zhong-Hua, Y. Numerical simulation of scouring funnel in front of bottom orifice. *J. Hydrodyn.* **2013**, *25*, 471–480. [CrossRef]
27. Mih, W.C.; Kabir, J. Impingement of water jets on non-uniform stream bed. *J. Hydraul. Eng.* **1983**, *109*, 536–548. [CrossRef]
28. Bao, J.; Zhou, T.; Huang, M.; Hou, Z.; Perkins, W.; Harding, S.; Titzler, S.; Hammond, G.; Ren, H.; Thorne, P.; et al. Modulating factors of hydrologic exchanges in a large-scale river reach: Insights from three-dimensional computational fluid dynamics simulations. *Hydrol. Process.* **2018**, *32*, 3446–3463. [CrossRef]
29. Ling, B.; Bao, J.; Oostrom, M.; Battiato, I.; Tartakovsky, A.M. Modeling variability in porescale multiphase flow experiments. *Adv. Water Resour.* **2017**, *105*, 29–38. [CrossRef]
30. Aderibigbe, O.; Rajaratnam, N. Effect of Sediment Gradation on Erosion by Plane Turbulent Wall Jets. *J. Hydraul. Eng.* **1998**, *124*, 1034–1042. [CrossRef]
31. Rajaratnam, N.; Mazurek, K.A. Erosion of Sand by Circular Impinging Water Jets with Small Tailwater. *J. Hydraul. Eng.* **2003**, *129*, 225–229. [CrossRef]
32. Ansari, S.A.; Kothyari, U.C.; Raju, K.G.R. Influence of Cohesion on Scour under Submerged Circular Vertical Jets. *J. Hydraul. Eng.* **2003**, *129*, 1014–1019. [CrossRef]
33. Ozan, A.Y.; Yüksel, Y. Simulation of a 3D submerged jet flow around a pile. *Ocean Eng.* **2010**, *37*, 819–832. [CrossRef]
34. Lei, G.U.; Yan, N.I.; Jiali, H.; Fusheng, N.I. Experimental study on the sand scouring by submerged vertical plane jet. *Yellow River* **2019**, *41*, 38–43.
35. Roulund, A.; Sumer, B.M.; Fredsøe, J.; Michelsen, J. Numerical and experimental investigation of flow and scour around a circular pile. *J. Fluid Mech.* **2005**, *534*, 351–401. [CrossRef]

Journal of
Marine Science and Engineering

Article

Energy, Economic and Environmental Effects of the Marine Diesel Engine Trigeneration Energy Systems

Ivan Gospić [1,*], Ivica Glavan [1], Igor Poljak [1] and Vedran Mrzljak [2]

[1] Department of Maritime Sciences, University of Zadar, Mihovila Pavlinovića 1, 23000 Zadar, Croatia; iglavan@unizd.hr (I.G.); ipoljak1@unizd.hr (I.P.)
[2] Faculty of Engineering, University of Rijeka, Vukovarska 58, 51000 Rijeka, Croatia; vedran.mrzljak@riteh.hr
* Correspondence: igospic@unizd.hr; Tel.: +385-91-919-4255

Abstract: The paper discusses the possibility of applying the trigeneration energy concept (cogeneration + absorption cooling) on diesel-powered refrigerated ships, based on systematic analyses of variable energy loads during the estimated life of the ship on a predefined navigation route. From a methodological point of view, mathematical modeling of predictable energy interactions of a ship with a realistic environment yields corresponding models of simultaneously occurring energy loads (propulsion, electrical and thermal), as well as the preferred trigenerational thermal effect (cooling and heating). Special emphasis is placed on the assessment of the upcoming total heat loads (refrigeration and heating) in live cargo air conditioning systems (unfrozen fruits and vegetables) as in ship accommodations. The obtained results indicate beneficiary energy, economic and environmental effects of the application of diesel engine trigeneration systems on ships intended for cargo transport whose storage temperatures range from −25 to 15 °C. Further analysis of trigeneration system application to the passenger ship air conditioning system indicates even greater achievable savings.

Keywords: trigeneration energy system; cogeneration; absorption cooling; heating and cooling output

Citation: Gospić, I.; Glavan, I.; Poljak, I.; Mrzljak, V. Energy, Economic and Environmental Effects of the Marine Diesel Engine Trigeneration Energy Systems. *J. Mar. Sci. Eng.* **2021**, *9*, 773. https://doi.org/10.3390/jmse9070773

Academic Editor: Tie Li

Received: 15 June 2021
Accepted: 13 July 2021
Published: 16 July 2021

Publisher's Note: MDPI stays neutral with regard to jurisdictional claims in published maps and institutional affiliations.

1. Introduction

Marine diesel engine trigeneration energy systems (MDETES) represent coupling diesel engine cogeneration systems with absorption cooling systems, which allow the use of a co-generation effect to balance the occurring overall heating loads on marine motor ships intended for the transport of moderately cooled cargoes, [1]. Modern marine four stroke engines' energy efficiency rate is in the range of about 49% to 50% at its nominal load, [2,3]. Two stroke engines' energy efficiency is slightly higher, at the gas mode about 53% [4,5]; however, on the diesel mode, energy efficiency is lower at about 52%. The calculation for the two-stroke engine is based on the LHV of the MDO 42.7 MJ/kg, and LHV of a typical LNG 48.0 MJ/kg, [6,7]. As the marine engines cannot convert all heat energy to power, the remaining exhaust heat energy may be utilized for the boiling process of the absorption cooling unit. The typical process utilization overview was given in [8,9]. The exhaust gas heat energy of marine engines [10,11] is not the only source of heat energy; waste cooling energy of the marine engines may be utilized for that purpose as well. The waste heat capacity potential of the typical stationery and marine engines is analyzed in [12,13], where it is concluded that this part of the energy also has potential for trigeneration purposes. The improvement of the waste heat recovery from marine diesel engines with the best fulfilment of the vessel needs in terms of mechanical, electric and thermal energies is analyzed in [14,15], where various solutions are proposed, with combined diesel engine, steam and gas turbines recovering part of the thermal energy of the diesel engine exhaust gas. The similar approach of feedwater regeneration to the boiler from the marine engines was presented in [16], which also concluded that waste

199

heat will beneficially contribute to the efficiency of waste heat recovery. In marine diesel engine cogeneration energy systems (MDECES), the propulsion of the main diesel engine (MDE) directly or indirectly drives the propeller (P), and the shaft electric generator (SEG), balancing both occurring propulsion and electrical load during the voyage, while the occurring heating load is balanced by the waste heat recovery steam generator (WHRSG). During both maneuver and standstill, the ship's occurring electric load is balanced by the diesel aggregates (DA), and the resulting heating load of the ship is balanced mainly by the fired steam generator (FSG) due to the insufficient availability of waste heat contained in the exhaust gases of the DA. The overview of the energy balance for the cruise ship and the chemical tanker is given in [17,18]. As the vessel is sailing, it is changing the local microclimate's environmental condition. The engine is not always operating in the steady condition but is changing load, which affects the absorption process. In order to reduce the negative effect on the absorption process, [19] carried out where proposed a solution to the amortization of the heat load changes with the small diameter tube bundle heat exchangers with large specific surface area. The changes of the local microclimate environment should not always affect the absorption process in the negative direction, according to the study under [20]; off design performance of the LiBr-H_2O, particularly in the lower generator or higher condenser temperature conditions, generates both higher COP and exergy efficiency. However, in this particular study the absorption process in the complex rolling and pitching environment of the ship is not considered.

Due to economic considerations, the ship's diesel engines are supplied with cheaper, heavy fuel (HFO, IFO380) [21,22]; therefore, the combustion gases must not substantially cool down because of the presence of sulfur oxides in them, which consequently limits the cogeneration plant heating effect. During navigation, unsteadiness of the cogeneration thermal effect is emphasized due to its dependence on emerging propulsion load (which is reflected through fluctuation of power and engine speed) and on the temperature of the surrounding air, which was discussed in [23–25]. On the other hand, the heating load of the ship has been determined by its purpose, respectively prescribed microclimate of the contained commodities compartments, and of the navigation route, so that it is continuously changing according to which climate zone the ship is in [26]. Apart from the need to balance the heating load, it is necessary to create an appropriate cooling effect for balancing the cooling load [27], which is in almost all modern ships balanced by the compression refrigeration plant (CRP); this significantly increases the electric load, while the heating effect of the installed cogeneration plant is being considerably unexploited [28]. By the use of an absorption refrigeration unit (ARU) with thermodynamic mixtures of water-lithium bromide (H_2O-LiBr) [29,30], and ammonia water (NH_3-H_2O) [31,32], and with negligible electrical load, the utilization of the cogeneration thermal output for a production appropriate cooling effect is allowed. The attainable lower temperature limit in the evaporator ammonium ARU (AARU) restricts its application on both accommodation air conditioning systems and commercial cargo that is being transported and stored in a moderate temperature range (−25 to 15 °C). On the other hand, an attainable lower temperature limit in the evaporator lithium bromide ARU (LBARU) restricts its application on the accommodation air conditioning systems. By coupling the DECES and ARU, the (MDETES) is created, which enables cost-effective, energy-efficient, and environmentally friendly balancing of the occurring heating loads of both air conditioning systems and moderate commercial refrigerated cargo by the use of the cogeneration heating effect. Reducing the electrical load automatically reduces the mechanical loads of the MDE during navigation, as well as of the DA during maneuvering and the ship's standstill in the terminals, which implies substantial fuel savings and consequently lower operating costs with the reduction of the majority of ecologically harmful effects (primarily CO_2 and SO_2 emissions) according to [33,34].

This paper continues with efforts in the rational usage of the waste heat from the marine diesel engines. It presents a comprehensive mathematic model of the proposed trigeneration system for reefer use. The concept interrelates the trajectory of a vessel and the wave conditions with a detailed described quasi-static model of the energetic, economic and environmental performance of a marine trigeneration unit installed onboard.

2. Defining Technically Possible Application Fields ARU's

Technically, the obtainable upper temperature of the absorption chiller cooker is $\vartheta_{cc} \sim 170\,°C$ (which is determined by the parameters of steam from the cogeneration plant), and the cooling sea temperature $\vartheta_{sd} \sim 32\,°C$. These two parameters determine the applicability of the absorption cooling with ammonia mixture on those cooling systems where the temperature of ammonia primary refrigerator in evaporator does not fall below $\vartheta_{se} \sim -35\,°C$ and considering the usual arrangement of the cooling system with $CaCl_2$ brine as secondary refrigerant, which corresponds to the storage temperature limited at the $\vartheta_{ca} \sim -25\,°C$. Such achievable lower temperature storage facilitates the transporting of a wide range of palletized (packed bulk) and liquid cargo, where the majority of commercial bulk palletized cargos comprise foodstuffs, transporting either frozen (dead) or unfrozen (live). The liquid cargos include nutritional liquids such as various fruit juices (e.g., orange juice), and industrial organic liquids such as some alkanes (paraffin), alkenes (olefins), alkyne (acetylene), alkanes (aldehydes), etc.

Among the frozen commodities, meat and fish dominate, and are generally stored at equal low temperature $\vartheta_{FM} \sim -18\,°C$, although the meat is often carried unfrozen in vacuum packs in the temperature range $\vartheta_{MV} \sim (-2\ \text{to}\ 10)\,°C$ depending on the type of meat and the duration of storage [35].

Micro-climate storage of the unfrozen (living) perishable products, among which prevail bananas, citruses, deciduous fruits and vegetables, is characterized by a relatively high relative humidity ($\varphi_{lp} \sim 0.8\ \text{to}\ 0.95$), moderately low temperature $\vartheta_{lp} \sim -2\ \text{to}\ 15\,°C$ and a relatively small proportion of fresh air (a small refreshment $g_0\ 1.5 \sim 2.5\%$). Nutritional liquids are refrigerated and held in separate tanks in an inert gas atmosphere, with the temperature depending on their respective sugar contents, for example, orange juice is held at $\vartheta_{OJ} \sim -7\,°C$, [36].

Typical industrial organic liquids are stored at appropriate saturation temperature $\vartheta_s\ (p_a)$ at normal atmospheric pressure p_a, and are as follows: paraffins (C_4H_{10} n-butane $-0.5\,°C$, isobutane $-11.7\,°C$, cyclobutane $10\,°C$), olefins (C_4H_8 butene $-6.6\,°C$, C_4H_6 cyclobutene $2.2\,°C$), acetylenes (propyne $-23.2\,°C$, butyne $7.7\,°C$ and aldehydes) methanal $-19\,°C$, ethanal $20.2\,°C$, etc., [37].

With cooling sea water of $\vartheta_{sd} \sim 32\,°C$ and with a heating sink of $\tilde{\vartheta}_s \sim 120\,°C$ (low-pressure saturation steam $\tilde{p}_s \sim 2$ bar), attainable temperature from mixture extracted water as primary refrigerator in the evaporator of lithium bromide ARU amounts to $\vartheta_{se} \sim 5\,°C$; hence, by the usual arrangement of the cooling system with water as secondary refrigerator in ships' accommodation air-conditioned systems, the conditioning air can be cooled down to a temperature of $\vartheta_{ac} \sim 13\,°C$, and consequently the application of this ARU to create the desired level of comfort is particularly suitable in air conditioning systems of passenger ships. Generally, there is a possibility that the same ship simultaneously carries different cargo stored in separate compartments with the appropriate prescribed microclimate. Taking into account ships' accommodation, balancing of the appearing cooling load of such a ship can be achieved either with two apart single-stage AARUs, one for "dead" and another for "live" products (which is expensive), or with a double-stage ARU for the both "dead" and "live" products, and with one lithium bromide LBARU for ships' accommodation.

In the case of the balancing of cooling load from "dead" product compartments, the overall occurring cooling load is absorbed by the circulation air, which is then cooled in the pertaining air coolers (AC) by the $CaCl_2$ brine, and furthermore, it is refrigerated in evaporator AARUs by the very rich ammonia water mixture. Transportation of "live" product is characterized, in addition to moderate temperature storage, by the removal of the products of their own metabolism followed by fresh air, and with continuous control of relative humidity air in compartments. Regulating the moisture content while balancing the ongoing cooling load and the continuous (or periodic) supply of fresh air, in addition to cooling it, also requires its heating in the air heater.

In the secondary cooling circuit of both cases, the secondary refrigerator $CaCl_2$ brine circulates by an electric motor driven cooling brine pump (CBP), while the compartment air is recirculated by an electric motor-driven recirculation air fan (RAF), and in transportation of "living" perishable goods the extraction of a part of the air saturated with products of their metabolism is ensured by a special extraction fan (EF). Although many times, depending on market conditions, the reefers simultaneously carry two or more different moderate refrigerated commercial cargoes to separate compartments, in order to define the appropriate cooling systems, it is assumed that during a voyage cycle reefer transport only one type of commercial cargo is transported. Here is an unambiguously defined sailing route. The transport of five different types of cargo is considered: dead-frozen meat (FM) and meat in vacuum packs (MVP), live perishable bananas (B), citruses (C—citrus fruits, e.g., oranges, lemons, grapefruits) and deciduous fruits (DF—kiwi, pears and apples). There is a possibility that, on the one hand, during the regime of navigation that the majority of occurring unsteady cooling and heating loads are balanced by MDETES, while on the other hand, during standstill reefer at final terminal destination, balancing of the corresponding generated cooling and heating loads is carried out by either CRP and FSG, respectively, or by installed absorption chillers that are in this case powered by heat flow produced in FSG. ARU cookers and HFO final heather (HFOFH) together with other ship's steam consumers (OSSC) determine the overall total unsteady heating load $\Phi_{hl}(t)$, which is balanced by the dry-saturated steam of $p_s{\sim}8$ bar produced in a single-pressure cogeneration system characterized with medium-pressure evaporator (MPE) and steam drum (MPD). In the case of the extended form of the cogeneration plant, whose application is based on the using of corrosion-resistant materials, additional main items are included, including a low-pressure steam drum (LPD) and low-pressure evaporator (LPE) that utilizes the remained waste heat from diesel motor exhausted gas after its passage through MPE. Simplified functional diagram Figure 1 illustrates the plain original MDETES, the application of which could be found on ships for transport of moderate refrigerated cargoes. Based on the submitted schemes, the following can be concluded. The reference design version of the ship's heating and cooling system is characterized by the required heating effect being produced by the combined steam generator (CSG = FSG + WHRSG), while the cooling effect is produced by a CRP. The introduction of the trigeneration concept offers a trigeneration design version characterized by the fact that during the entire transport cycle the occurring cooling load is balanced by the appropriate ARUs, while by the WHRSG unbalanced amount of the overall unsteady heating load is balanced by the FSG. In order to achieve sufficient redundancy, it is recommended in the case of application of the trigeneration concept to install the CRP of the corresponding cooling effect.

Description of the Trigeneration Scheme

The enclosed Figure 1 illustrates a simplified scheme of generic MDETES. The functional-interactive connection of the involved subsystems with the flows of the involved media—engine exhaust, water vapor, heavy fuel oil, cooling freshwater, cooling brine, cooling sea, and air of accommodation air conditioning system (AACS) and air of cargo storage air conditioning systems (CACS)—does not illustrates interactive links within the building subsystems of the involved absorption devices (AARU and LBARU).

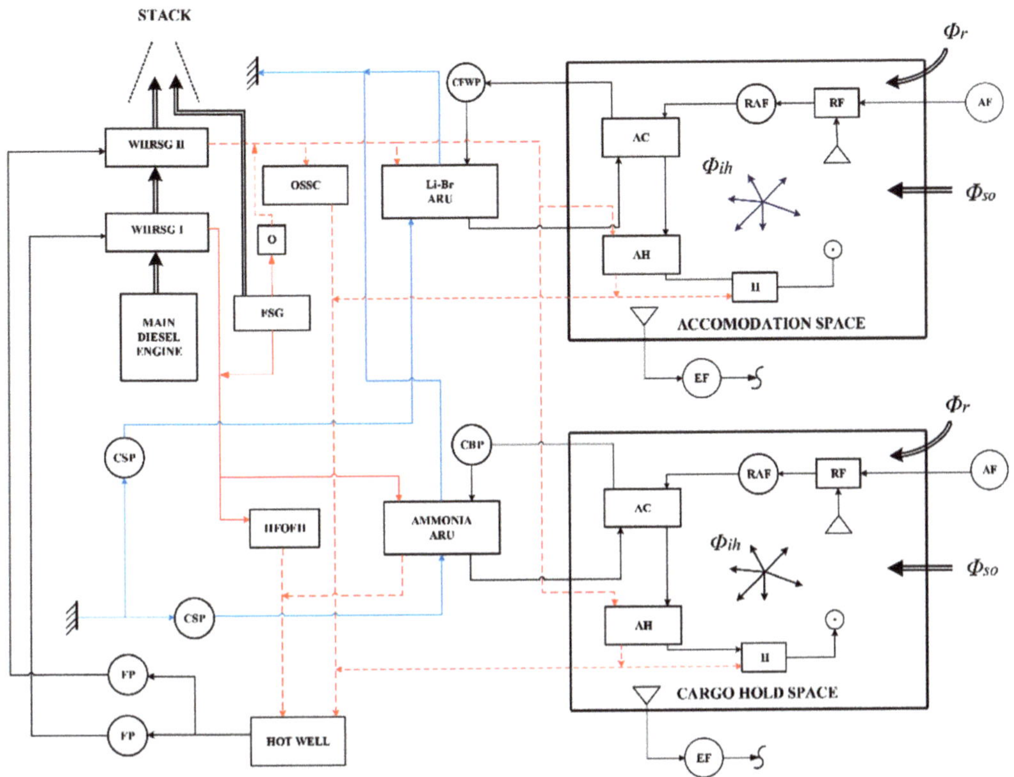

Figure 1. The trigeneration scheme.

During navigation, the two-pressure WHRSG is driven by the exhaust gases of the MDE, which simultaneously produces medium-pressure and low-pressure water vapor to balance the corresponding unsteady heat loads—HFOFH and AARU with medium-pressure steam, LBARU, AHa, AHc, and OSSC (Other Ship's Steam Consumers, such as heaters in HFO S&TS (HFO Storage and Treatment System), as well in HPW (Hot Portable Water)) with low-pressure steam—whereas in cases of insufficient cogeneration effect, the unbalanced part of the occurring heat load (including the heat load of the heavy fuel heater of the fired boiler (FHFSG)) is balanced with medium-pressure steam generated by the fired boiler. In both cases, the medium-pressure and low-pressure condensate are transferred to the hot well (HW) by the appropriate condensate lines. In the air conditioning systems of housing and storage, the extraction of the precisely prescribed part of the metabolic products of saturated air is carried out by extraction fans (EFa and EFc). The amount of the fresh air replacing extraction saturated air and that is mixing with the recirculated saturated is regulated by the regulation flap (RF), immediately in front of recirculation air fans (RAFa and RAFc). The mass flow of this mixture, its temperature, and relative humidity at the outlet of the associated conditioning units, must lie in the prescribed range of comfort parameters to ensure the balance both of humidity and the resulting overall heat load of the air-conditioned space. That is achieved by an appropriate distribution system and by appropriate cooling and/or heating, as well as possible humidification of the same. The unsteady heat loads in the air coolers (ACa and ACc) are balanced by the secondary coolers (fresh water in AACS and brine in CACS), while unsteady heat loads in the associated air heaters (AHa and AHc) are balanced by the low-pressure steam. The secondary coolers are cooled in the evaporators of the respective ARUs. The total heat loads there are increased by the thermal gains of the pertaining conductor systems. The unsteady cooling effect's generation in ARU evaporators causes the occurrence of unsteady heat loads of contained heat exchangers: cookers that are balanced by water vapor from the cogeneration system, as well absorbers and condensers that are balanced with the cooling sea supplied by the electrically driven cooling brine pump (CBP) of low power rate. In practice, by applying the trigeneration concept, the overall thermal load of the AACS and CACS is balanced by generating the appropriate cogeneration heat output. When transporting "dead" products, there is no need for air conditioning of the storage air, so despite the significantly lower storage temperatures than for "living" products, the thermal load of the storage of "living" products is much higher. This is primarily due to the appearance of increased infiltration heat gains inherent to refreshing. Practically, when maintaining the prescribed micro-climatic conditions of the storage space, a slight change in the intensity of refreshing significantly changes the occurring heat load of the air conditioning system, which may lead to the need for a short-term (but permissible), slight reduction of refreshing in conditions of temporary insufficiency of cogeneration. When modeling MDETES, this scenario was also taken into account, which is confirmed by the results of the simulations of the occurring heat load during navigation for the set refreshing intensities.

The Figure 1 symbols are; AF—Air Filter, AC—Air Cooler, AH—Air Heater, CBP—Cooling Brine Pump, CSP—Cooling Sea Pump, H—Humidifier, HFOFH—Heavy Fuel Oil Final Heater, OSSC—Other ship's Steam Consumer, RF—Regulation Flap, RAF—Recirculation Air Fan, FP—Feed Pump, O—Orifice, WHRSG—Waste Heat Recovery Steam Generator, FSG—Fire Steam Generator, EF—Extraction Fan, CFWP—Cooling Fresh Water Pump.

3. Mathematical Modeling Methodology of Energy Interactions of Ship and Environment during Navigation

In order to calculate the performance of the proposed system, models of each respective component are built. Modeling and simulation of the complex physical system are made in Wolfram Mathematica environment.

During the navigation on the predefined navigation route, the ship's energy system balances the occurring unsteady loads which are fluctuating in a wide range of values due to the action of unsteady and intermittent excitation of the occurring characteristic scalar and vector quantities of the environment, as illustrated in Figure 2. The contained sizes are: vectors—$\vec{v}_{sc}(\varphi_r, t_Y)$ and $\vec{v}_w(\varphi_r, t_Y)$ are sea current and wind velocity, respectively (these vectors are aligned in the tangential plane of the sphere), $\vec{q}_{si}(\varphi_r, t_Y)$ is vector of the solar irradiance; scalar—$\vartheta_{a_Y}(\varphi_r, t_Y)$, $\Delta\vartheta_{a_d}(\varphi_r, t_Y)$ and $\Delta\vartheta_{a_m}(\varphi_r, t_Y)$ are yearly average daily air temperatures (DAT), yearly average temperature differences between max and min DAT, monthly average temperature differences between yearly average DAT, respectively; $\chi_s(\mu_r, t_Y)$—the saturation level of the moist air; $\vartheta_{sd}(\varphi_r, t_Y)$—sea temperature at the depth d measured from the free surface; $\Gamma_c(\varphi_r, t_Y)$ and $\Psi_c(\varphi_r, t_Y)$—sky coverage with the clouds and the attenuation of the sun's radiation due to clouds, respectively; $H_s(\varphi_r, t_Y)$ and $T_s(\varphi_r, t_Y)$—significant wave height and period, respectively.

By available reliably statistically processed data of the occurring environment values over a defined navigation route [38], it is possible to create appropriate mathematical formulations that express the dependence of these quantities on calendar time t_y and geographical position with given either latitude φ or longitude μ. The mathematical model for each defined navigation route is a precise mathematical explicit approximate functional dependence between latitude and longitude—$\varphi(\mu)$ or $\mu(\varphi)$, respectively—either in the form of a polynomial or a trigonometric order with sine terms, as follows:

$$\mu_r(\varphi_r) = \sum_i^n a_i\varphi_r^i \text{ or } \mu_r(\varphi_r) = \sum_i^n [a_i \sin(i\varphi_r + \varepsilon_i)], \tag{1}$$

while in the case of sailing on an orthodrome, $\mu_r(\varphi_r)$ is defined by the following formula:

$$\mu_r(\varphi_r) = \mu_A + \arcsin\{\sin(\mu_B - \mu_A)\cot(\varphi_B - \varphi_A)\tan[\varphi_r(t) - \varphi_A]\}. \tag{2}$$

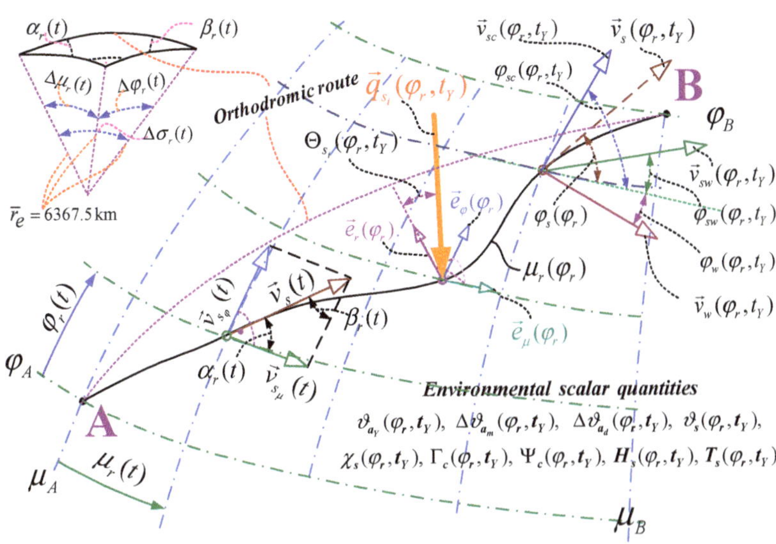

Figure 2. The navigation route through realistic surroundings.

Based on that, any emerging j-th scalar or vector item of the environment that interacts with the ship's power system on the navigation route can be represented in the form of the product of trigonometric series of sine functions depending on t_Y and φ_r as follows:

$$S_j(t_Y,\varphi_r) = \sum_{i_t}^{n_t}\sum_{i_\varphi}^{n_\varphi}[A_{i_t}\sin(i_t\omega_t t_Y + \varepsilon_{i_t})]\left[A_{i_\varphi}\sin\left(i_\varphi\varphi_r + \varepsilon_{i_\varphi}\right)\right] \tag{3}$$

Thus, for example, based on the obtained statistical data for the characteristic ambient air temperatures, it is possible to create a mathematical model for the atmospheric air temperature on a certain navigation route, as follows:

$$\vartheta_a(\varphi_r,t_Y) = \left\{ \begin{array}{l} \vartheta_{a_y}(\varphi_r,t_Y) + \frac{1}{2}\Delta\vartheta_{a_m}(\varphi_r,t_Y)\sin\left[\omega_{sm}t_Y + \gamma_{sm}(\varphi_r,t_Y)\right] \\ -\frac{1}{2}\Delta\vartheta_{a_d}(\varphi_r,t_Y)\cos\left[\omega_{sd}t_Y + \gamma_{sd}(\varphi_r,t_Y)\right] \end{array} \right\} \tag{4}$$

For the fluctuation period of the $\vartheta_{a_m}(\varphi_r,t_Y)$, it seems appropriate to take half an average synodic month $\overline{\tau}_{sm} = 29.55\,\overline{\tau}_{sd}$, where the length of the mean solar day is $\overline{\tau}_{sd} = 24\,\mathrm{h}$, from which it follows for an unsteady circular frequency $\omega_{sm} = \pi/\overline{\tau}_{sm}$. The fluctuation period for $\Delta\vartheta_{a_d}(\varphi_r,t_Y)$ can be obtained by the fact that the daily temperature minimum occurs just after dawn, whose unsteady phase shift is $\gamma_{sd}(\varphi_r,t_Y)$, and which possesses the apparent circular frequency $\omega_{sd} = 2\pi/\overline{\tau}_{sd}$. Using this methodology, models of characteristic air temperatures were developed, which are illustrated in the Figure 3. Similarly, Figure 4 illustrates wind speed and significant wave height.

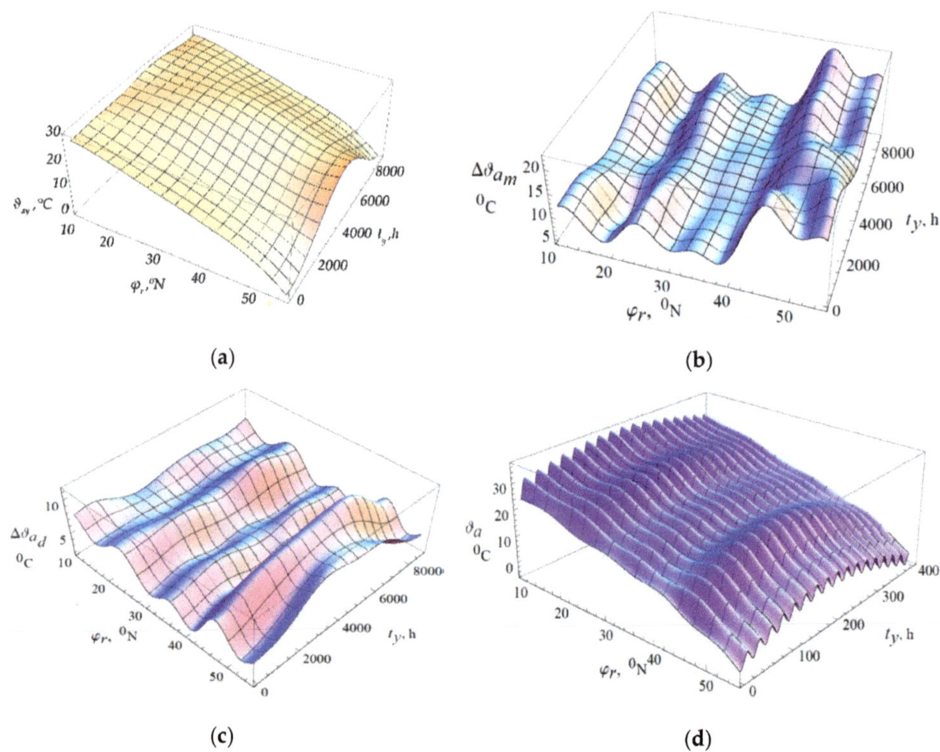

Figure 3. Air temperature models en route: (**a**) $\vartheta_{a_Y}(\varphi_r, t_Y)$, (**b**) $\Delta\vartheta_{a_m}(\varphi_r, t_Y)$, (**c**) $\Delta\vartheta_{a_d}(\varphi_r, t_Y)$, (**d**) $\vartheta_{a_d}(\varphi_r, t_Y)$.

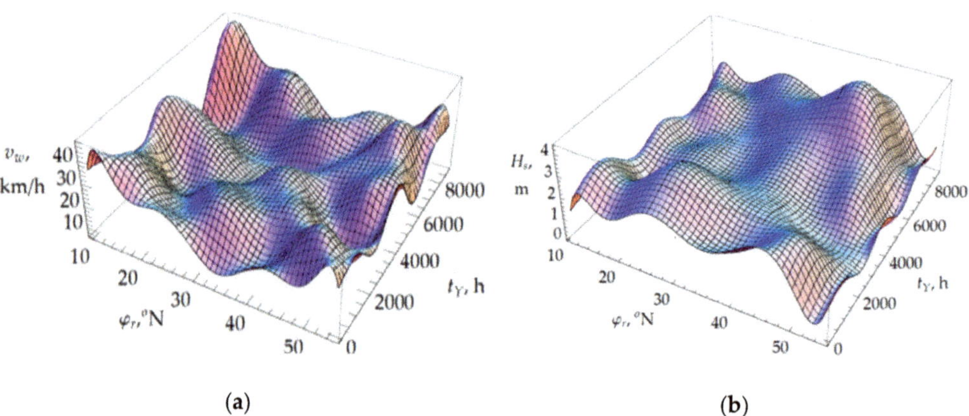

Figure 4. (**a**) wind velocity modules $v_w(\varphi_r, t_Y)$ and (**b**) significant wave high $H_s(\varphi_r, t_Y)$ en route.

Ignoring the geoid shape due to the flatness of the sphere, the navigation route can be defined with sufficient accuracy by a position vector which, with a defined dependence $\mu_\mathrm{r}(\phi_r)$, taking into account the radius of the sphere $r_e = 6376.5$ km, becomes:

$$\vec{r}_r(\varphi_r) = r_e\left\{\cos\varphi_r\cos[\mu_r(\varphi_r)]\,\vec{i} + \cos\varphi_r\sin[\mu_r(\varphi_r)]\,\vec{j} + \sin\varphi_r\,\vec{k}\right\}, \tag{5}$$

According to, the sailing speed can be defined as follows:

$$\vec{v}_s(t) = \frac{d\vec{r}_r[\varphi_r(t)]}{dt} = \vec{v}_s[\varphi_r(t), \dot{\varphi}_r(t)]; \quad \dot{\varphi}_r(t) = \frac{d\varphi_r(t)}{dt}, \tag{6}$$

The overall navigational resistance increased by the amount of propulsion reduction for a given ship's loading condition $T_{lo} = R_{tot}/(1 - \hat{t})$, which is charged to the propeller, except for the navigation speed $\vec{v}_s$ and the continuous increase in resistance due to hull pollution and fouling, depends on the occurring vector quantities: wind $\vec{v}_w$, sea currents $\vec{v}_{sc}$, and waves $\vec{v}_{sw}$ [39,40], which according to (3) are dependent on $t_Y = t + \tau_0$ and φ_r. The propeller-generated thrust force T_p required to balance this load and the required torque of the propeller Q_p, in addition to its speed n or wing pitch p, depend on the average sea inflow speed into the propeller disk $v_a = v_s(1 - w)$, which depends on the modulus of sailing speed v_s and the mean coefficient of wake w, whose functional dependence as well as ship's thrust deduction coefficient $\hat{t}$ are defined by the expression:

$$\begin{aligned} w &= w\left(\vec{v}_s, \vec{v}_w, \vec{v}_{sc}, \vec{v}_{sw}\right) \\ \hat{t} &= \hat{t}\left(\vec{v}_s, \vec{v}_w, \vec{v}_{sc}, \vec{v}_{sw}\right) \end{aligned} \tag{7}$$

Taking into account the above mathematical formulations, the quasi-static characteristics of thrust and torque take on generalized functional dependencies:

$$\begin{aligned} T_p &= T_p(n, v_a) = T_p[n, \varphi_r(t), \dot{\varphi}_r(t), t, \tau_0] \\ Q_p &= Q_p(n, v_a) = Q_p[n, \varphi_r(t), \dot{\varphi}_r(t), t, \tau_0] \end{aligned} \tag{8}$$

The generalized functional dependence of the overall quasi-static thrust load (taking into account the total time spent in the service after the docking $\tau_s = \tau_0$) is formulated by the expression:

$$T_{lo} = T_{lo}\left(\vec{v}_s, \vec{v}_w, \vec{v}_{sc}, \vec{v}_{sw}, t_Y, \tau_s\right) = T_{lo}[\varphi_r(t), \dot{\varphi}_r(t), t, \tau_0], \tag{9}$$

The actual torque $Q_{de} = Q_{de}(k_F, n)$ of the MDE which balances the oncoming propulsion load increased by the counter friction torque of the propeller shafting $Q_{cs} = Q_{cs}(n)$, except for n, depends on the motor load factor k_F which is on the fixed geometry of the propeller (FPP, p = const.), the only control variable which ensures (with constant fluctuation of n and v_s) safe, reliable and seakeeping acceptable navigation.

Ignoring the emerging capacitances within the DEPS (diesel engine propulsion system), to conduct credible navigation simulations through a realistic environment, it is necessary to set up an appropriate system of dynamic equilibrium of thrust and torque equations as follows:

$$m_s = \frac{dv_s[\varphi_r(t), \dot{\varphi}_r(t)]}{dt} = T_p[\varphi_r(t), \dot{\varphi}_r(t), t, \tau_0] - T_{lo}[\varphi_r(t), \dot{\varphi}_r(t), t, \tau_0], \tag{10}$$

$$I_p = \frac{\pi}{30}\frac{dn(t)}{dt} = Q_{de}[k_F(t), n(t)] - Q_{cs}[n(t)] - Q_p[n, \varphi_r(t), \dot{\varphi}_r(t), t, \tau_0], \tag{11}$$

where: m_s is the mass of the loaded ship increased by the mass of the surrounding water affected by the movement of the ship (about 10% of the mass of the ship), and I_p is the polar moment of inertia of all rotating masses of the propulsion system increased by the inertia of the propelled sea mass.

This system of nonlinear differential equations (2nd order $\varphi_r(t)$ and 1st order $n(t)$), solvable exclusively numerically based on the prescribed initial conditions $[\varphi_r(0), \dot{\varphi}_r(0), n(0)]$, gives unknown time dependencies $\varphi_{r_{AB}}(t)$ and $n_{AB}(t)$ during navigation between destinations A and B, sailing duration τ_{AB}. Furthermore, by the numerical processing of $\varphi_{r_{AB}}(t)$, $\dot{\varphi}_{r_{AB}}(t)$ is obtained and according to (6) unsteady sailing speed $v_{s_{AB}}(t)$ is calculated; additionally, based on both the $n_{AB}(t)$ and SMCR (Service Maximum Continuous Rating) of MDE, the time dependency control variable $k_{F_{AB}}(t)$ is also obtained. In principle, the simulation of navigation from B to A is the same, except that the prescribed initial parameters and conditions are different, and thus the functional dependences of unsteady quantities on time $\varphi_{r_{AB}}(t)$, $n_{BA}(t)$, $v_{s_{AB}}(t)$, $k_{F_{BA}}(t)$ are obtained with the duration of navigation $\tau_{BA}(t)$, which are illustrated in Figure 5.

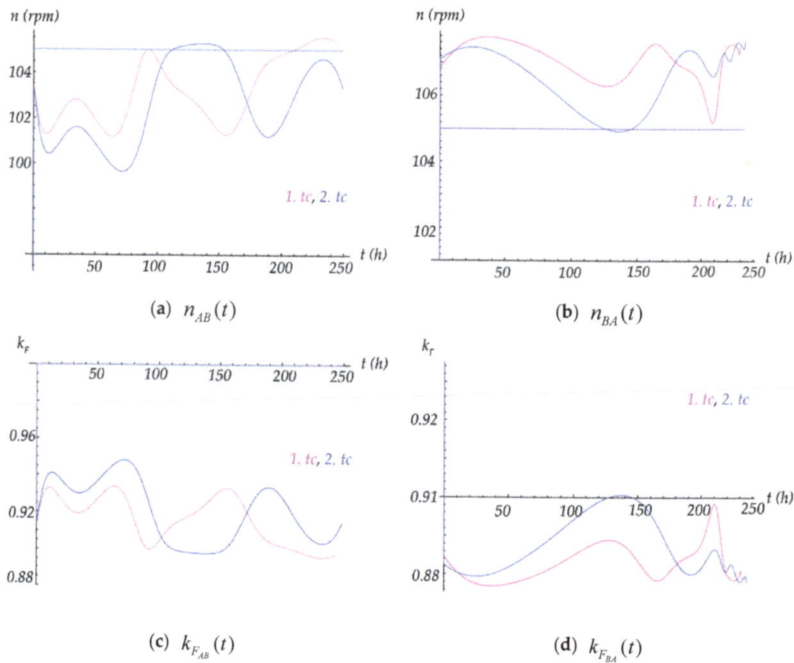

Figure 5. Unsteady balanced items of MDE for navigation in 1st and 2nd transport cycle.

With the estimated time intervals of stay in the final destinations τ_A and τ_B, the total duration of the kth transport cycle is: $\tau_{tc_k} = \tau_{Ak} + \tau_{ABk} + \tau_{Bk} + \tau_{BAk}$, so when simulating navigation in the $(k+1)$'s transport cycle, the time parameter τ_0 is obtained:

$$\tau_{0(A)k} = \sum_{k=1}^{n_k-1} \tau_{tc(k-1)}, \quad \tau_{0(AB)k} = \tau_{0(A)k} + \tau_{0(AB)k}, \quad \tau_{0(B)k} = \tau_{0(A)k} + \tau_{0(AB)k}, \quad \tau_{0(BA)k} = \tau_{0(A)k} + \tau_{Ak} + \tau_{ABk} + \tau_{Bk} \quad (12)$$

By using models for air and sea temperature (according to Equation (3)) and ship sailing speeds during the transport cycle (obtained from simulated navigation through realistic surroundings) the unsteady profiles of air and sea temperature are obtained as illustrated in Figure 6.

3.1. Quasi-Static Effect of Cogeneration System Scheme

Taking into account predictable values of mass flow $\dot{m}_{eg}$ and temperature $\vartheta_{eg}(t)$ of MDE exhaust gases at the SMCR (Service Maximum Continues Rating), and by estimating techno-economically acceptable minimum temperature difference $\Delta\vartheta_{min}$ at the WHRSG outlet, the main cogeneration design parameters can be determined, such as exhaust gas outlet temperature and heat exchanger area A. Its required area A is determined for the selected configuration, which is interactively related to the overall heat transfer coefficient k.

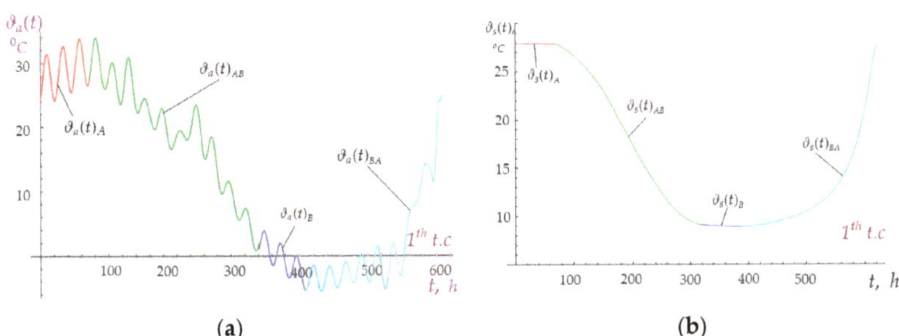

Figure 6. (a) Air $\vartheta_a(t)$ and (b) sea temperature $\vartheta_{sd}(t)$ through one transport cycle.

During navigation, the fluctuation $\dot{m}_{eg}$ and $\vartheta_{eg}(t)$ at the WHRSG inlet, with the determined temperature of the generated steam ϑ_s and the selected configuration of the overall flow surface A, leads to volatility $k(t)$, so based on the energy balance and the applicable mathematical formulation for heat exchange via mean logarithmic temperature difference (MLTD-$\Delta\vartheta_m$), unsteady cogeneration heat effect is obtained:

$$\Phi_{co}(t) = C_{eg}(t)\left[\vartheta_{eg}(t) - \vartheta_s\right]\left\{1 - \exp\left[-\frac{K(t)}{C_{eg}(t)}\right]\right\}, \tag{13}$$

where: $C_{eg}(t) = \dot{m}_{eg}(t)c_p(t)$ is the heat capacity of exhaust gases, and $K(t) = k_i(t)A_i$ is the exchanger heat transfer capability.

The total unsteady overall heat transfer coefficient $k_i(t)$ corresponding to A_i is defined by the expression:

$$k_i(t) = \left[\frac{1}{\alpha_s(t)} + \frac{r_i}{\lambda_t}\ln\left(\frac{r_e}{r_i}\right) + \frac{r_i}{r_e}\frac{1}{\alpha_{eg}(t)}\right]^{-1}, \tag{14}$$

where the appropriate heat transfer coefficients are $\alpha_s(t)$ and $\alpha_{eg}(t)$ on both the steam and exhaust side, and $\alpha_{eg}(t)$ is markedly dependent on the oncoming mean temperature $\overline{\vartheta}_{eg}(t)$ and on the mean flow rate through the exchanger section $\overline{v}_{eg}(t)$.

Taking into account the functional dependence of the emerging operating parameters of the diesel engine $k_F(t)$ and $n(t)$, as well as the ambient air temperature $\vartheta_a(t)$ during navigation, their involvement in (10) easily results in the emerging thermal effects of the cogeneration system $\Phi_{co_{AB}}(t)$ and $\Phi_{co_{BA}}(t)$.

When using a two-pressure cogeneration system, the steam production process takes place in such a way that even when $\Phi_{hl}(t) < \Phi_{co}(t) < \Phi_{HL}(t)$, all available flue gas heat flow is used to produce medium-pressure steam, which results in the flue gas temperature at the inlet to the low-pressure evaporator being exactly equal to the flue gas temperature outlet of the medium-pressure evaporator $\overline{\vartheta}_{eg_i}(t) = \overline{\vartheta}_{eg_o}(t)$:

$$\vartheta_{eg}(t) = \vartheta_s + \left[\vartheta_{eg}(t) - \vartheta_s\right]\exp\left[-\frac{K(t)}{C_{eg}(t)}\right], \tag{15}$$

Thus the available heat output of the low-pressure evaporator is obtained:

$$\tilde{\Phi}_{co}(t) = C_{eg}(t)\left\{1 - \exp\left[-\frac{\tilde{K}(t)}{C_{eg}(t)}\right]\right\}\left\{\vartheta_s - \tilde{\vartheta}_s + \left[\vartheta_{eg}(t) - \vartheta_s\right]\left[1 - \exp\left(-\frac{K(t)}{C_{eg}(t)}\right)\right]\right\}, \tag{16}$$

where $\tilde{K}(t) = \tilde{k}_i(t)\tilde{A}_i$ is the heat transfer capability of the low-pressure evaporator, the flow surface $\tilde{A}_i$ and the total heat transfer coefficient $\tilde{k}_i(t)$. Using mathematical models expressed in formulas (13) and (16), the unsteady MP cogeneration effects $\Phi_{co_{AB}}(t)$ and $\Phi_{co_{BA}}(t)$ (during navigation; from A to B and from B to A, respectively), as well unsteady

LP cogeneration effects $\widetilde{\Phi}_{co_{AB}}(t)$ and $\widetilde{\Phi}_{co_{BA}}(t)$ (during navigation; from *A* to *B* and from *B* to *A*, respectively) for 1st and 2nd transport cycles were calculated and shown in Figure 7.

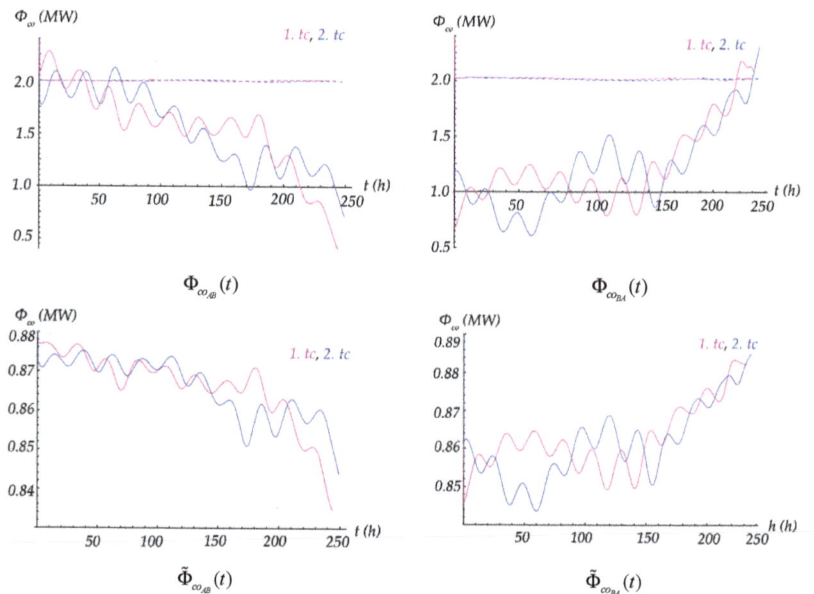

Figure 7. Unsteady cogeneration effects for navigation in first and second transport cycle.

3.2. Quasi-Static Heating Loads Scheme

The total unsteady heat load of the ship's power system during navigation between the final destinations A and B $\Phi_{HL}(t) = \Phi_{hl}(t) + \widetilde{\Phi}_{hl}(t)$ consists of an MP heat load balanced with MP steam $\Phi_{hl}(t)$ and an LP heat load balanced with LP steam $\widetilde{\Phi}_{hl}(t)$, defined by the corresponding terms as follows:

$$\Phi_{hl}(t) = \Phi_{CAA}(t) + \Phi_{FH_{HFO}}(t), \tag{17}$$

$$\widetilde{\Phi}_{hl}(t) = \widetilde{\Phi}_{CLBA}(t) + \widetilde{\Phi}_{AHa}(t) + \widetilde{\Phi}_{AHc}(t) + \widetilde{\Phi}_{OSSC}(t), \tag{18}$$

where unsteady heat loads are contained as follows: $\Phi_{CAA}(t)$ represents AARU cookers; $\Phi_{FH_{HFO}}(t)$ represents HFOFH; $\widetilde{\Phi}_{CLBA}(t)$ represents LBARU cookers; $\widetilde{\Phi}_{AHa}(t)$ and $\widetilde{\Phi}_{AHc}(t)$ represent AH in the AACS and CACS, respectively; and $\widetilde{\Phi}_{OSSC}(t)$ represents OSSC heaters. Unsteady heat loads of AARU and LBARU cookers are defined by the terms:

$$\Phi_{CAA}(t) = \frac{\Phi_{CEc}(t)}{COP_{AA}(t)}, \quad \Phi_{CLBA}(t) = \frac{\Phi_{CEa}(t)}{COP_{LBA}(t)}, \tag{19}$$

where $\Phi_{CEc}(t)$ and $\Phi_{CEa}(t)$ are cooling effects of AARU-a and LBARU, respectively, and $COP_{AA}(t)$ and $COP_{LBA}(t)$ are their coefficients of performance (COP). $COP_{AA}(t)$ and $COP_{LBA}(t)$ are functionally dependent on the equilibrium pressures in the evaporator $p_{CE}(t)$, and on the unsteady temperature of the cooling sea water $\vartheta_{cs} = \vartheta_{sd}(t)$, which limits the evaporation range for a certain mixture when there is the steady heating source of the cooker (water vapor from the cogeneration system). As ϑ_{cs} is lower, the possible evaporating range is higher, which increases COP, as illustrated in Figure 8.

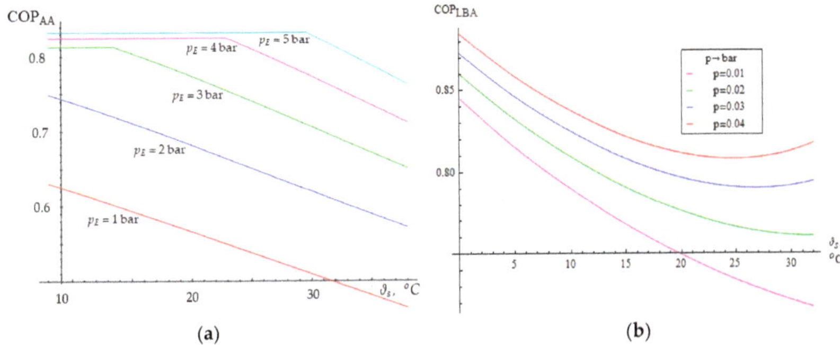

Figure 8. Quasi-static characteristics (**a**) COP_{AA} (p_{CE}, ϑ_{cs}) and (**b**) COP_{LBA} (p_{CE}, ϑ_{cs}).

Unsteady cooling effects corresponding to the occurring heat loads of AARU and LBARU evaporators are:

$$\Phi_{CEc}(t) = \sum_{i=1}^{n_s} \Phi_{ACs_i}(t) + P_{cbp} + \Phi_{bcs}(t), \tag{20}$$

$$\Phi_{CEa}(t) = \Phi_{ACa}(t) + P_{cwp} + \Phi_{wcs}(t). \tag{21}$$

where: $\Phi_{ACs_i}(t)$ and $\Phi_{ACs_i}(t)$ are the heat loads of air coolers in climatization systems of cargo spaces and accommodation, respectively; P_{cbp} and P_{cfwp} are the mechanical powers of the cooling brain pumps (for cargo space) and cooling fresh water pump (for accommodation), respectively; and $\Phi_{fwcs}(t)$ and $\Phi_{bcs}(t)$ are the heating incomes of the conducting systems from pertaining surroundings.

The heat loads of both climatization system (storage space and accommodation) depend on both predefined microclimatic comfort parameters (mainly air temperature and relative humidity), and total sensible and infiltration heating loads. In principle there are two emerging characteristics operating scenarios, either humidification or dehumidification of the wet air mixture as a result of mixing of the recirculation air from climatization space and fresh air from surroundings (Figure 9).

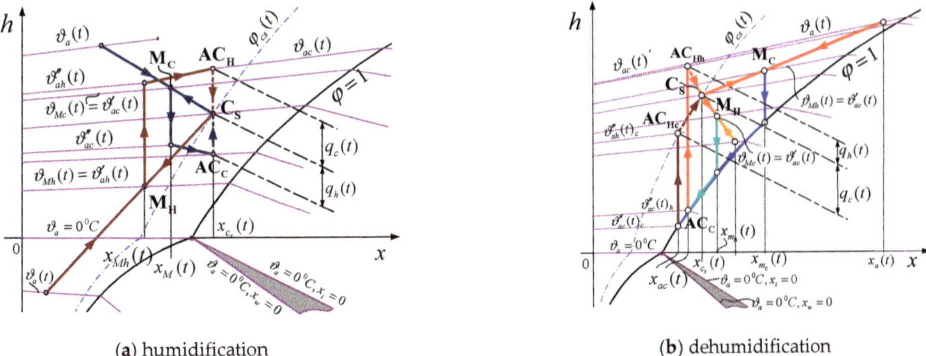

(**a**) humidification

(**b**) dehumidification

Figure 9. The air conditioning processes for the ship's spaces in a Mollier h-x diagram.

Consequently, heat loads of air coolers are defined according to the corresponding formulas for cases of humidification and dehumidification as follows:

Humidification and cooling, if $x_{cs}(t) - x_{Mc}(t) > 0$, and $\Phi_{AC}(t)_{(1)} < 0$,

$$\Phi_{AC}(t)_{(1)} = \dot{m}_{sa}\{h_{cs}(t) - h_{Mc}(t) - [x_{cs}(t) - x_{Mc}(t)]c_w\vartheta_w(t)\} + \Phi_{hls}(t) - P_{RF} - \Phi_{ds}(t), \quad (22)$$

Dehumidification and heating, if $x_{ac}(t) - x_{Mc}(t) < 0$, $x_{ac}(t) = x_{cs}(t) - \dot{m}_{we}/\dot{m}_{sa}$,

$$\Phi_{AC}(t)_{(2)} = \dot{m}_{sa}\left[h_{dp}(t) - h_{Mc}(t)\right] - P_{RF} - \Phi_{ds}(t), \quad (23)$$

where moisture contents are $x(t)$ and specific enthalpies are $h(t)$ (per kg dry air); $x_{cs}(t)$ and $h_{cs}(t)$ are for air of air conditioned space; $x_{Mc}(t)$, $x_{Mh}(t)$ and $h_{Mc}(t)$, $h_{Mh}(t)$ are for wet air mixture obtained by mixing of the recirculated air and fresh environmental air, in the corresponding cases when cooling or heating is carried out; $x_{dp}(t)$ and $h_{dp}(t)$ are for saturated wet air after excessive moisture extraction (from the condition that is $x_{dp}(t) = x_{ac}(t)$); and $\vartheta_w(t)$ is temperature of water for humidification.

Specific enthalpies of wet air when $\varphi \leq 1$ are defined by the following formula:

$$h_i(t) = c_{p_a}\vartheta_i(t) + x_i(t)\left[r_0 + c_{p_s}\vartheta_i(t)\right], \quad (24)$$

where: c_{p_a}, c_{p_s} and c_w are specific heat capacities at p = const. of the dry air, water steam and water, respectively; r_0 is specific evaporation heat of water steam at $0\,°C$; and $\vartheta_i(t)$ is wet air temperature.

Further specific enthalpies and moisture contents of the wet air mixtures are defined by the corresponding expressions as follows:

$$h_M(t) = g_{fa}h_{fa}(t) + g_{ra}h_{ra}(t), \quad (25)$$

$$x_M(t) = g_{fa}x_{fa}(t) + g_{ra}x_{ra}(t), \text{ where } x_{ra}(t) = x_{cs}(t) \quad (26)$$

and according to (24), (25) and (26) the temperature of the mixture is defined as follows:

$$\vartheta_M(t) = \frac{h_M(t) - r_0\,x_M(t)}{c_{p_a} + c_{p_s}x_M(t)} = \frac{g_{fa}h_{fa}(t) + g_{ra}h_{ra}(t) - r_0\left[g_{fa}x_{fa}(t) + g_{ra}x_{ra}(t)\right]}{c_{p_a} + c_{p_s}\left[g_{fa}x_{fa}(t) + g_{ra}x_{ra}(t)\right]}, \quad (27)$$

where: $g_{fa} = \dot{m}_{fa}/\dot{m}_{sa}$ and $g_{ra} = \dot{m}_{ra}/\dot{m}_{sa}$ are mass fractions of the fresh air and recirculated air, respectively; and $\dot{m}_{fa}, \dot{m}_{ra}, \dot{m}_{sa} = \dot{m}_{fa} + \dot{m}_{ra}$ are mass flows of the fresh air, recirculated air and air mixture, respectively.

Finally, there are: P_{RAF}—power of recirculating fan electromotor, $\Phi_{ds}(t)$—heat gains of the air distribution ducting system, and $\Phi_{hls}(t)$—sensible heating loads of the climatizated space.

In the occurring air conditioning processes of both air conditioning systems (for cargo space and accommodation), it is necessary that the air is heated, apart from in the case of humidification plus cooling (Figure 9a); hence, heating loads of air heaters are defined as follows:

Humidification and heating, if $x_{cs}(t) - x_{Mh}(t) > 0$, and $\Phi_{AH}(t)_{(1)} > 0$),

$$\Phi_{AH}(t)_{(1)} = \dot{m}_{sa}\{h_{cs}(t) - h_{Mh}(t) - [x_{cs}(t) - x_{Mh}(t)]c_w\vartheta_w(t)\} + \Phi_{hls}(t) - P_{RF} - \Phi_{ds}(t), \quad (28)$$

Dehumidification and heating, if $x_{dcs}(t) - x_{Mcs}(t) < 0$ and $\Phi_{AH}(t)_{(2)} > 0$,

$$\Phi_{AH}(t)_{(2)} = \dot{m}_{sa}\left[h_{cs}(t) - h_{dp}(t)\right] + \Phi_{hls}(t), \quad (29)$$

3.3. The Sensible Heating Load

This load involves heat flows exchanging between the air conditioned space and surroundings $\Phi_{sh}(t)$, and internal heat gains $\Phi_{ih}(t)$ including the respiration heat flow

$\Phi_r(t)$, either from live products (cargo hold space) or people (accommodation), as well from contained energized equipment such as lighting, etc. For defining the $\Phi_{sh}(t)$, a concept of the quasi-static thermal network is used for both cargo hold space and accommodation, illustrated for the latter in Figure 10. By implementation of this concept, sensible heat load of a ship's air conditioned space is defined by the formula:

$$\Phi_{sh}(t) = \sum_{k=1}^{n_a} \left[\frac{\vartheta_a(t) - \vartheta_{sa}(t) - R_{k_o} a_k A_k q_{si_k}(t)}{R_{k_o} + R_{k_i}} \right] + \sum_{p=1}^{n_p} \left[\frac{\vartheta_a(t) - \vartheta_{ap}(t)}{R_{Tp}} \right] + \sum_{sw=1}^{n_s} \left[\frac{\vartheta_{sw}(t) - \vartheta_{sa}(t)}{R_{Tsw}} \right], \tag{30}$$

where contained heat resistances are as follows:

$$R_{k_O} = \frac{1}{A_k}\left[\frac{1}{\alpha_{k_O}} + \sum_{i=1}^{n_{ki}}\left(\frac{\delta_i}{\lambda_i}\right)\right], \quad R_{k_I} = \frac{1}{A_k \alpha_{k_I}}, \quad R_{Tp} = \frac{1}{A_p}\left[\frac{1}{\alpha_{pI}} + \sum_{i=1}^{n_{pi}}\left(\frac{\delta_i}{\lambda_i}\right) + \frac{1}{\alpha_{pO}}\right], \quad R_{Tsw} = \frac{1}{A_{sw}}\left[\frac{1}{\alpha_{swI}} + \sum_{i=1}^{n_{swi}}\left(\frac{\delta_i}{\lambda_i}\right) + \frac{1}{\alpha_{swO}}\right]. \tag{31}$$

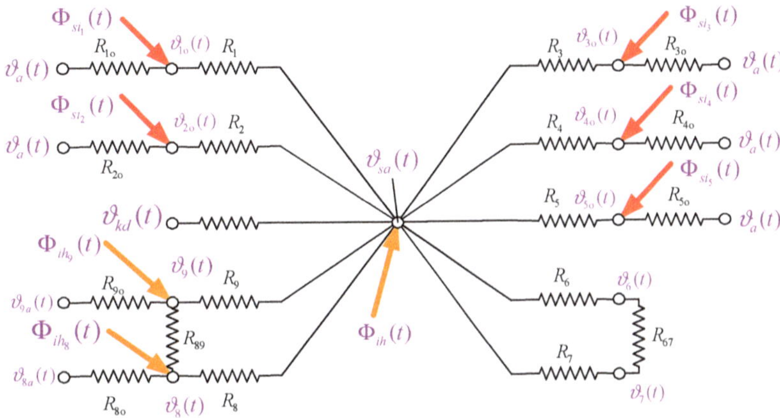

Figure 10. Simplified quasi-static thermal network of the accommodation.

Involved items in formulas (30) and (31) are: α_{k_O}, α_{k_I}, α_{p_O}, α_{p_I} and α_{sw_O}, α_{sw_I}, the convective heat transfer coefficients on the outer (indices o) and inner (indices I) flat surfaces of ship's enclosures that are in interacting with; environmental air (indices k), surrounding air non air conditioned ship's compartments (indices p), and sea (indices s_w); A_k, A_p and A_{sw} are areas of the ship's flat surfaces of the enclosures indexed by k, p and sw, respectively; and δ_i and λ_i are thicknesses and heat conductivity of the involved multilayer enclosures (n_{ki}, n_{pi} and n_{swi} are pertaining layers' number enclosures indexed, respectively, by k, p and sw), respectively.

Further, $\vartheta_{ap}(t)$ is temperature of the pth non-air conditioned ship's compartment and $\vartheta_{sw}(t)$ is sea temperature on the place corresponding to sea depth at the center of s_wth flat surface wetting by sea water, and a_k is absorbance of the kth ship's flat surface exposed to solar irradiation, the intensity of which is displayed as $q_{si_k}(t)$.

3.4. Solar Irradiation

For any kth ship's flat surface exposed to solar radiation intensity of the overall acting solar irradiation, $\Phi_{si_k}(t)$ is defined by the following formula:

$$\Phi_{si_k}(t) = A_k q_{si_k}(t) = A_k \left[q_d(t) \cos \phi_k(t) + q_{dif}(t)(1 + \cos \eta_k)/2 + q_r(t)(1 - \cos \eta_k)/2 \right], \tag{32}$$

where: $q_d(t)$, $q_{dif}(t)$ and $q_r(t)$ are intensities of the solar irradiation directing normally on the sphere tangential surface (Figure 11), diffuse sky irradiation and overall reflected solar irradiation by the sea surface, respectively. These components of irradiation are defined by the corresponding expressions as follows:

$$q_{si_k}(t) = q_{d_e}(t) \exp[-\Psi_c(t)\, B(t)\, \csc\Theta(t)], \tag{33}$$

$$q_{dif}(t) = q_d(t)\, D(t_Y), . \tag{34}$$

$$q_r(t) = r_{ss}(t)\big[q_d(t)\sin\Theta(t) + q_{dif}(t)\big], \tag{35}$$

where $D(t_Y) = 0.0904 - 0.04116\cos[\omega_G(t_Y - t_o)]$ is the diffuse irradiation parameter depending on changeable amounts of the moisture and dust particles in the atmosphere [41], $q_{dif}(t)$ and $q_r(t)$ are the intensities, and $r_s(t) = 1 - 0.95\sin\Theta(t)$ is the reflectivity of the sea surface.

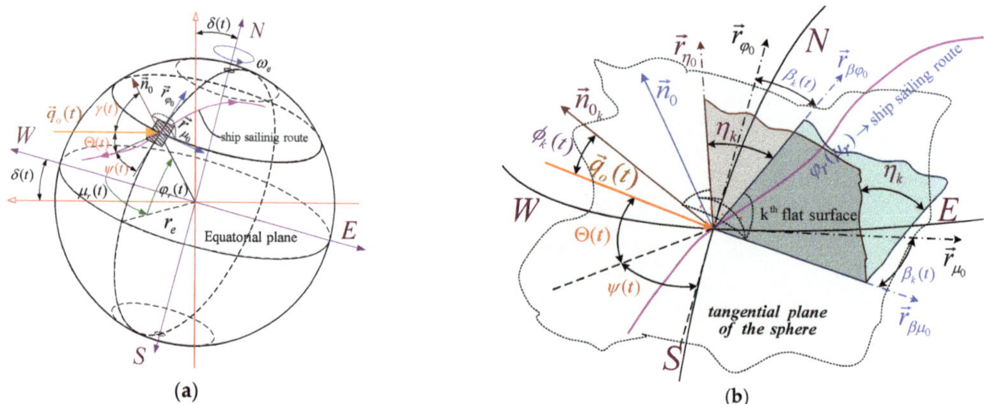

Figure 11. Main parameters of the solar geometry on the sailing route; (**a**) for tangential sphere surface and (**b**) for arbitrary place *k*th ship's flat surface.

By using the solar constant $q_S = 1.373\ \text{kW/m}^2$, the direct extraterrestrial irradiation is defined by approximated expression $\tilde{q}_{d_e} = q_S[1 + 0.033\cos(\omega_Y t_Y)]$ [42], or more precisely by the formula:

$$q_{d_e}(t) = \frac{\sigma T_s^4 \bar{r}_S^2}{[R_{SE}(t) - \bar{r}_S - \bar{r}_a]^2}, \tag{36}$$

where $\bar{r}_S = 6.923\cdot 10^5$ km is the mean Sun chromosphere radius, $\bar{r}_a = 6467.5$ km is the mean radius of the Earth's atmosphere, $\sigma T_s^4(\bar{r}_S/\overline{R}_{SE})^2 = q_S$, $\sigma = 5.67\cdot 10^{-8}\,\text{W/(m}^2\text{K}^4)$ is the Stefan–Boltzmann constant, $T_S \approx 5791$ K is the absolute chromosphere temperature, and module of radii vector of the Earth's path around of the Sun is:

$$R_{SE}(t) = \frac{a(1 - \in^2)}{1 + \in\cos[\Omega(t_Y, t_0)]}, \tag{37}$$

where $a = 149.5\cdot 10^6$ km are the big semi-axes and is the $\in = 0.017$ eccentricity of the elliptical path, and finally the revolution angle of the Earth around of Sun is defined, according to [43], by the formula:

$$\Omega(t_Y, t_0) = \omega_Y(t_Y - t_0) + \sum_{i=1}^{2} k_i \sin[2\omega_Y(t_Y - t_0)], \tag{38}$$

where $k_1 = 0.033985$ and $k_2 = 3.61128 \cdot 10^{-4}$ are contained constants, $\omega_Y = 2\pi/\tau_Y$ is the mean angular velocity of Earth's revolution around the Sun, and $\tau_Y \approx 8766\,\text{h}$ is the revolution period.

The atmosphere attenuation factor is defined by $B(t_Y) = 0.17164 - 0.034686 \cos[\omega_G(t_Y - t_o)]$ [40], while irradiation attenuation produced by the cloudiness depends on the factor $\Psi_c(\varphi_r, t_Y)$.

That is simulated by applying formula (3) on the obtainable data from isonephs maps (line connecting the places with equal mean cloudiness), which is illustrated in Figure 12a and by use results for navigation trough realistic surroundings, the unsteady attenuation factor while transport cycle is obtained as shown in Figure 12b.

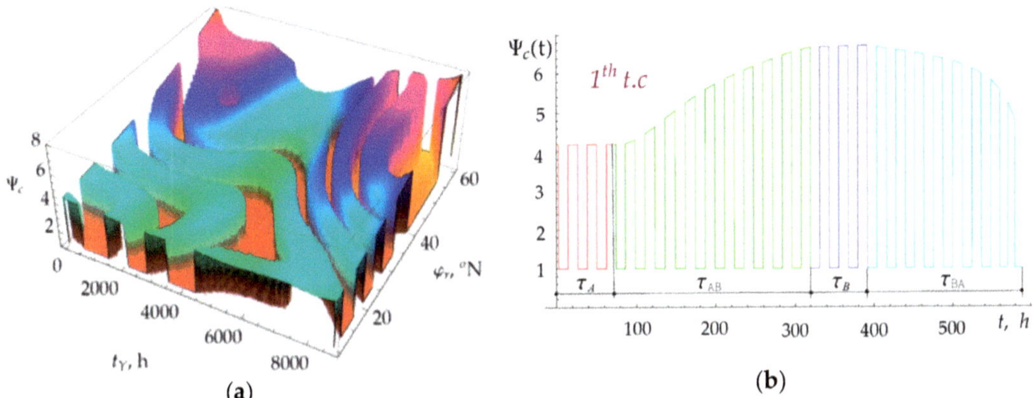

Figure 12. Equivalent attenuation of the irradiation by the cloudiness: (a) $\Psi_c(\varphi_r, t_Y)$ and (b) $\Psi_c(t)_{1.tc}$.

Sun high $\sin\Theta(t)$ is defined by the scalar product of unity vectors, the normal of the tangential surface of the sphere $\vec{n}_0(t)$ and acting Sun radiation $\vec{q}_0(t)$ (Figure 11):

$$\sin\Theta(t) = \vec{n}_0(t) \cdot \vec{q}_0(t), \tag{39}$$

where unity vectors are involved as follows:

$$\vec{n}_0(t) = \cos\varphi(t)\cos\mu(t)\,\vec{i} + \cos\varphi(t)\sin\mu(t)\,\vec{j} + \sin\varphi(t)\,\vec{k}, \tag{40}$$

$$\vec{q}_0(t) = \cos\delta(t)\,\vec{i} + \sin\delta(t)\,\vec{k}, \tag{41}$$

where $\delta(t_Y)$ is the Sun's inclination according to:

$$\delta(t_Y) = \delta_0 \cos[\Omega(t_Y, t_0) - \Omega_{\delta_0}] = \delta_0 \cos\left\{\omega_Y(t_Y - t_0 - \tilde{\tau}_{\delta_0}) + \sum_{i=1}^{2} k_i \sin[i\,\omega_Y(t_Y - t_0)]\right\}. \tag{42}$$

Further, $\cos\phi_k(t)$ is defined by formula:

$$\cos\phi_k(t) = \vec{n}_{k_0}(t) \cdot \vec{q}_0(t), \tag{43}$$

where $\vec{n}_{k_0}(t)$ is the unity normal of kth arbitrarily placed ship's flat rectangular surface, defined by vector's product of the pertaining unity base vectors as follows:

$$\begin{aligned} \vec{n}_{k_0}(t) &= \vec{r}_{\beta_0}(t) \times \vec{r}_{\eta_0}(t), \quad \vec{r}_{\beta_0}(t) = \cos\beta_k\,\vec{r}_{\mu_0}(t) - \sin\beta_k\,\vec{r}_{\varphi_0}(t), \\ \vec{r}_{\eta_0}(t) &= \cos\eta_k\sin\beta_k\,\vec{r}_{\mu_0}(t) + \cos\eta_k\cos\beta_k\,\vec{r}_{\varphi_0}(t) + \sin\eta_k\,\vec{n}_0(t) \end{aligned} \tag{44}$$

In this formula, contained orts are:

$$\vec{r}_{\mu_0}(t) \;=\; \frac{1}{\left|\vec{r}_\mu\right|}\,\frac{\partial \vec{n}_0(t)}{\partial \mu}, \quad \vec{r}_{\varphi_0}(t) \;=\; \frac{1}{\left|\vec{r}_\varphi\right|}\,\frac{\partial \vec{n}_0(t)}{\partial \varphi}. \tag{45}$$

Further, contained angles β_k and η_k present azimuth and elevation of a kth ship's flat surfaces, respectively, as shown in Figure 11b.

Other parameters involved in above expressions are: $\delta_0 = -23.45^0$—min. inclination for north hemisphere, $t_0 = 68\,\text{h}$—time shift for perihelion, $\widetilde{\tau}_{\delta_0} = -230\,\text{h}$—time shift for winter solstice to which corresponds angle shift $\Omega_{\delta_0} = \Omega_{\delta_0}(\widetilde{\tau}_{\delta_0}) = -12.660$; finally, of the course in modelling through a transport sequence, ts timing relations $t_Y = t + \tau_{0ts}$ must be taken into account. Figure 13 illustrates overall solar irradiations for the (a) navigation A to B, and (b) navigation B to A.

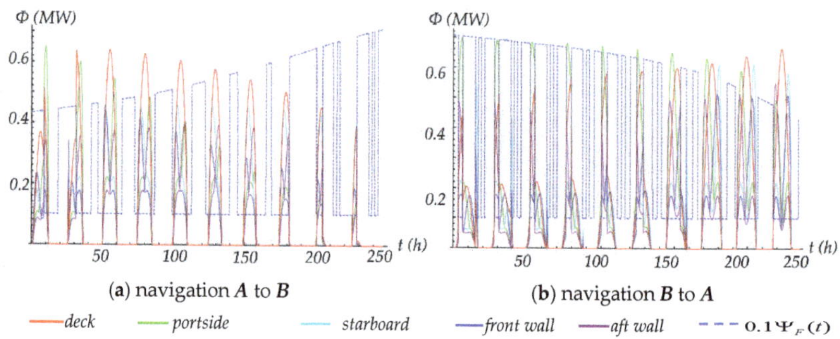

Figure 13. Overall solar irradiation for the ship's accommodation enclosures during navigation through first transport cycle.

Finally, by applying the above-mentioned for characteristic quasi-static ship's energy items during the navigation through any transport cycle, the example illustrated by Figure 14 for unsteady heat load is obtained.

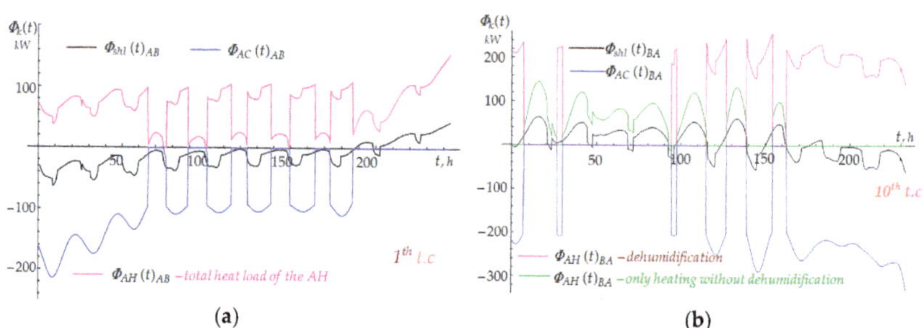

Figure 14. Characteristics of heat loads for ship's accommodation during navigation: (**a**) from A to B through 1st transport cycle and (**b**) during navigation from B to A through 10th transport cycle.

Taking into account that all design parameters of the involved heat exchangers are predetermined on the extreme base values of pertaining heat loads, such as heat exchange areas, overall heat transfer coefficients, mean logarithmic temperature, etc., and assuming that heat capacities of the involved cooling (heating) media (fresh water and brain), as well as heat transfer capabilities of exchangers, are unchangeable during navigation, it can be possible to determine the corresponding quasi-static balanced temperatures on the side cooling media contained in LBARU and AARU.

As an example, Figure 15 illustrates design and unsteady temperatures of the involved cooling media in the air conditioned system of the accommodation, while expressions (46) and (47) define corresponding heat transfer capability and capacities of conditioned air and cooling fresh water:

$$K_{ex} = k_{ex} A_{ex} = \frac{|\Phi_{ex}|_{max}}{\Delta \vartheta m_d} = \frac{|\Phi_{ex}|_{max}}{\delta_{1d} - \delta_{2d}} \ln\left(\frac{\delta_{1d}}{\delta_{2d}}\right), \tag{46}$$

$$C_{ac}(t) = \frac{\Phi_{AC}(t)}{\vartheta'_{ac}(t) - \vartheta''_{ac}(t)}, \quad C_{cw} = \frac{\Phi_{AC}(t)}{\vartheta''_{sw}(t) - \vartheta'_{sw}(t)}, \tag{47}$$

where δ_{1d} and δ_{2d} are design temperature differences on the ends of the counter flow exchanger.

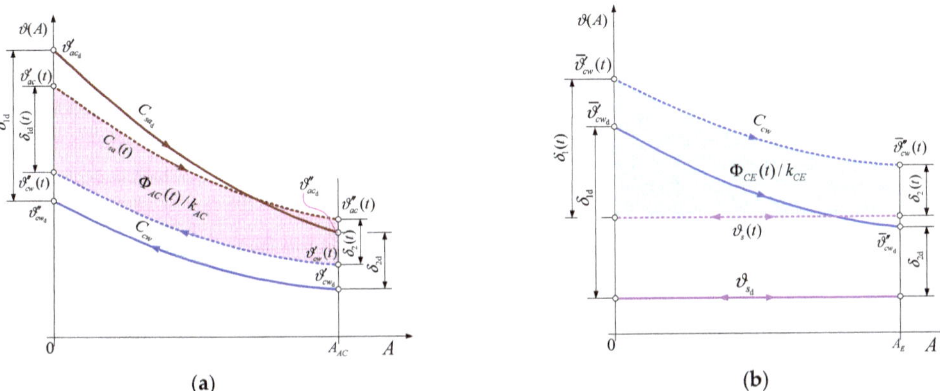

Figure 15. Quasi-static balanced temperatures for (**a**) conditioned air and cooling water in AC and (**b**) cooling water and cooling steam in evaporator of LBARU-a.

Further, for characteristics of unsteady balancing temperatures the corresponding expressions are determined:

The inlet temperature of the cooling fresh water in AC is:

$$\vartheta'_{cw}(t) = \vartheta'_{ac}(t) + \frac{\Phi_{AC}(t)}{C_{cw}}\left\{1 - \exp\left[K_{AC}\left(\frac{1}{C_{cw}} - \frac{1}{C_{ac}(t)}\right)\right]\right\}^{-1}, \tag{48}$$

The outlet temperature of the cooling fresh water in AC is:

$$\vartheta''_{cw}(t) = \vartheta'_{ac}(t) + \frac{\Phi_{AC}(t)}{C_{cw}}\exp\left[K_{AC}\left(\frac{1}{C_{cw}} - \frac{1}{C_{ac}(t)}\right)\right]\left\{1 - \exp\left[K_{AC}\left(\frac{1}{C_{cw}} - \frac{1}{C_{ac}(t)}\right)\right]\right\}^{-1}, \tag{49}$$

The temperature of the cooling media in the cooling evaporator is:

$$\vartheta_s(t) = \tilde{\vartheta}'_{cw}(t) - \frac{\Phi_{CE}(t)}{C_{cw}}\exp\left(\frac{K_{CE}}{C_{cw}}\right)\left[1 - \exp\left(\frac{K_{CE}}{C_{cw}}\right)\right]^{-1}, \tag{50}$$

The temperature of the heating media (LP water steam) in the air heater is:

$$\vartheta_s(t) = \vartheta'_{ah}(t) + \frac{\Phi_{AH}(t)}{C_{sa}}\left[1 - \exp\left(\frac{K_{Ah}}{C_{sa}}\right)\right]^{-1}. \tag{51}$$

By adding equal amounts of the heat gains of the cooling media conducting system $\Phi_{CS}(t) = P_{cp} + \Phi_{cs}(t)$ on the inlet and outlet of the cooling evaporator, $\tilde{\vartheta}'_{cw}(t) = \vartheta''_{cw}(t) + 0.5\,\Phi_{CS}(t)\,\tilde{\vartheta}''_{cw}(t) = \vartheta'_{cw}(t) + 0.5\,\Phi_{CS}(t)$ are obtained for the inlet and outlet temperatures of the secondary cooling media (fresh water or brine) in the cooling evaporator. By using above mentioned quasi-static balanced temperature of the involved media in AACS are obtained, as illustrated in Figure 16.

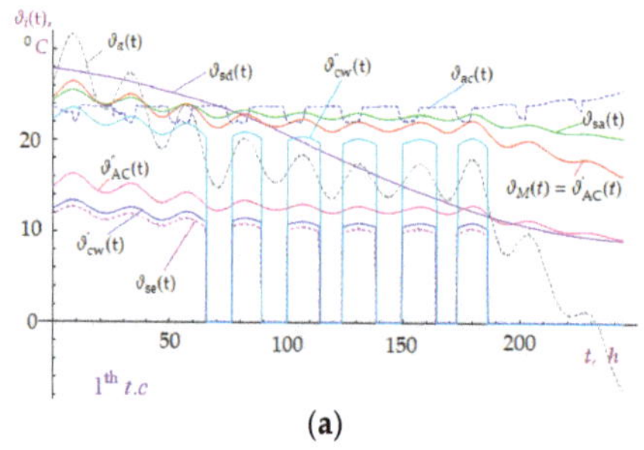

(a)

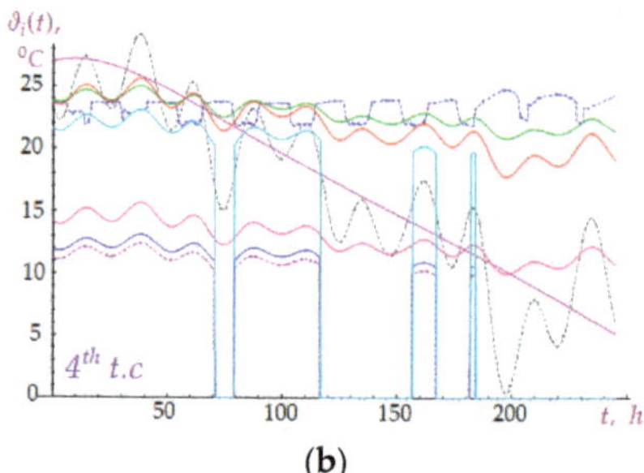

(b)

Figure 16. The characteristics of quasi-static balanced temperatures of involved process's media in the ship's accommodation climatization system; (**a**) for 1st transport cycle and (**b**) for the 4th transport cycle.

4. Preliminary Economic, Energy, and Environmental Indicators of the Application of the Trigeneration Concept

By applying the appropriate original mathematical model for unsteady thermal loads of the ship and the unsteady cogeneration heating effect, the techno-energy applicability of the trigeneration concept on the ships intended for the transport of moderate refrigerated bulk "live" and "dead" commodities is examined. In this sense, a one-year reefer service through a simplified orthodromic navigation route between terminal *A* (corresponding to Marcaibo, Venezuela) and terminal *B* (corresponding to an estuary of the river Laba, Hamburg) for reefer which main particulars are given in Table 1 is simulated. The simulation includes 14 transport cycles; both navigations loaded reefer and steaming in the ballast, where the continuous change of energy effects and energy loads that are balanced are taking the place. The application of the double-stage cogeneration system that balances both medium-pressure (MP) heating load (thermal load resulting from AARU cooker and the HFOFH) and low-pressure (LP) heating load (thermal load that results from all other consumers of heat energy including LBARU cooker) is considered for the typical reefer as follows:

Table 1. The model ship principal particulars.

Principal Particulars	
Length, overall	162.3 m
Length, between perpendiculars	150 m
Breadth, middle	23.4 m
Depth, middle	13.2 m
Draught, scantling middle	9 m
Draught, ballast condition	6.3 m
Deadweight	13,390 tones
Hold Capacity	16,999.1 m^3
Main Engine MAN B&W S60MC-C	
Output MCR	13.56 MW at 105 min^{-1}
Service speed	17 knots

4.1. Energy Sufficiency

Combined display of total unsteady heating load of a loaded ship with characteristic bulk cargo (Banana—B, Citrus—C, deciduous fruit—DF, frozen meat—FM), and unsteady cogeneration heating effect, for two characteristic transport cycles, are illustrated in Figure 17. It is shown that the MP cogeneration heating effect is sufficient to balance the generated unsteady MP heating loads in the transport of all above-mentioned commodities.

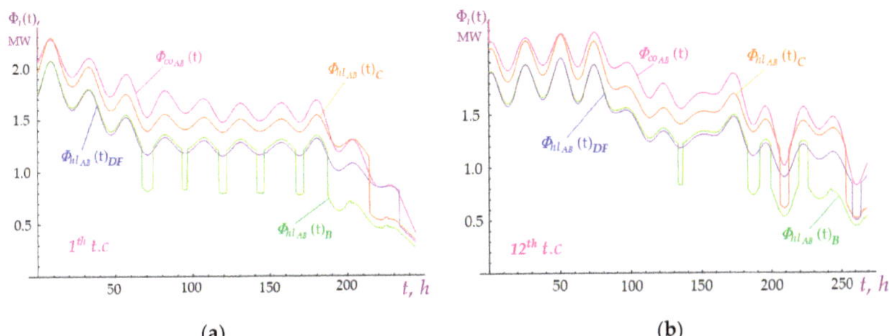

Figure 17. MP heating load $\Phi_{hl_{AB}}(t)$ and heating effect of MP cogeneration $\Phi_{co_{AB}}(t)$ for various commodities during 1st and 12th transport cycle for navigation (**a**) from *A* to *B*, and (**b**) *B* to *A*.

Possible, short-term insufficiency of the double-stage cogeneration effect in balancing the generated heating load can easily be remedied by the short-term changing of microclimate parameters (relatively small increase in relative humidity, or a small reduction in the proportion of fresh air g_{fa}), as illustrated in Figure 18.

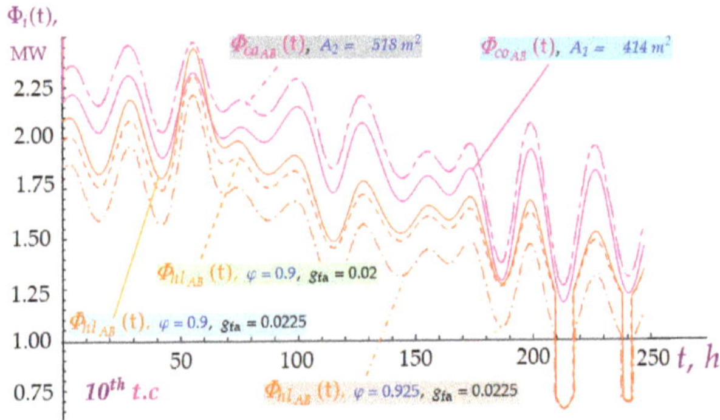

Figure 18. Cogeneration effect $\Phi_{co_{AB}}(t)$ for different evaporator area and heating loads $\Phi_{hl_{AB}}(t)_C$ for different citrus storage microclimatic parameters φ_a and g_{fa}, during 10th transport cycle.

The total LP unsteady heating load of the reefer transporting characteristic commodities for certain transport cycles is obtained by summation of the contained basic LP heating loads, as illustrated in Figure 19.

The corresponding total (low-pressure and high-pressure) heating loads for the ballast voyage when maintained storage temperature of cargo was predetermined by the product to be transported in the opposite direction, for the most demanding heating loads (frozen meat—FM, deciduous fruit—DF, orange—O, and bananas—B), is shown in Figure 20. In addition, Figure 5 shows a common view unsteady heating effect of the medium-pressure cogeneration system and the characteristic load caused by the overall thermal load of the ship, which shows energy sufficiency of the high-pressure cogeneration system in balancing the overall heating load of the ship. During practically any condition in the ballast voyage, there is no need for activating the low-pressure cogeneration system.

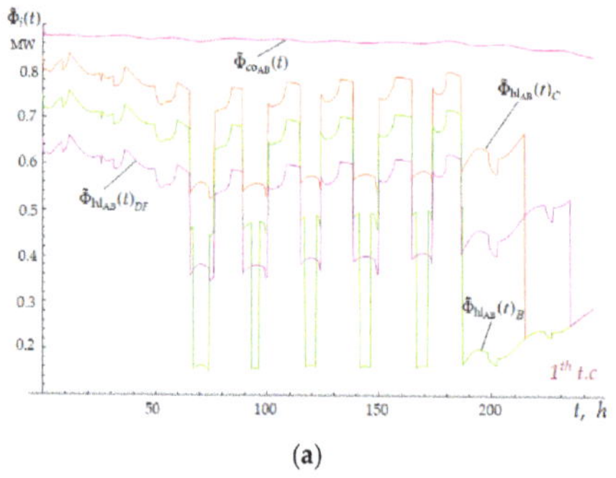

(a)

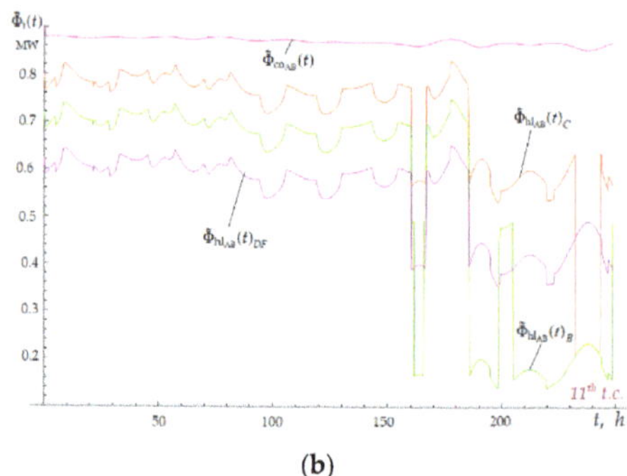

(b)

Figure 19. Heating effect of LP cogeneration $\widetilde{\Phi}_{co_{AB}}(t)$ and LP heating load $\widetilde{\Phi}_{hl_{AB}}(t)$ for various commodities during 1st and 11th transport cycle for navigation from **A** to **B** for (**a**) 1st and (**b**) 11th transport cycle.

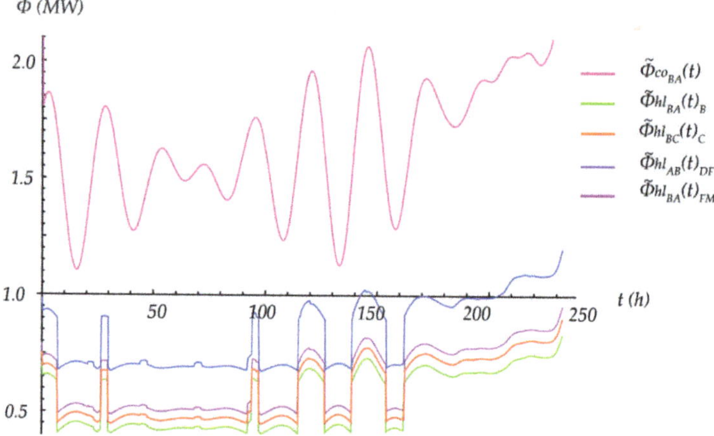

Figure 20. Unsteady cogeneration heating effect and total unsteady low-pressure heating loads for products **B**, **C**, **DF** and **FM** during the ballast voyage, from **B** to **A** for 10th transport cycle.

4.2. Economic Impacts

Furthermore, in brief, the simplified economic review of the positive economic effects of applying DETES based on two-pressure cogeneration systems for reefer ships is given.

In this regard, primarily referring to the estimated scenario of the commercial engagements of a ship, we determine fuel savings compared to competitive design solutions with CRPs in the CACS and in the AACS. Fuel saving corresponds to the fuel consumption of the CRP in the CACS during the steaming in ballast for several characteristic transport cycles, illustrated in Figure 21, while saving fuel for the CRP in AACS for several characteristic transport cycles, illustrated Figure 22.

Fuel consumption accounting for the operation of the compressor plant to balance the occurring storage cooling load during navigation of a loaded ship for two characteristic transport cycles, through a one-year period, is illustrated in Figure 23.

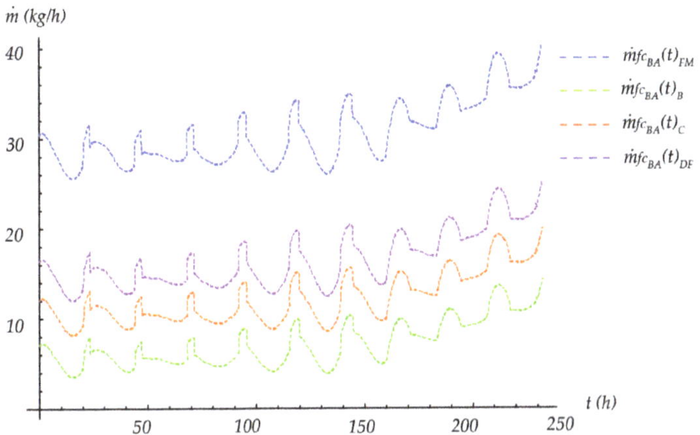

Figure 21. Fuel savings (mass flow kg/h) in the CACS for products *B*, *C*, *DF* and *FM* during navigation of unloaded reefer from *B* to *A* for 10th transport cycles.

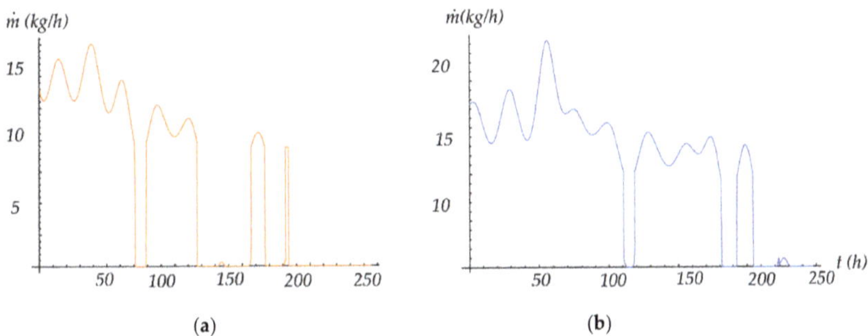

Figure 22. Fuel savings in air conditioning systems of the ship's accommodation for 1th and 10th transport cycles during navigation of loaded ship from *A* to *B* for; (**a**) 1st and (**b**) 10th transport cycle.

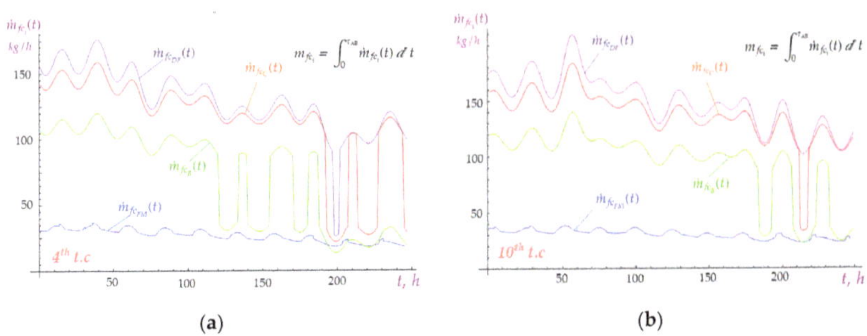

Figure 23. Fuel savings in maintenance of the prescribed microclimate in the storage space for various products during navigation of loaded reefer from *A* to *B* for; (**a**) 4th and (**b**) 10th transport cycles.

By the numerical integration of fuel consumption (mass flow) over the navigation period, the following results are obtained: average fuel savings for the AACS over one transport cycle and one year of service which consists of 14 transport cycles, as well as the total savings for the 25-year ship lifetime period (SLT), were taken into account alongside the yearly trend price of low sulfur heavy fuel oil (LSHFO, containing sulfur

$s \leq 0.5\,\%,\ mas.$), $\overline{p}_{2021} \approx 500$ US\$/t [44], giving a total saving of FSAC = 1.3146 mil. US\$. In this case, in which the reefer in one direction navigates loaded with typical product, and the second direction navigates in ballast, the obtained results of the possible combinations of transport over one year of service, by the numerical integration of the curves of mass fuel consumption, are listed in Table 2.

Table 2. Fuel savings for a combination laden voyage from *A* to *B* and a ballast voyage (*BL*) from *B* to *A*, where: *t.c.*—transport cycle, *B*—bananas, *C*—citruses, *DC*—deciduous fruits, *FM*—frozen meat.

t.c.	*B* + *BL*	*C* + *BL*	*DF* + *BL*	*FM* + *BL*
1. t. c. *A-B* + *B-A*	19.406 + 0.855	28.298 + 1.932	31.373 + 2.835	6.500 + 5.977
2. t. c. *A-B* + *B-A*	16.275 + 0.875	29.400 + 1.988	31.923 + 2.917	6.606 + 6.035
3. t. c. *A-B* + *B-A*	17.912 + 0.925	30.600 + 2.005	33.410 + 2.922	7.021 + 6.340
4. t. c. *A-B* + *B-A*	17.545 + 0.962	28.237 + 2.037	31.840 + 2.942	6.615 + 6.062
5. t. c. *A-B* + *B-A*	17.418 + 1.212	27.33 + 2.237	30.949 + 3.242	6.609 + 6.662
6. t. c. *A-B* + *B-A*	21.299 + 1.454	31.513 + 2.655	34.052 + 3.482	6.797 + 6.934
7. t. c. *A-B* + *B-A*	24.680 + 1.695	33.776 + 2.855	36.721 + 3.861	7.125 + 7.132
8. t. c. *A-B* + *B-A*	23.225 + 1.705	33.507 + 2.915	36.306 + 3.902	7.222 + 7.144
9. t. c. *A-B* + *B-A*	23.449 + 1.695	33.740 + 2.895	36.369 + 3.875	7.309 + 7.115
10. t. c. *A-B* + *B-A*	23.459 + 1.663	33.781 + 2.825	36.920 + 3.830	7.555 + 7.109
11. t. c. *A-B* + *B-A*	22.979 + 0.972	33.223 + 2.107	36.528 + 2.992	7.486 + 6.102
12. t. c. *A-B* + *B-A*	23.065 + 0.912	34.043 + 2.011	37.744 + 2.900	7.795 + 6.058
13. t. c. *A-B* + *B-A*	19.778 + 0.885	30.785 + 1.996	33.561 + 2.911	6.916 + 6.095
14. t. c. *A-B* + *B-A*	19.350 + 0.865	28.220 + 1.944	31.651 + 2.856	6.702 + 5.985
	Σ = 306.515 tones	Σ = 468.855 tones	Σ = 524.814 tones	Σ = 189.068 tones
T_{SLT} = 25 years p = 500 US\$/ton	7662.875 tones	11,721.34 tones	1312.35 tones	4736.7 tones
T_{SLT} = 25 years	3.832 mil. US\$	5.861 mil. US\$	6.560 mil. US\$	2.368 mil. US\$

In the economically more favorable case, which is the case of a complete engagement of the ship in transporting of moderate refrigerated commodities, the various possible combinations of transport are illustrated in Table 3.

Table 3. Savings for various combinations of cargo in the event of total transport engagement.

t.c.	*B*	*C*	*DF*	*FM*
1. t. c. *A-B*	19.406	28.298	31.373	6.500
2. t. c. *A-B*	16.275	29.400	31.923	6.606
3. t. c. *A-B*	17.912	30.600	33.410	7.021
4. t. c. *A-B*	17.545	28.237	31.840	6.615
5. t. c. *A-B*	17.418	27.33	30.949	6.609
6. t. c. *A-B*	21.299	31.513	34.052	6.797
7. t. c. *A-B*	24.680	33.776	36.721	7.125
8. t. c. *A-B*	23.225	33.507	36.306	7.222
9. t. c. *A-B*	23.449	33.740	36.369	7.309
10. t. c. *A-B*	23.459	33.781	36.920	7.555
11. t. c. *A-B*	22.979	33.223	36.528	7.486
12. t. c. *A-B*	23.065	34.043	37.744	7.795
13. t. c. *A-B*	19.778	30.785	33.561	6.916
14. t. c. *A-B*	19.350	28.220	31.651	6.702
	Σ = 289.840 tones	Σ = 436.453 tones	Σ = 479.340 tones	Σ = 98.318 tones
p = 500 US\$/ton		1 year	TLP 25 year	
Combination 1	*B* + *C*	346.146 US\$	9.079 mil. US\$	
Combination 2	*B* + *DF*	384.594 US\$	9.615 mil. US\$	
Combination 3	*B* + *FM*	194.079 US\$	4.852 mil. US\$	Average savings
Combination 4	*C* + *DF*	457.901 US\$	11.448 mil. US\$	AS = 8.15 mil. US\$
Combination 5	*C* + *FM*	267.386 US\$	6.685 mil. US\$	
Combination 6	*FM* + *DF*	288.832 US\$	7.221 mil. US\$	

Taking into account the current specific price [45], CRP, ARU and the CP, which are $p_{CRP} = 137,000$ US\$/MWce, $p_{ARU} = 411,000$ US\$/MWce, and $p_{CP} = 120,000$ US\$/MWhe, respectively, in the economically most unfavorable case when the 100% redundancy cooling-heating system of the ship is required, both the CRP and ARU rated cooling effects are installed, and for the same reasons FSG has an extra heating effect installed, the following net cost-benefit equation is obtained:

$$P_{EB} = p_{LSHFO} \cdot m_{FS} - I_{AA} - I_{LBA} - I_{FSG}, \tag{52}$$

where, I_{AA}, I_{LBA} and I_{FSG} are the present investment values of AARU, and LBARU and FSG, respectively. Contained investment values are defined by the following terms:

$$I_{ARU} = p_{ARU} \cdot \Phi_{CE}, \quad \Phi_{CE} = \Phi_{AAce} + \Phi_{LBAce}, \tag{53}$$

$$I_{FSG} = p_{FSG} \cdot (\Phi_{TC} - \Phi_{CRP}), \tag{54}$$

where are the nominal cooling effects of the AARU and LBARU; $\Phi_{AAce} = 1.85$ MW and $\Phi_{LBAce} = 0.35$ MW, correspondingly, while the nominal cogeneration heating effects are; for trigeneration energy concept (DECES+ARU) $\Phi_{TC} = 3$ MW, and for conventional energy concept (DECES+CRP) $\Phi_{CRP} = 1$ MW.

Following the above, the total, present, additional investment cost of $I_{TOT} = 1.045$ mil. US\$ is obtained, which is less than the economic fuel savings achieved by applying LBARU in the AACS (FSAC = 1.3146 mil. US\$). Accordingly, for economic gains, the fuel savings m_{FS} obtained by applying AARU in CACS for various transport combinations through the 25-year economic life of the ship, alongside the same savings that are contained in Tables 2 and 3, are illustrated in Figure 24.

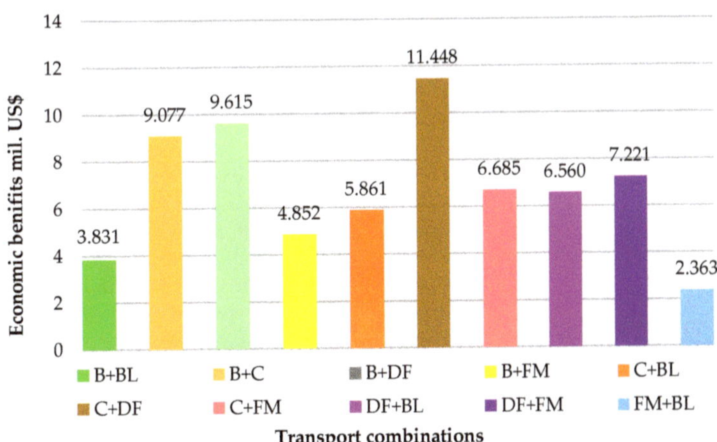

Figure 24. Estimated economic benefits for the characteristic combination of transport over the 25-year ship's lifetime.

Based on the above it can be concluded that the economic effects of the application of trigeneration energy systems on motor ships intended for the transport of moderately refrigerated commodities are appreciable. By the extrapolating of techno-economic parameters (the price of heavy fuel and reefers' hold capacity), the value of preliminary economic benefits would be increased. For example, for a reefer twice the capacity of that considered, at the same price of fuel, profits would be nearly doubled, while at a price of fuel at least twice that of the current (which is realistic to expect after a five-year period), economic profit for the considered ship would be double, but for a ship with double the capacity it would increase nearly fourfold.

4.3. Environmental Impact

Most environmentally harmful substances are manifested through the emission of environmentally harmful combustion gases: CO_2, CO, SO_2, SO_3, and various nitrogen oxides NO_x. By reducing fuel consumption, these emissions are almost proportionally reduced, but also in the assumed complete combustion of fuel reduced mass amounts of CO_2 and SO_2 will be easily achieved based on of the following terms:

$$\Delta m_{CO_2} = c \cdot \frac{M_{CO_2}}{M_C} \cdot m_F \cdot \tau_{SLT}, \qquad \Delta m_{SO_2} = s \cdot \frac{M_{SO_2}}{M_S} \cdot m_F \cdot \tau_{SLT} \qquad (55)$$

where $c = 0.84\ \mathrm{kg_c/kg_f}$; $s = 0.005\ \mathrm{kg_s/kg_f}$ are the average concentrations of carbon and sulfur in heavy fuel, respectively; $M_{CO_2} = 44$, $M_C = 12$, $M_S = 32$ and $M_{SO_2} = 64$ kg/kmol are appropriate molar mass; m_F is annual fuel savings; and τ_{SLT} is the ship's economic lifetime.

The calculation results for the 25-year ship's economic lifetime are given in the following Table 4. The application of trigeneration energy systems on ships intended for transporting moderately cooled products considerably contributes to reducing environmentally negative effects.

Table 4. Emissions reduction for various combinations of cargo in the event of total transport engagement.

	Transport Combination	m_F [tones]	Δm_{CO_2} [tones]	Δm_{SO_2} [tones]
1.	*B + BL*	306.515	23,601.655	76.629
2.	*C + BL*	468.855	36,101.835	117.214
3.	*DF + BL*	524.814	40,410.678	131.204
4.	*FM + BL*	189.068	14,558.236	47.267
5.	*B + C*	726.293	55,924.561	181.573
6.	*B + DF*	769.187	59,227.399	192.297
7.	*B + FM.*	388.158	29,888.166	97.039
8.	*C + DF*	915.800	70,516.600	228.950
9.	*C + FM*	534.771	41,177.367	133.693
10.	*DF + FM*	577.665	44,480.205	144.416

This confirms the adequacy of diesel engine trigeneration energy systems on motor reefers in balancing the occurring overall heating load during the voyage, as well as the respectable positive economic, energy, and environmental effects of its application. Therefore, the conclusion is that the application of diesel engine trigeneration energy systems on ships intended for the transport of moderately refrigerated cargo is appreciably economically beneficial, and environmentally acceptable. In addition to the considered types of ships, trigeneration energy systems can also be applied on modern passenger ships, where the necessary energy for ensuring a high level of comfort is very high and often exceeds the power of the ship's propulsion engines.

5. Conclusions

By applying mathematical models of characteristic environment sizes (based on available WMO data), which interact with the ship system and with developed mathematical models of quasi-static characteristics of energy components involved with the diesel engine trigeneration energy system, we created models of unsteady energy balancing during the ship's characteristic operating intervals. The concept interrelates the trajectory of a vessel and the wave conditions with a detailed described quasi-static model of the energetic, economic and environmental performance of a marine trigeneration unit installed on-board. The developed mathematical models of quasi-static characteristics for single-stage AARUs and LBARUs, enable modelling of unsteady thermal loads of their cookers, absorbers and condensers. Furthermore, the models enable the management of absorption cooling processes in the most energy-efficient way within the emerging unsteady environment.

Within the defined technically possible area of application of absorption cooling, with the application of LBARU for balancing the refrigeration load in the AACS, the application of AARU in CACS is tested for several characteristic moderately refrigerated cargoes: frozen meat, not frozen vacuum packed meat, bananas, citrus and deciduous fruits, where it is shown that on the increasing of the thermal loads, the most influenced variables are: higher intensity of refreshing, lower relative humidity and to a lesser extent the respiratory heat flux of the transported commodity. This implies that the total refrigeration loads when transporting live cargo at the prescribed storage temperature are almost an order of magnitude larger than the corresponding sensible heat loads. Applying these models over a one-year period of service of the ship, which includes 14 repeated transport cycles, it was shown that the trigeneration system is energy sufficient to balance the overall unsteady heat load of ships intended for the transport of moderately refrigerated "dead" and "live" cargo. Quasi-static characteristics of the (CRP) enable the determination of the appropriate fuel consumption that falls on its drive when it balances the unsteady refrigeration load during the navigation.

By integrating the quasi-static fuel consumption curves of the CRP over the intervals of navigation routes for all contained 14 transport cycles, and summing the obtained values, one-year fuel consumption is obtained, i.e., one-year fuel savings. Based on one-year fuel savings, for the estimated economic life of the ship, taking into account the current fuel price as well as additional investment costs of trigeneration and cogeneration plant, significant economic savings were obtained for certain transport combinations, which fully confirm the economically positive effects of the trigeneration concept on ships intended for the transport of moderately refrigerated cargo.

Finally, using the mean reference values for the composition of the LSHFO, and assuming complete combustion, no negligible amounts of reductions in greenhouse gas emissions, especially CO_2, were obtained. In conclusion, according to the above, the energy, economic and environmental effects of the application of diesel engine trigeneration energy systems on ships intended for the transport of moderately refrigerated cargo are significantly positive. In addition, the trigeneration energy concept can be applied on modern passenger ships, where the energy amounts required to ensure a high degree of comfort are significant, and often exceed the installation power of propulsion diesel engines.

Author Contributions: Conceptualization, I.G. (Ivan Gospić) and I.P.; methodology, I.G. (Ivan Gospić); software, I.G. (Ivica Glavan); validation, I.G. (Ivan Gospić), I.P. and V.M.; formal analysis, I.G. (Ivan Gospić); investigation, I.G. (Ivan Gospić); resources, I.G. (Ivica Glavan); data curation, I.G. (Ivica Glavan); writing—original draft preparation, I.G. (Ivan Gospić); writing—review and editing, I.P.; visualization, I.G. (Ivica Glavan); supervision, V.M.; project administration, I.P.; funding acquisition, V.M. All authors have read and agreed to the published version of the manuscript.

Funding: This research received no external funding.

Institutional Review Board Statement: Not applicable.

Informed Consent Statement: Not applicable.

Data Availability Statement: Not applicable.

Conflicts of Interest: The authors declare no conflict of interest.

Abbreviations

Nomenclature
Latin symbols

A	area, m^2
B	atmosphere attenuation factor
C	flow stream capacity, W/K
I_p	polar moment inertia of rotating mass, kg m^2

D	diffusely irradiation parameter,
I	investment cost, US$
K	heat transfer capability, W/K
P	power, W
R	total thermal resistance, m^{2}K/W
R_{SE}	module of radii vector between Sun and Earth, m
Q_p	propeller torque, Nm
T	absolute temperature, K
T_p	propeller trust, N
a	big semi-axes of Earth eliptic path, km
c	specific heat capacity, J/(kgK)
g	mass fractions a component in the mixture
h	specific enthalpy, J/kg
$\in$	eccentricity of Earth elliptical path
k	overall heat transfer coefficient, W/m^{2}K
k_F	diesel engine load factor
m	mass, kg
$\dot{m}$	mass flow rate, kg/h
n	rotational speed, rev/min
r	radius, m
r_0	specific evaporation heat of water steam at 0 °C, J/kg
q	heat flow intensity, W/m^2
t	time, s
$\hat{t}$	trust deduction coefficient
x	moisture content in wet air,
v	Speed, velocity, m/s
w	wake coefficient

Greek symbols

Γ	sky coverageness with the clouds
$\Delta\vartheta_m$	mean logaritmic tempearture difference, K
Θ	Sun hight angle, °
Φ	heat flow, W
Ψ	cloudiness attenuation factor
Ω	revolution angle of the Earth around of Sun, °
α	convective heat transfer coefficient, W/(m^{2}K)
β	azimuth of ship's flat surface, °
δ	Sun inclination, °
δ_i	thickness of the ith layer of the multilayer wall, m
η	elevation of ship's flat surface, °
ϑ	temperature, °C
λ	heat conductivity, W/(mK)
μ_r	geographic longitude (on navigation route), °
φ_r	geographic latitude (on navigation route), °
φ	relative humidity of wet air
ϕ	angle between Solar irradiation and normal vector of a flat ship's surface, °
χ	the saturation level of the moist air
ψ	Sun azimuth, °
σ	Stefan-Boltzmann constant, W/(m^{2}K^4)
τ	time interval, s
ω	angular speed, rad/s

Abbreviations

AACS	accommodation air conditioning system
AARU	ammonia absorption refrigeration unit
ARU	absorption refrigeration unit
AC	air cooler
AF	air filter

AH	air heater
B	bananas
C	citruses (citrus fruits)
CACS	cargo storage air conditioning system
CBP	cooling brine pump
CFWP	cooling fresh water pump
COP	coefficient of performance
CRP	compression refrigeration plant
CSP	cooling sea pump
DA	diesel aggregate
DAT	daily air temperatures
DEPS	diesel engine propulsion system
DF	deciduous fruits
EF	extraction fan
FM	frozen meat
FPP	fixed pitch propeller
FSAC	fuel savings of the accommodation climatization
FSG	fired steam generator
HFO	heavy fuel oil
HFOFH	heavy fuel oil final heater
LBARU	lithium-bromide absorption refrigeration unit
LHV	lower heating value
LNG	liquefied natural gas
LP	low pressure
LSHFO	low sulphur heavy fuel oil
MCR	maximum continuous rating
MDE	main diesel engine
MDECES	marine diesel engine cogeneration energy system
MDETES	marine diesel engine trigeneration energy system
MP	medium pressure
MVP	meat in vacuum packs
OSSC	other ship's steam consumers
P	propeller
RAF	recirculation air fan
RF	regulation flap
SEG	shaft electric generator
SMCR	service maximum continuous rating
WHRSG	waste heat recovery steam generator
t.c.	transport cycle

Subscripts

a	accommodation, atmospheric air
ac	air conditioned
c	cargo space
ca	cargo space air
ce	cooling effect
cbp	cooling brine pump
$cfwp$	cooling freshwater pump
co	cogeneration
cw	cooling water
d	direct solar irradiation
dif	diffusely solar irradiation
ds	ducting system
d_e	direct extraterrestrial irradiation
ex	heat exchanger
fa	fresh air
fc	fuel consumption
$fwcs$	freshwater conducting system

hl	heating load
lp	live product
r	reflected irradiation
ra	recirculating air
s	saturation, sailing
sc	sea current
sd	seawater on the sea depth
se	evaporator cooling steam
si	solar irradiation
sw	seawater, seawaves
w	wind
AA	ammonia ARU
AB	navigation from *A* to *B*
AC	air cooler
AH	air heather
BA	navigation from *B* to *A*
CE	cooling evaporator, total cooling effect
CS	conducting system
CRP	compressor refrigeration plant
HL	total heating load
FS	fuel saving
FSG	fired steam generator
LBA	lithium-bromide ARU
LP	low pressure
M	mixture
MP	medium pressure
SLT	ship lifetime
TC	trigeneration concept
Y	year, timing

Superscripts

$'$	heat exchanger inlet
$''$	heat exchanger outlet
$\sim$	physical items related on the low-pressure system
$-$	mean value

References

1. Gospić, I.; Donjerković, P. The Application of the Tri-generation Systems on Offshore Vessels. In Proceedings of the 4th International Symposium Power and Process Plant, Dubrovnik, Croatia, 22 May 2000.
2. Improving Efficiency. Available online: https://www.wartsila.com/sustainability/innovating-for-sustainability/improving-efficiency (accessed on 3 May 2021).
3. Casisi, M.; Pinamonti, P.; Reini, M. Increasing the Energy Efficiency of an Internal Combustion Engine for Ship Propulsion with Bottom ORCs. *Appl. Sci.* **2020**, *10*, 6919. [CrossRef]
4. Low-Speed Engines 2021. Available online: https://www.wingd.com/en/documents/general/brochures/wingd-low-speed-engines-booklet-2021/ (accessed on 3 May 2021).
5. Efficiency of MAN B&W Two-Stroke Engines. Available online: https://man-es.com/docs/default-source/energy-storage/5510-0200-02_man-bw-two-stroke-engines_v4_lt_web.pdf?sfvrsn=c0d11a61_10 (accessed on 3 May 2021).
6. Zamiatina, N. Comparative Overview of Marine Fuel Quality on Diesel Engine Operation. *Procedia Eng.* **2016**, *134*, 157–164. [CrossRef]
7. Tana, R.; Durub, O. Prapisala Thepsithard, Assessment of relative fuel cost for dual fuel marine engines along major Asian container shipping routes. *Transp. Res. E* **2020**, *140*, 102004. [CrossRef]
8. Moussawi, H.A.; Fardouna, F. Hasna Louahliab, Selection based on differences between cogeneration and trigeneration in. various prime mover technologies. *Renew. Sustain. Energy Rev.* **2017**, *74*, 491–511. [CrossRef]
9. Al-Sulaiman, F.A.; Hamdullahpur, F.; Dincer, I. Trigeneration: A comprehensive review based on prime movers. *Int. J. Energy Res.* **2011**, *35*, 233–258. [CrossRef]
10. Yao, Z.M.; Qian, Z.Q.; Li, R.E.H. Energy efficiency analysis of marine high-powered medium-speed diesel engine base on energy balance and exergy. *Energy J.* **2019**, *176*, 991–1006. [CrossRef]
11. Dere, C.; Deniz, C. Effect analysis on energy efficiency enhancement of controlled cylinder liner temperatures in marine diesel engines with model based approach. *Energy Convers. Manag.* **2020**, *220*, 113015. [CrossRef]

12. Cavalcanti, E.J.C. Energy, exergy and exergoenvironmental analyses on gas-diesel fuel marine engine used for trigeneration system. *Appl. Therm. Eng.* **2021**, *184*, 116211. [CrossRef]
13. Marty, P.; Hétet, J.-F.; Chalet, D.; Corrignan, P. Exergy Analysis of Complex Ship Energy Systems. *Entropy* **2016**, *18*, 127. [CrossRef]
14. Benvenuto, G.; Trucco, A.; Campora, U. Optimization of waste heat recovery from the exhaust gas of marine diesel engines. Proceedings of the Institution of Mechanical Engineers. *Part M J. Eng. Marit. Environ.* **2016**, *230*, 83–94. [CrossRef]
15. Altosole, M.; Benvenuto, G.; Zaccone, R.; Campora, U. Comparison of Saturated and Superheated Steam Plants for Waste-Heat Recovery of Dual-Fuel Marine Engines. *Energies* **2020**, *13*, 985. [CrossRef]
16. Theotokatos, G.; Livanos, G. Techno-economical analysis of single pressure exhaust gas waste heat recovery systems in marine propulsion plants. Proceedings of the Institution of Mechanical Engineers. *Part M J. Eng. Marit. Environ.* **2013**, *227*, 83–97.
17. Baldi, F.; Ahlgren, F.; Nguyen, T.-V.; Thern, M.; Andersson, K. Energy and Exergy Analysis of a Cruise Ship. *Energies* **2018**, *11*, 2508. [CrossRef]
18. Baldi, F.; Johnson, H.; Gabrielii, C.; Andersson, K. Energy and exergy analysis of ship energy systems—The case study of a chemical tanker. *Int. J. Thermodyn.* **2015**, *18*, 82–93. [CrossRef]
19. Du, S.; Wang, R.Z.; Chen, X. Development and experimental study of an ammonia water absorption refrigeration prototype driven by diesel engine exhaust heat. *Energy* **2017**, *130*, 420–432. [CrossRef]
20. Ren, J.; Qian, Z.; Yao, Z.; Gan, N.; Zhang, Y. Thermodynamic Evaluation of LiCl-H_2O and LiBr-H_2O Absorption Refrigeration Systems Based on a Novel Model and Algorithm. *Energies* **2019**, *12*, 3037. [CrossRef]
21. Oil Monster. Available online: https://www.oilmonster.com/ (accessed on 3 May 2021).
22. BIX Bunker Index. Available online: https://www.bunkerindex.com/ (accessed on 3 May 2021).
23. Vallianou, V.A.; Frangopoulos, C.A. Dynamic Operation Optimization of a Trigeneration System. *Int. J. Thermodyn.* **2012**, *15*, 239–247. [CrossRef]
24. Glavan, I.; Prelec, Z. The analysis of trigeneration energy systems and selection of the best option based on criteria of ghg emission, cost and efficiency. *Eng. Rev.* **2012**, *32*, 131–139.
25. Frangopoulos, C.A. Developments, Trends, and Challenges in Optimization of Ship Energy Systems. *Appl. Sci.* **2020**, *10*, 4639. [CrossRef]
26. Gospić, I.; Boras, I.; Mravak, Z. Low-Temperature Ship Operations, Brodogradnja: Teorija i Praksa Brodogradnje i Pomorske Tehnike. 2011, Volume 62. Available online: https://hrcak.srce.hr/75662 (accessed on 3 May 2021).
27. Lugo-Villalba, R.A.; Guerra, M.Á.; López, B.S. Calculation of marine air conditioning systems based on energy savings. *Ciencia Y Tecnología Buques* **2017**, *11*, 103–117. [CrossRef]
28. Yang, S.; Vargas, J.; Ordonez, J. Ship HVAC System Analysis and Optimization Tool. In Proceedings of the 2019 IEEE Electric Ship Technologies Symposium, Arlington, VA, USA, 14–16 August 2019. [CrossRef]
29. Mărcuș, G.; Lungu, C.I. Partial load efficiency analysis of a CCHP plant with RICE and H2O-LiBr absorption chiller. In Proceedings of the CLIMA 2019 Congress, Bucharest, Romania, 26–29 May 2019. [CrossRef]
30. Han, B.; Li, W.; Li, M.; Liu, L.; Song, J. Study on Libr/H2O absorption cooling system based on enhanced geothermal system for data center. *Energy Rep.* **2020**, *6*, 1090–1098. [CrossRef]
31. Lima, A.A.S.; Leite, G.d.N.P.; Ochoa, A.A.V.; Santos, C.A.C.d.; Costa, J.A.P.d.; Michima, P.S.A.; Caldas, A.M.A. Absorption Refrigeration Systems Based on Ammonia as Refrigerant Using Different Absorbents: Review and Applications. *Energies* **2021**, *14*, 48. [CrossRef]
32. Ouadha, A.; El-Gotni, Y. Integration of an ammonia-water absorption refrigeration system with a marine Diesel engine: A thermodynamic study. *Proc. Comput. Sci.* **2013**, *19*, 754–761. [CrossRef]
33. IMO. Prevention of Air Pollution from Ships. International Maritime Organization (IMO). 2017. Available online: https://www.imo.org/en/OurWork/Environment/Pages/Air-Pollution.aspx (accessed on 3 May 2021).
34. Ammar, N.R.; Seddiek, I.S. Eco-environmental analysis of ship emission control methods: Case study RO-RO cargo vessel. *Ocean Eng.* **2017**, *137*, 166–173. [CrossRef]
35. Food Handling, Storage and Preparation. Maritime Labour Convention, 2006 (MLC, 2006). Available online: https://webcache.googleusercontent.com/search?q=cache:Nbx7BJ33CHIJ:https://www.ilo.org/dyn/normlex/en/f%3Fp%3D1000:53:::NO:53:P53_FILE_ID:3132751+&cd=1&hl=en&ct=clnk&gl=kr (accessed on 3 May 2021).
36. Panama Maritime Authority Department of Seafarers, Instructions for Handling, Storage and Preparation of Food Under the Maritime Labour Convention, 2006 (Rule 3.2 Rule 3.2). Available online: http://webcache.googleusercontent.com/search?q=cache:udhN1MNh1ggJ:www.ilo.org/dyn/normlex/en/f%3Fp%3D1000:53::::53:P53_FILE_ID:3130264+&cd=1&hl=en&ct=clnk&gl=kr (accessed on 3 May 2021).
37. Lemmon, E.W.; Bell, I.H.; Huber, M.L.; McLinden, M.O. *REFerence fluid PROPerties 10*, Applied Chemicals and Materials Division National Institute of Standards and Technology: Boulder, CO, USA.
38. WMO-World Meteorological Organization. Available online: https://public.wmo.int/en (accessed on 3 May 2021).
39. Martić, I.; Degiuli, N.; Farkas, A.; Gospić, I. Evaluation of the Effect of Container Ship Characteristics on Added Resistance in Waves. *J. Mar. Sci. Eng.* **2020**, *8*, 696. [CrossRef]
40. Farkas, A.; Degiuli, N.; Martić, I. Assessment of hydrodynamic characteristics of a full-scale ship at different draughts. *Ocean Eng.* **2018**, *156*, 135–152. [CrossRef]

41. Thomas, A.P.; Thekaekara, M.P. Experimental and Theoretical Studies on Solar Energy for Energy Conversion. In Proceedings of the Joint Conference, Winnipeg, MB, Canada, 15–20 August 1976.
42. Kreider, J.F. *Rabi A: Heating and Cooling of Buildings, Design for Efficiency*; McGraw-Hill, Inc.: New York, NY, USA, 1994.
43. Gospić, I. Modeliranje Brodskih Dizelmotornih Trigeneracijskih Energetskih Sustava. Ph.D. Thesis, University of Zagreb, Zagreb, Croatia, 2008.
44. Global Average Bunker Price. Available online: https://shipandbunker.com/prices/av/global/av-glb-global-average-bunker-price#MGO (accessed on 7 June 2021).
45. Summerheat. EU Intelligent Energy Europe Programme EIE-06-194. April 2009. Available online: https://www.euroheat.org/wp-content/uploads/2016/04/SUMMERHEAT_Report.pdf (accessed on 7 June 2021).

MDPI

St. Alban-Anlage 66

4052 Basel

Switzerland

Tel. +41 61 683 77 34

Fax +41 61 302 89 18

www.mdpi.com

Journal of Marine Science and Engineering Editorial Office

E-mail: jmse@mdpi.com

www.mdpi.com/journal/jmse